Malin Falkenmark is currently with the SRC, Stockholm University, and the Stockholm International Water Institute (SIWI). She is a future-oriented scientist and a pioneer of interdisciplinary environmental and water research. Her deepest interests lie in the linkages between land/water/ecosystems, particularly as they relate to water scarcity, regional similarities and differences, and their policy implications. She has introduced three broadly used concepts: the water scarcity indicator, the concepts of green and blue water, and hydrosolidarity. Professor Falkenmark has held several high level posts on international boards and committees, and has received several international awards, including the Rachel Carson Prize and the Volvo Environmental Prize.

Carl Folke is Professor in Natural Resource Management, founder and Science Director of the SRC, and Director of the Beijer Institute of Ecological Economics of the Royal Swedish Academy of Sciences. He is a leading reseacher on social–ecological systems and resilience, sustainability science and ecological economics, and co-founder of the Resilience Alliance. Dr Folke's work focuses on the essential role of life-support ecosystems, ecosystem services and biodiversity and their governance, emphasising that people are part of and fundamentally dependent on the capacity of the biosphere to sustain development. He has published some 180 research papers, numerous book chapters and 12 books. He serves as scientific advisor to international research institutes, organisations and other actors in science, policy and practice.

Mats Lannerstad is a Research Fellow at the SEI and a Scientist at the International Livestock Research Institute (ILRI). His research is centred on natural resource use in the global food system, with particular focus on freshwater requirements for food security on multiple scales and pathways towards a sustainable intensification in food production. Two key areas are the dietary change towards more animal-source foods following rising affluence and urbanisation, and the linkages between livestock production and the use of different water sources. Dr Lannerstad has authored 15 peer-reviewed publications, three books/book chapters and numerous research reports.

Jennie Barron is a Research Leader at SEI and a researcher at the SRC. Her research applies a systems approach to farm and landscape agricultural development trajectories, to improve productivity, livelihoods, food security and ecosystems services, and to advise on opportunities for action and investment. She works in local, national and international partnerships in sub-Saharan Africa and South Asia. Dr Barron has written or co-authored more than 25 peer-reviewed publications, and developed various outreach communications and research for development and policy at local, national and international levels.

Elin Enfors is a Researcher at the SRC. Her main interest concerns pathways to transformation in agro-ecosystems, with a special focus on how ecosystem services generated in these systems can contribute to poverty alleviation. She has a background in systems ecology, with extensive experience of field research from Africa, and a keen interest in participatory research.

Line Gordon is an Associate Professor and Deputy Science Director at the SRC and leads the Landscape research theme. Her research centres on interactions among freshwater resources, ecosystem services and food production, with a focus on how resilience thinking can enable better management of these resources. Dr. Gordon has published over 20 peer-reviewed papers and book chapters in a wide range of journals. She has extensive international experience, and is conducting transdisciplinary research with a strong interest in the interface of science and society.

Jens Heinke is a Research Fellow and doctoral candidate at Potsdam Institute for Climate Impact Research (PIK), and a Scientist at the ILRI. He primarily works with the Lund-Potsdam-Jena dynamic global vegetation model (LPJ), as both a developer and analyst. His research is focussed on global change impacts on the water cycle with an emphasis on water resources, food production and livestock production. He also works on mapping and understanding the flows and linkages in the global food system, and is the author or co-author of more than 20 scientific papers and several research reports.

Holger Hoff is Senior Research Fellow at the SEI and PIK. His research focus is on climate impacts and adaptation in water, agriculture and ecosystems, and integrated water and natural resource management, across scales up to the global level. He works primarily in the Middle East and North Africa, sub-Saharan Africa and South Asia regions, and at PIK he currently works on sustainability boundaries of water and land use. Mr Hoff co-coordinates the Nexus Initiative at SEI, using WEAP and other nexus tools, and is the author of more than 30 peer-reviewed publications, several book chapters and numerous research reports and policy papers.

Claudia Pahl-Wostl is Professor for resources management and Director of the Institute for Environmental Systems Research at the University of Osnabrück, Germany. Her major research interests are adaptive governance and management of water resources, social and societal learning and their role in transformation processes towards sustainability, global water governance and multi-level governance systems and conceptual and methodological frameworks to analyse social–ecological systems. She has authored numerous papers in peer-reviewed journals, chapters in edited books, policy briefs and popular reports, and has also edited three books and twelve special issues in peer reviewed journals.

Water Resilience for Human Prosperity

Johan Rockström
Stockholm Resilience Centre, Stockholm University

Malin Falkenmark
Stockholm Resilience Centre, Stockholm University and Stockholm International Water Institute

Carl Folke
Stockholm Resilience Centre, Stockholm University and the Beijer Institute of Ecological Economics of the Royal Swedish Academy of Sciences

Mats Lannerstad
Stockholm Environment Institute and International Livestock Research Institute

Jennie Barron
Stockholm Environment Institute

Elin Enfors
Stockholm Resilience Centre, Stockholm University

Line Gordon
Stockholm Resilience Centre, Stockholm University

Jens Heinke
Potsdam Institute for Climate Impact Research and International Livestock Research Institute

Holger Hoff
Stockholm Environment Institute and Potsdam Institute for Climate Impact Research

Claudia Pahl-Wostl
Institute for Environmental Systems Research at the University of Osnabrück

Shaftesbury Road, Cambridge CB2 8EA, United Kingdom

One Liberty Plaza, 20th Floor, New York, NY 10006, USA

477 Williamstown Road, Port Melbourne, VIC 3207, Australia

314–321, 3rd Floor, Plot 3, Splendor Forum, Jasola District Centre, New Delhi – 110025, India

103 Penang Road, #05–06/07, Visioncrest Commercial, Singapore 238467

Cambridge University Press is part of Cambridge University Press & Assessment, a department of the University of Cambridge.

We share the University's mission to contribute to society through the pursuit of education, learning and research at the highest international levels of excellence.

www.cambridge.org
Information on this title: www.cambridge.org/9781107024199

First published 2014

A catalogue record for this publication is available from the British Library

Library of Congress Cataloging-in-Publication data
Water resilience for human prosperity / Johan Rockström, Malin Falkenmark, Carl Foilke
pages cm
ISBN 978-1-107-02419-9 (Hardback)
1. Water–History. 2. Water supply–History. 3. Drinking water–History. I. Rockström, Johan.
GB659.6.W38 2014
333.91–dc23 2013028561

ISBN 978-1-107-02419-9 Hardback

Additional resources for this publication at www.cambridge.org/9781107024199

Contents

Part II – Living in a human-dominated world

Part IV – Governance and pathways

Contributors

These authors contributed the informative boxes on particular issues found throughout this volume.

Göran Berndes
Chalmers University of Technology

Petra Döll
University of Frankfurt

Ellen M. Douglas
University of Massachusetts, Boston

Ruud J. van der Ent
Delft University of Technology

Lance Gunderson
Emory University

Elke Herrfahrdt-Pähle
German Development Institute

Eloise Kendy
The Nature Conservancy

Yuanhong Li
Gansu Research Institute for Water Conservancy (GRIWAC)

Michael E. McClain
UNESCO-IHE Institute for Water Education

Denis Mpairwe
Makerere University

Jay O'Keeffe
Rhodes University

Donald Peden
International Livestock Research Institute (ILRI)

Marcela Quintero
International Centre for Tropical Agriculture (CIAT)

Wilhelm Ripl
Technical University of Berlin (Professor Emeritus)

Hubert H.G. Savenije
Delft University of Technology

Bridget Scanlon
University of Texas at Austin

Maja Schlüter
Stockholm Resilience Centre

Jan Sendzimir
University of Natural Resources and Applied Life Science, Vienna (BOKU)

Will Steffen
The Australian National University and Climate Commissioner, Australia

Alain Vidal
CGIAR Challenge Program on Water and Food

Brian Walker
Commonwealth Scientific Industrial Research Organisation (CSIRO)

Qiang Zhu
Gansu Research Institute for Water Conservancy (GRIWAC)

Preface

Why yet another book on water? Partial thinking and sectoral approaches have dominated resource and environmental management for too long, and this is also true for freshwater. Perspectives are rapidly changing, however, expanding on the conventional perception of freshwater as 'blue water' – a natural resource to be extracted from rivers and groundwater for households, industry, irrigation and economic production. Integrated water resource management (IWRM), although still predominantly concerned with the blue water branch of the water cycle, has extended the focus to interacting sectors in catchments. More recently, water vapour or 'green water' has increased focus in the policy arena on issues such as rainfed agriculture. The role of freshwater in ecosystem services, both terrestrial and aquatic, is now on the agenda, as well as work on their water trade-offs or in relation to water-related tipping points in dynamic landscapes. New approaches are emerging, such as adaptive water governance of landscapes and catchments.

The biosphere – the sphere of life – is the living part of the outermost layer of our rocky planet – the part of the Earth's crust, oceans and atmosphere where life dwells. It is the global life-support system that integrates all living beings and their relationships. Life on Earth interacts in myriad ways with the chemistry of the atmosphere, the circulation of the oceans and the water cycle, including solid water in polar and permafrost regions, to form favourable conditions for life on Earth. People and societies are integrated parts of the biosphere, dependent on its functioning and life support.

Water plays a key role in the operation of the biosphere, from the level of the cell to the dynamics of the atmosphere. The water cycle functions as the bloodstream of the biosphere. Like any organism, humans have evolved with water, benefitting from its many functions and the mineral salts it carries. Water is required for soil formation and is critical to the production of the food we eat. Continents are connected by rainfall patterns, it provides climate-regulating services and plays a central role in extreme events such as floods, storms and droughts. On the blue planet, the water cycle is clearly essential to our existence and a precondition for our evolution.

It is now apparent that humanity has become a major force in the dynamics of the biosphere, shaping it not only locally and regionally but also globally, and leaving a significant imprint on the operation of the biosphere as a whole. Drivers of change such as rising human numbers, urbanisation, migration patterns, emerging markets, the diffusion of new technologies and social innovations can combine with sudden events such as floods, fires, pandemics, rapid shifts in fuel prices and volatile financial markets to trigger tipping points. The global social–ecological system is complex and dynamic, and subject to unexpected, often rapid, changes – not as exceptions but increasingly as the rule. Such changes play out in cascading fashion in a world where everyone is in everyone else's backyard. Thresholds and tipping points are now part of the furniture.

This new situation – the Anthropocene – calls for a fundamental shift in perspectives and world views, reconnecting development and progress to the capacity of the biosphere and its water cycle to sustain society and prosperity. This reconnection is linked to the insight that humanity has been prospering from a stability that is exceptional in the history of the Earth. The past 10 000 years, the Holocene geological epoch, was an era during which agriculture and human civilisations emerged and flourished. Many take the favourable Holocene conditions for granted. In our view, a greater appreciation is needed of water as part of biosphere dynamics and resilience. Hence the call for a broader water perspective that connects the local with the global. Resilience, in the way we approach it, is about persistence in the face of change, having the capacity to continually adapt to complex dynamics, and to develop in order to get out of traps and even transform and shift into new development pathways. The capacity of the biosphere and the water bloodstream sets the framework for such pathways – the planetary boundaries for prosperous societal development.

It is in this context that we have written this book – to take on the challenge of expanding mindsets towards water as the bloodstream of the biosphere of which people are an embedded part. In the globally interconnected world, humanity is critically dependent on the capacity of the biosphere to support our way of life, and the way we have organised societies, technologies and economies. The water bloodstream approach is not just an ethical stand. It is about prosperity and ultimately about survival. It is also about biosphere stewardship and innovation for sustainable development for humanity.

We have written this book in search of a deeper understanding of the new water dynamics in the globally integrated system of people and nature, to put forward new conceptual systems, perspectives, hypotheses and findings. We believe that science has a responsibility to search for a better understanding of the new challenges facing humanity, and to explore pathways for a sustainable world. We describe and analyse the role of water in the biosphere and how it relates to human actions and well-being from the global to the local levels, and we introduce new concepts such as water resilience and water stewardship in the new Anthropocene era. Striving for water stewardship and a resilient biosphere is not about preserving the status quo or circumventing change. It is about having the capacity to deal with change, turning crises into opportunities and shifting into sustainable pathways.

Resilience thinking encourages us to anticipate, experiment, adapt and transform. Water resilience and water stewardship are about strengthening the resilience of social–ecological systems to deal with changing conditions, and finding ways to live in prosperity in the Anthropocene era. This will require an appreciation of the critical role of water in the operation of the biosphere for human well-being. We hope this book will inspire people in this direction.

Carl Folke, Malin Falkenmark and Johan Rockström

Introduction to the book

Scope of the book

This book aims at synthesising our current state of knowledge and probing the key area of how recent insights from social–ecological systems and resilience research influence our understanding of water resource governance and management in a world subject to rapid global environmental change. It advances a proposed new framework on 'water resilience' as an integral part of sustainable water resource management. We have a focus on ecosystem services in productive landscapes, especially food production (and bioresources), seen from the perspective of land, water, ecosystem interactions and resilience building. Focus is on water resources from local to global scale, exploring dynamic interactions between sectors, components of the Earth system and scales. The book will therefore only briefly address water quality issues. The water resource focus of the book includes water flows from the local water balance to the global hydrological cycle – i.e. the governance and management of precipitation, vapour flows, as well as surface and sub-surface runoff flows and resources. It is, furthermore, global in scope, even though a particular focus is set on the regions of the world facing the most challenging future in terms of water resource scarcity and water resilience challenges related to current and future global environmental change. This means that a particular focus is given to the semi-arid and dry sub-humid tropical savannah regions of the world.

The water and ecosystems focus of the book, places the emphasis on the relations between freshwater and the living systems in the biosphere. The book thus takes as a starting point the role of water resources in the generation of ecosystem functions and services from terrestrial and aquatic ecosystems, and how these define the resilience of ecosystems; how human interactions with water impact on ecosystem and resilience; and how innovative water governance and management principles can be applied to human challenges in an era of rapid global changes. In essence we attempt to advance a social–ecological systems approach to water resilience for human prosperity in the Anthropocene.

The book thereby does not focus on water in marine ecosystems, and does not explore the important role of water for resource use (e.g. in mining) nor for domestic and urban water supply and water for industrial purposes. This said, the book obviously takes an integrated perspective on the trade-offs between water use for living systems and other resource and social uses. Our special focus on water and food in a changing world is justified by the fact that no human sector consumes so much freshwater as bioresources for food, energy and biomass, which raises, apart from trade-offs between different water needs, the challenge of how to build water resilient *food production* in the world.

Target audience

The book is targeted at graduate/post-graduate students, water resource professionals and senior water planners, and is therefore a book targeting higher education, which can also inform key water professionals in different sectors from agriculture and environment to industry and river basin planning.

We allow ourselves to be relatively detailed and in-depth, and quite technical where needed, while trying to reach a broader professional audience. We want to explain and give examples related to complex issues ranging from vapour shift, water-induced regime shifts, moisture feedback, water resilience, etc. The text is interspersed with a set of *boxes*, authored by invited water scientists looking deeper into a number of issues discussed or referred to in the main text.

Book sections

The book is divided into four parts.

Part I. A new perspective

Chapter 1 is an overview and framing chapter on the emerging challenge of water resilience in the Anthropocene. It explains the crucial roles played by water in the life-support systems on Earth in an era of rapid global and regional change. It discusses different disturbance regimes and the emerging threats and dilemmas, and highlights potential thresholds of critical concern. It explains the core roles of water in sustaining a desired 'Holocene-like' state on Planet Earth, and the risk for human-induced water thresholds. It also explains three core roles of water for resilience. The chapter furthermore highlights the central role of water partitioning changes, motivating special focus on foreseeable future land-use alterations, in particular future human use of bioresources, especially implications of feeding a growing humanity.

Part II. Living in a human-dominated world

Chapter 2 offers an overview of past human alterations to the Earth system and the main drivers of change. It highlights climate change in particular, as it interacts profoundly with the planet's global water cycle. It demonstrates the socially driven connectivity between different global regions, and human-generated impacts on the Earth System. It stresses that humanity is now living in the new Anthropocene, a new geological epoch where humanity constitutes a quasi-geological force of planetary change, at risk of and approaching various water-related tipping points.

Chapter 3 analyses the options for safe global pathways towards sustainable water development and the dangers to be avoided in the form of water-related thresholds, rigidity and poverty traps. It addresses water's involvement in abrupt, unexpected regime shifts in social–ecological systems. Resilience is characterised by the existence of reinforcing processes and stabilising feedbacks. Water's many different roles in the life-support system mean that it is profoundly involved in the processes of and responses to regime shifts, as both a state variable and a control variable.

Chapter 4 examines human dependence on the global water system (GWS), and the role of water as the bloodstream of the biosphere. It highlights human-generated changes in the system, including a number of remote water-related connections between regions (so-called teleconnections) such as trade-related virtual water flows. Resilience-related changes are summarised including land-use change and its implications for green–blue water partitioning; climate change, noting that aridification can reduce resilience to droughts; growing water demands; and groundwater overexploitation. It stresses that basin closure represents a critical threshold beyond which new processes and interactions are triggered.

Part III. Food production globally: in hotspot regions and in the landscape

Chapter 5 analyses the challenge of feeding a growing humanity from a water perspective. It describes the growing food demand up to 2050, considering population increase, average per capita food supply levels, and changed composition of animal and vegetal source foods in food supply. Country-level assessments of food water requirements are given for different scenarios, including climate change, irrigation development, water productivity improvements, alternative dietary options and reduced food losses. The chapter highlights the need for large-scale virtual water transfer through expanded food trade. Food supply is examined from a dynamic perspective in terms of the ability to cope with shocks and change, and the adaptability and social–ecological resilience required.

Chapter 6 analyses the large and rising social–ecological challenge in the water-poor savannah zone with rapidly increasing populations and demands for water. What are the implications of food supply efforts, and the implementation difficulties in these regional hotspot regions? The chapter clarifies that, contrary to popular beliefs, this zone has a substantial (and untapped) agro-hydrological potential. Rather than facing absolute lack in water, the challenge is the huge fluctuations in rainfall and the large amount of water lost to the farming system through evaporation, runoff and drainage. Water resilience strategies involve practices for dry-spell mitigation, using, e.g. water harvesting systems.

Chapter 7 focuses on basin-level challenges and the meso-scale perspective, which is where land-use changes can aggregate and affect ecosystem services, and consequently livelihood and development opportunities, and ecosystem sustainability. Agriculture, which is itself an ecosystem service provider, is

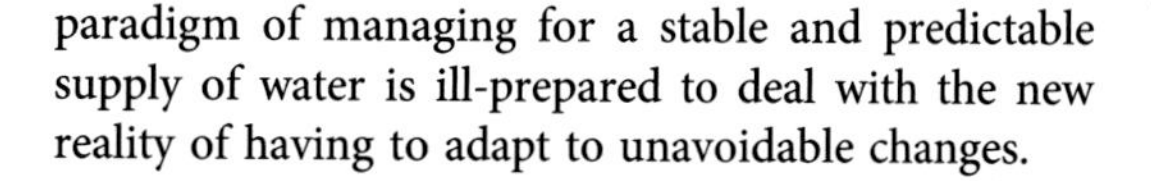

primarily a way to manage particular benefits from ecosystems, but other ecosystem services may be affected. Three landscapes are analysed in terms of landscape multifunctionality, exemplifying common development trends and emerging upstream–downstream conflicts of interest.

Part IV. Governance and pathways

Chapter 8 addresses the challenge of water governance of landscapes and basins for resilience, sustainability and human well-being. Integrated governance of land and water resources for the generation of ecosystem services, safeguarding development and avoiding crossing critical thresholds, is at the heart of this analysis, integrating global dynamics, the necessity of water governance that 'safeguards rainfall and wetness in landscapes'. We focus on the blue–green water partitioning, the blue–green trade-off between upstream and downstream activities and strategies for stabilising moisture feedbacks (the source of future rainfall). The chapter highlights governance challenges and transformations needed.

Chapter 9 concludes by addressing insights and pathways for a world transition towards sustainability by adopting a social–ecological systems approach to IWRM. It describes the evolution of water governance and management from a largely blue water focused paradigm from the early 1970s until the early 1990s, which has served humanity quite well in a world of relative water abundance, but which, now, under pressures of growing human demands, water use and the recognition of shifts in water supply and risks of thresholds due to global environmental change, necessitates a new integrated green–blue water paradigm. It summarises the new insights in terms of what we have learnt on water and resilience, and highlights the grand global challenge of feeding a world population within a safe operating space of planetary boundaries. It also notes that the current water governance paradigm of managing for a stable and predictable supply of water is ill-prepared to deal with the new reality of having to adapt to unavoidable changes.

The four-step resilience chain

As is noted in the Preface, we have written this book in search of a deeper understanding of the new water dynamics in the globally integrated system of people and nature, and to put forward new conceptual systems, perspectives, hypotheses and findings. We believe that science has a responsibility to search for a better understanding of the new challenges facing humanity, and to explore pathways for a sustainable world. All the different chapters analyse the role of water in the biosphere, and how it relates human actions and well-being to the global to local levels. New concepts are introduced, such as water resilience and water stewardship.

Striving for sustainable water stewardship and a resilient biosphere is not about preserving the status quo or circumventing change. It is about having the capacity to deal with change, turning crises into opportunities and shifting on to sustainable pathways. Special emphasis is put on the world's most water-dependent sector – agriculture.

Governing and managing water for resilience encompasses a range of actions from mitigation to resilience building, adaptation and transformation. The figure below shows these actions in a schematic way, indicating the interconnected challenges facing global water resource management. The range of actions along this 'change continuum' includes mitigation to reduce human pressures on the Earth System, building the resilience of Earth System components, adaptation to materialised responses and transformation after regime shifts in social–ecological systems.

Every chapter opens with a short *resilience-oriented ingress*, clarifying how it relates to the above sequence of stages in resilience thinking. The reader

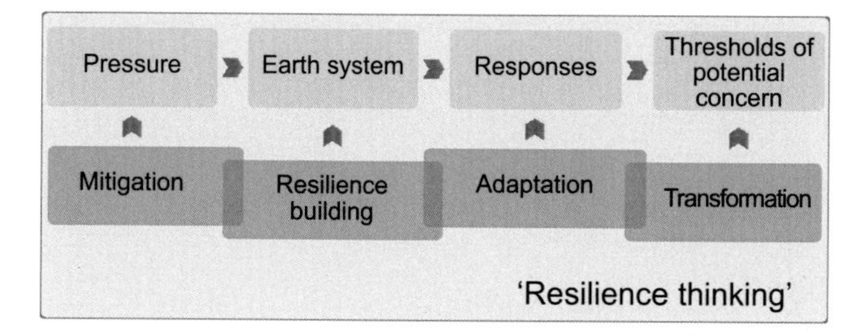

Figure I.1 The challenges facing the world in the Anthropocene, from a range of pressures to impacts on the Earth System, influence responses from societies and the possibilities of passing thresholds that change social–ecological systems. The chain of actions to build resilience for global sustainability includes mitigation, local resilience building, adaptation and transformation.

will find key components of the different steps in the resilience chain exemplified in the different chapters.

Pressure	Chapters 2, 4, 5, 6: drivers of change, land-use change, feeding humanity, water supply, energy supply, industrial production, urbanisation, technological development and international trade
Earth system	Chapters 2 and 4: land productivity, Holocene equilibrium, GWS, ecosystem functions, biodiversity
Responses	Chapters 2, 5, 6: land degradation, CO_2-enrichment, water stress, aquifer overexploitation, virtual water flows, megafires, traps, migration, famine
Thresholds/ tipping points	Chapters 1, 3, 4, 9: desertification, savannisation, salinisation, monsoon weakening, basin closure, aquifer depletion, thresholds of potential concern
Regime shifts	Chapters 1 and 3: Anthropocene dynamics, ecosystem shifts, unproductive land, biodiversity loss, poverty traps, rigidity traps
Resilience building	Chapters 1, 3, 6, 7, 8, 9: land stewardship, moisture feedback, balanced water uses, secured partitioning, environmental flow, planetary boundaries, vital ecosystem functions and services, upstream–downstream trade-offs

Authorship

This book is a result of a joint collaborative effort among all authors. The book was written by synthesising recent research, advancing new insights through a series of author workshops and a distributed responsibility for different chapters among co-authors. The lead authorship responsibility was shared as follows: Chapter 1, lead author Johan Rockström; Chapter 2, lead author Holger Hoff; Chapter 3, lead author Line Gordon; Chapter 4, lead author Holger Hoff; Chapter 5, co-lead authors Mats Lannerstad and Malin Falkenmark, data modelling and analysis Jens Heinke; Chapter 6, lead author Elin Enfors; Chapter 7, lead author Jennie Barron; Chapter 8, co-lead authors Carl Folke and Claudia Pahl-Wostl; Chapter 9, lead author Johan Rockström. Johan Rockström and Carl Folke led the effort together with Malin Falkenmark in distilling key messages and structuring the line of argument on water-related resilience thinking throughout the book.

Production staff

Mats Lannerstad
Book project manager

Jens Heinke
Modelling and data analysis

Hugo Ahlenius, Nordpil
Illustrations and production coordination

Andrew Mash
Language and sub-editing

Acknowledgements

We wish to thank colleagues at the Stockholm Resilience Centre (SRC), Beijer Institute of Ecological Economics of the Royal Swedish Academy of Sciences, the Potsdam Institute for Climate Impact Research and the Stockholm Environment Institute (SEI) for the stimulating discussions during the advancement of this book.

We are grateful to several donors and projects, without which this book could not have been written. Core funding was provided by Mistra, through its support to the Stockholm Resilience Centre, and from the Swedish Research Council Formas, through its Centre of Excellence grant to SRC and SEI. We also appreciate the support from the Kjell and Märta Beijer Foundation, which enabled the engagement of Beijer Institute colleagues, the Ebba och Sven Schwartz Stiftelse and the Swedish International Development Cooperation Agency (SIDA).

The authors greatly appreciate the contributions made by colleagues in the form of the informative boxes on particular issues found throughout this volume: Göran Berndes, Petra Döll, Ellen M. Douglas, Ruud van der Ent, Lance Gunderson, Elke Herrfahrdt-Pähle, Eloise Kendy, Yuanhong Li, Michael E. McClain, Denis Mpairwe, Jay O'Keeffe, Donald Peden, Marcela Quintero, Wilhelm Ripl, Hubert Savenije, Bridget Scanlon, Maja Schlüter, Jan Sendzimir, Will Steffen, Alain Vidal, Brian Walker and Qiang Zhu.

Many research colleagues have contributed significantly to our thinking in this book. We thank you all collectively, but would like to particularly mention Patrick Keys who assisted in development of several figures and the conceptual thinking; Howard Cambridge, SEI, ran the Nariale water balances in SWAT; Kausal Garg, ICRISAT, kindly developed additional maps for the Kothapally case study, and Philippe Cecchi, IRD UMR G-eau, has provided invaluable understanding of the Nariale catchment development, including providing land-use data over time. Charles Batchelor, independent consultant, kindly shared the case data from Rajasthan, India. Thanks to Dieter Gerten, the leading scientist at the Potsdam Institute on the LPJ model (used for all global estimates of green–blue water flows), and to several water scientists for dialogues stimulating and triggering the writing of this book (in particular, Hubert Savenije, Sandra Postel, Tony Allan, Charles Vörösmarty, Joseph Alcamo, Gretchen Daily, Suhas Wani, Theib Oweis, David Molden and Vladimir Smakhtin).

We greatly appreciate the enthusiasm and deep commitment to the realisation of this book shown by all these contributors.

Special thanks to Hugo Ahlenius for preparing all the figures and tables and leading the final stretch in organising the manuscript for submission, Mats Lannerstad for the management of this highly interactive book project, and to the language editor Andrew Mash, for your invaluable contributions. Without your tireless efforts we would not have been able to finalise this book.

A new perspective

The rainforest of Borneo is one of the greatest biodiversity hotspots in the world, but is under threat from deforestation and conversion to oil palm plantations. This leads to a drier open landscape with higher risk of severe forest fires, and lower moisture feedback, with implications for local regional rainfall.

The role played by water in the biosphere

The future of humanity will depend on our capacity to govern and manage water in ways that build resilience in an era of rapid global change and growing indications of large-scale, undesirable risks caused by the unsustainable exploitation of ecosystems. We define this strategic domain of global sustainability as 'water resilience', i.e. the role of water in achieving social–ecological resilience in support of sustainable development in the world. The chapter presents the new conceptual framework for reconnecting our societies to the biosphere and introduces the focus of the book: freshwater and the living systems of the biosphere.

1.1 The fundamental role of water in sustaining life on Earth

Water is understood, and has been for centuries, as a fundamental component of human well-being and socio-economic development. This insight dates back to the ancient water civilisations in human history, ranging from the Mesopotamian irrigation societies of the early years of the Holocene geological era, some 8000 years ago, to the great water-engineering feats of the Egyptian, Maya, Chinese and Roman empires, all the way through to sophisticated local contemporary water societies such as the Bali water temples and the intricate Dutch water-control boards. Nonetheless, there is ample evidence to suggest that we have reached a new situation in which our current way of governing and managing freshwater is becoming obsolete in relation to the social and environmental challenges facing humanity in the coming 50 years.

1.1.1 Water as a strategic agent in building resilience in our societies

We need to rethink how we govern and manage freshwater resources. The purpose of this book is to present a new freshwater paradigm, which originates from several strands of scientific advances and empirical developments. These include, in short, the insights that:

1. humanity may be approaching a point of planetary water overshoot, i.e. that water use will exceed sustainable boundaries at the planetary scale;
2. the social and economic demands for freshwater, driven predominantly by food production, exceed what we can sustainably supply with current policies and practices;
3. we have reached a globalised phase of sustainability in which accelerated global environmental change – ranging from climate change to loss of biodiversity – undermines the ability of the planet to supply freshwater in a way that is conducive to human development on a planet which will have at least nine billion people by 2050;
4. there is increasing evidence of the role of water in sustaining the resilience of ecosystems and thereby access to and the use of natural capital, which is key to our development and our ability to avoid undesirable, rapid and irreversible tipping points in social and ecological systems;
5. water needs to be actively governed in order to sustain terrestrial and aquatic ecosystem functions and services as well as the direct supply of freshwater for societies;
6. water is intimately linked essentially to all other biophysical processes on the planet that regulate the functioning of the Earth System and thereby its ability to support human development – from the generation of biomass to the regulation of climate; and
7. water, at any given location in the world, can no longer be governed and managed without an active understanding of the drivers of and impacts on other spatial and temporal scales, from the local to the global, for now and the decades and centuries to come.

What emerges is a deeper understanding of the fundamental role played by water in sustaining the living biosphere on Earth (Falkenmark and Chapman, 1989), as well as the role of water as a key driver of change (IPCC, 2007) and as a strategic agent in building resilience in our societies (Falkenmark and Folke, 2003). The above insights are currently maturing in the scientific world, at a time of increased social, economic and ecological turbulence across the globe. We are seeing increasing evidence of abrupt changes and undesirable social feedback – of financial and social crises moving from the local to the regional and the global scales – but there is also evidence of a similar globalisation of environmental change, generating unexpected interactions between social and ecological changes, posing new challenges for human development. Water plays a central role in this new era of social, economic and ecological globalisation.

This book is written in the context of the new water-related turbulence in the world. We address the question of how the governance and management of water need to change to enable a transition to global sustainability that meets human needs for water and water-dependent ecosystem services (such as food, medicine and bioenergy) while building resilience to unavoidable change.

The new social–ecological global water complex is central to our understanding of the threats to human development, and key to identifying new ways of transforming water governance and management in a direction that supports human development in the future. It is an interesting and challenging paradox that as we realise the growing and excessive human pressures on finite freshwater resources on the planet, we also better understand our total dependence on freshwater not only for food, industry and the domestic water supply, but also for the generation of fundamental or essential ecological functions and services in terrestrial and aquatic ecosystems, which in turn form the basis for social–ecological resilience.

Box 1.1 Water: the mysterious basis of life

Wilhelm Ripl, Technical University of Berlin (Professor Emeritus)

One dazzling property of water – its ability to determine the lifespan of structures at several different levels of organisation and of the water cycle – is linked to its chemical dissipative property: the partial dissociation of water into protons and electrons. Its molecular structure, with an angle of about 105° between the two hydrogen atoms and the oxygen atom, helps to make water a polar agent and the most abundant solvent on Earth for salts and even organic compounds. When salts are dissolved, anions and cations are evenly distributed in the solvent water. Water molecules show paramagnetic properties, and ionic solutions show electric properties.

Water reacts with carbon dioxide to form carbohydrate radicals that in turn combine to polymerise into glucose. This reaction system reacts with nitrogen ions and continues reacting to give long-chained fibre structures of, e.g. cellulose or, say, starch. Preferentially, molecules containing carbon and nitrogen form long-chained complex molecules with hydrogen and oxygen.

Water plays crucial roles as a transportation, reaction and cooling medium in self-organising ecosystems. According to Odum (1969), ecosystem development is a two-phase process. In the establishment phase, pioneer plants prove to be the most efficient at utilising nutrients and water for rapid reproduction, and thus cover the available empty surface or space. When the available space is filled and limitations occur, a change in strategy is forced by negative feedback. The maintenance phase takes over in which assemblages that develop matter-recycling capabilities, characteristic of 'mature' ecosystems, appear. In the maintenance phase, ecosystems are analogous to organisms designed in a cellular way, and are known as dissipative ecological units (DEUs). Five functionally defined components are necessary to form such a unit (see Figure 1.b1.1): (1) three types of organism – green plants (primary producers), bacteria and fungi (decomposers) and the food chain (all kind of animals as grazers and predators of all kinds of organisms, opening up space for growth and reproduction, and keeping the system efficient); and (2) two non-living components – dead organic debris, which serves as a stock of energy, nutrients and minerals, and water, which serves as the cooling, transportation and reactive medium.

The green plant usually has a double function: as a source of energy for all kinds of organisms and as an active water pump, sucking water through the roots to the leaves and maintaining the process of transpiration which is coupled to the photosynthetic processes in the root zones by way of feedback. Evapotranspiration coupled with net production reduces the amount of water in the soil zone capillaries and gives air access to the debris layer, enhancing the activity of decomposers and providing

Box 1.1 (*cont.*)

mineralised nutrients for the production process just at the time when they are needed – and almost without energy, nutrient or mineral losses.

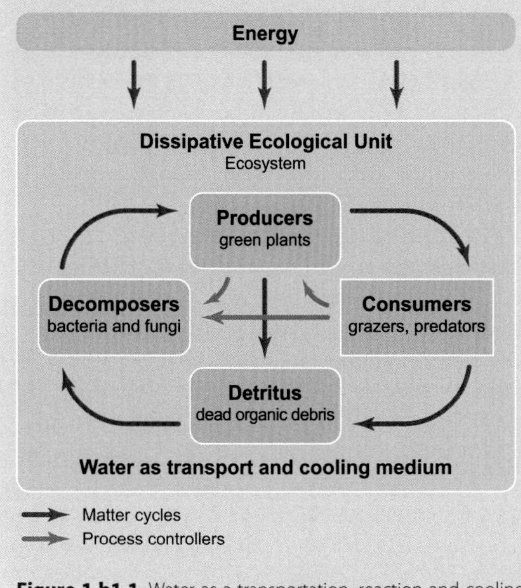

Figure 1.b1.1 Water as a transportation, reaction and cooling medium, and the Dissipative Ecological Unit (DEU) (Ripl and Hildmann, 2000).

Water is not only the bloodstream of the landscape – as has been stated insightfully over the decades – it is also the bloodstream of human societies and of the human enterprise on Earth. It is, of course, late in the day to be deepening and widening our understanding of the role of water in human development. We have reached a precarious situation for humanity, with large and growing demands for water in the world, occurring in an era of rapid global environmental change, which could trigger a serious undermining of freshwater availability in the future. This is a challenging predicament. In a situation where we face a tougher and tougher battle to secure freshwater for immediate human needs in an increasingly populated world, we now need to incorporate into our thinking, policy and practice the need to secure and be active stewards of freshwater flows to sustain ecosystem services and resilience outside the traditional water sector. We thus face the urgency of meeting rapidly rising social needs for water, while simultaneously safeguarding an increasingly large proportion of freshwater resources for ecosystem services and resilience. This includes freshwater to sustain biodiversity, carbon sequestration in soils and the ability of landscapes to buffer storm flows to avoid disastrous floods. Simply put, at a time when we may be running short of freshwater to sustain the traditional 'economic water sectors' – agriculture, industry and domestic needs – we are increasingly realising the need to secure water for ecosystems not for their preservation, but for our ability to prosper in the future.

1.1.2 Towards sustainable stewardship of freshwater on a planet with finite resources

Significant advances on several scientific fronts have contributed to the insight that we need to develop a new integrated paradigm on the governance and management of water resources. Several components of a new approach to water and development either already exist or are advancing rapidly, including, for example, the management of environmental water flow in aquatic ecosystems, the integration of climate change impacts on water resources at 'governing' scales (e.g. river basins) and advances in frameworks for managing land-based ecosystems to regulate water flows (e.g. managing forests and the spatial configuration of rivers and wetlands).

There have also been significant scientific advances across disciplines that converge towards a social–ecological approach to integrated land and water resource management from the local to the global scale. These include, for example, water resources research focused more actively on the role of evapotranspiration in generating ecosystem functions and services (Enfors, 2013; Jansson *et al.*, 1999; Liquete *et al.*, 2011; Rockström *et al.*, 1999), a growing recognition of water-induced thresholds (Gordon *et al.*, 2008), the role of freshwater in terrestrial ecosystems (Poff *et al.*, 1997; Poff and Zimmerman, 2009; Richter *et al.*, 1997), research on IWRM, broadened out to a stronger focus on land and water (de Vries and de Boer, 2010; Duda, 2003), a stronger integration of water and land management into Earth System science (Canadell *et al.*, 2007; Lambin and Geist, 2006) particularly on climate change and water (Alcamo and Henrichs, 2002; Bogardi *et al.*, 2012), a deeper focus on the role of water in complex system dynamics and tipping points in ecosystems (Scheffer *et al.*, 2001), advances in institutional research focused on the need for adaptive co-management across scales to address the complex and intertwined challenges of water, ecosystems and development (Pahl-Wostl *et al.*, 2012) and managing common pool resources and property rights (Cole and Ostrom, 2011).

Our task in this book is to integrate these strands in order to put forward a social–ecological resilience-based approach to the way we understand and approach water resources and human development in an era of rapid global change. To achieve this, water resource governance and management needs its own paradigm shift, away from a focus on securing freshwater supply and minimising the negative impacts on freshwater resources, to a broader perspective on freshwater in social–ecological systems for ecosystem services and resilience, in order to avoid undesirable feedback and promote adaptive capacity and desirable transformations.

The purpose is not only to develop a new approach to sustainable water use. It is also to establish the foundations for an approach to water governance and management which recognises that the expected increase in water turbulence in the world, triggered by abrupt shifts in water flows, and rising risks of water-related regional or global traumas, arising from shifts in the stability and functioning of the biosphere, cannot be addressed with truncated approaches to water.

We build this approach on the different strands of advances across the social and natural sciences related to water resources. This is not a blueprint. It should rather be seen as an effort to draw logical conclusions from the latest science on water resources and how water relates to global change, resilience, ecosystem services and development. From these, we propose a framework – to be further explored and advanced by scientists, in education and by water professionals – for how to proceed towards the sustainable stewardship of freshwater on a planet with finite resources, which has entered a new geological epoch – the Anthropocene, or the global phase of human pressure on Planet Earth.

1.2 Water in the era of the great acceleration of human enterprise

The water situation is becoming increasingly precarious. The number of hungry people in the world remains stubbornly high, approaching a staggering 1 billion (FAO, 2012a). The planet is more or less committed to another 2 billion inhabitants by 2050 (UN DESA, 2011). Just to meet the basic food requirements of the currently malnourished and the unavoidable growth in the world's population will require an increase in world food production of a staggering 50–70% (Godfray *et al.*, 2010; McIntyre *et al.*, 2009; Tilman *et al.*, 2011). Furthermore, the world is experiencing unprecedented levels of growth in the new global middle class, predicted to rise from less than 2 billion to over 4 billion in the next 30 years (Kharas, 2010), driven by developments in China, South Asia and Latin America in particular. This causes rapid shifts away from vegetable-based diets to water-greedy meat-dominated diets.

This increasing wealth and demand for food is potentially the most dramatic trend in terms of the impact on freshwater resources in the future. Food production – and particularly livestock production in terms of the amount of water input per unit calorie output – is by far the largest direct water-consuming sector in society. To produce – using current agricultural practice – an adequate diet for an adult requires in the order of 1300 m^3 of freshwater per person per year, which equates to 3.5 to 4 m^3 of freshwater per person per day, or 80 to 90% of total freshwater use per person, i.e. some five times higher than water for domestic and industrial purposes. It is estimated that annual world food production consumes 5100 km^3/year of freshwater (see Chapter 5 in this volume). Recent assessments of the additional water required by 2050 for all forms of food production to lift people out of hunger and feed the growing world population amount to 1500–4000 km^3 per year (McIntyre *et al.*, 2009). This on a planet where a significant number of the larger rivers in the world, such as the Colorado river, the Rio Grande and the Yellow river, run dry before they reach the ocean, due to the overuse of freshwater, primarily for irrigation (Molle and Wester, 2009).

Despite remarkable success stories in terms of the accelerated exploitation of freshwater on Earth through massive water-engineering feats –the world has over 6000 km^3 of storage capacity for water behind large dams – we know that approximately half the world population of ~7 billion people already faces various degrees of water scarcity, with 30% facing severe stress (Kummu *et al.*, 2010). As shown in Figure 1.1 various indicators of water infrastructure show an accelerated pace of expansion over the past 100 years. Despite this remarkable expansion, which has come at a very high cost for the environment, the world is far from keeping pace with growing human water shortages, and the overuse of runoff water is

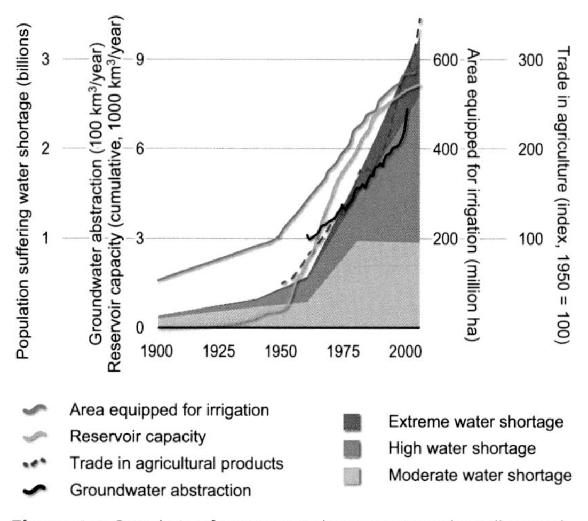

Figure 1.1 Population facing water shortage, using the Falkenmark index of average per capita blue water availability in each country. < 500 m^3/year means absolute or extreme water shortage; 500–1000 m^3/year per capita is severe or high water shortage; and 1000–1700 m^3/ year per capita is moderate water shortage. The lines provide several indicators of water infrastructure development in the world, ranging from reservoirs to abstraction of groundwater (adapted from Kummu *et al.*, 2010, based on Chao *et al.*, 2008; Freydank and Siebert, 2008; Wada *et al.*, 2010; WTO, 2012).

undermining the ability of aquatic freshwater ecosystems to fulfil crucial ecosystem functions and services (Millennium Ecosystem Assessment, 2005). This is an absolutely critical point. Furthermore, human-induced climate change undermines the availability of freshwater in many regions, and will increase the frequency and magnitude of extreme events such as droughts and floods (IPCC, 2007).

To these water-related drivers must now be added the interacting complexity of social and environmental systems on the local to the global scale, which directly influence the availability and use of freshwater. This requires a deeper integration of the human dimensions of water resource development, the role of freshwater in global change processes – ranging from climate change to biodiversity loss and global trade flows – and the role of water in building social and ecological resilience to various shocks and disturbances.

1.2.1 Humans as an integrated part of the Earth System

The availability of water at the river basin, regional and global levels has long been regarded as predefined – a given, predictable volume that oscillates within a relatively narrow natural variability, which can be estimated based on empirical measurements of runoff in major rivers. Moreover, this is thought to change only slowly, in incremental and therefore controllable ways. This is the basis of water resource governance and management. This water-centric, 'mono-scale' approach, always depicted by stable water resource data – as hydrographs of base flow and total runoff for the major river basins world – fails to recognise that the availability of water changes rapidly according to a broad set of social–ecological drivers. These range from rates of local land degradation or regional deforestation in rainforests, which affect moisture feedback and therefore rainfall patterns, to the impacts on water availability of climate change and air pollution.

A key message of this book is that humans are an integral part of the Earth System, and water availability for human needs is intimately connected to dynamic changes in the biophysical systems of the Earth. Water cannot therefore be dealt with in isolation from land management, climate mitigation, ecosystem stewardship, air pollution abatement and other key social and economic activity that exploits natural capital at different scales. Water determines the outcome of essentially all the processes on Earth that matter for human well-being, from food production to the temperature of the atmosphere. In turn, water is affected by essentially all the environmental processes influenced by humans, from loss of biodiversity to emissions of greenhouse gases.

This is what makes the governance and management of freshwater so challenging – well beyond the conventional, truncated approach to water resource management and development. Water is a determining factor behind the functioning, structure and stability of the biophysical systems on Earth, all of which are being put under tremendous stress by human beings. Water does this by regulating the climate system – as the most important greenhouse gas. It is a prerequisite for all biomass growth and therefore all living species on Earth, and also sets the pace of the global cycles of carbon, nitrogen and phosphorus, is an agent for transportation and a solvent of chemical compounds. At the same time, water delivers human well-being, beyond the usual focus on the conventional domestic, industrial and agricultural sectors of water supply, by generating ecosystem functions and services from terrestrial and aquatic water use and providing the basis for social–ecological resilience.

This means, paradoxically, that a book on integrated water resources logically starts outside of the hydrosphere, by focusing on the trends and dynamics, in relation to water, of the other systems on Earth, covering the biosphere, the atmosphere, the stratosphere and the cryosphere.

1.2.2 Acceleration of human enterprise since the 1950s

What makes the current global water situation so different from the past is that the Earth has entered a new geological era, the Anthropocene, in which humanity constitutes the major driving force of planetary change (Crutzen, 2002a, b; Crutzen and Steffen, 2003; Crutzen and Stoermer, 2000). There is evidence to suggest that humanity has become the driving force behind abrupt changes on a par with or even greater than (particularly in terms of pace) natural geophysical forces, such as the position of the planet in relation to the Sun, and geological events, such as earthquakes and volcanic eruptions. There is also evidence that we have reached a point of human-induced global ecological overshoot on a planet with finite resources (Ewing *et al.*, 2010; Loh *et al.*, 2005).

These two facts together completely change the agenda for sustainable development. The United Nations Conference on Environment and Development (UNCED) process in Rio in 1992 made tremendous progress in pursuit of sustainable development through the conceptual advances in 'Our Common Future' (Brundtland, 1987) and the development and publication of its implementation plan, Agenda 21, in 1992 saw the birth of the three leading United Nations conventions on the environment (the United Nations Framework Convention on Climate Change (UNFCCC), the United Nations Convention to Combat Desertification (UNCCD) and the Convention on Biodiversity). Nonetheless, the dominant way in which nations choose to address environmental problems remains embedded in a sectoral and largely local to regional approach, aimed at minimising environmental impacts while securing social and economic development. It is important to remember that the sectoral approach to sustainable development that evolved out of the UNCED process also had a big influence on the water resource agenda. Thus, framed by the UNCED process, the Dublin principles for IWRM were born in 1991 (International Conference on Water and the Environment, 1992). The IWRM was framed around a set of basic principles: water has an economic value and its competing uses should be recognised as an economic good; water should be recognised as a scarce and vulnerable resource; the need for participatory approaches to water resource management; and the key role played by women in water management. These focused on managing runoff for societal purposes, i.e. a heavily runoff or blue water-centric approach to the governance and management of water, and one moreover that is disconnected from cross-scale interactions.

Now the situation has changed. Major scientific advances since 1992 clearly show that we are not only in the Anthropocene era, but have reached a globalised phase of sustainable development in which the aggregate effects of all local uses of natural capital directly influence environment processes at the regional to global scales. Water is no exception. That this globalised phase of humanity affects our life-support systems – of which water is the most fundamental – was recognised in the run-up to the United Nations Rio+20 Earth Summit, which took place in Rio de Janeiro in June 2012, where the world gathered 20 years after the UNCED conference and 40 years after the Stockholm conference on environment and development. The UN Secretary-General, Ban Ki-Moon, established a Global Sustainability Panel, which delivered a pre-Rio+20 report, 'Resilient people, resilient planet: a future worth choosing', which was the 'Brundtland equivalent' of the 2012 summit. The report concluded that the current world development paradigm is not sustainable, and that the risk of unacceptable tipping points in the Earth System, due to human-induced global environmental change, needs to be recognised and integrated into our development paradigm. The Rio+20 summit decided to transform the Millennium Development Goals (MDGs) into a set of global Sustainable Development Goals (SDGs), with the aim of addressing social and economic development in the context of global sustainability.

The most important evidence of humanity having entered the Anthropocene epoch was published in 2004 in a synthesis of Earth System science by the International Geo-Biosphere Program (IBGP) (Steffen *et al.*, 2004). We refer specifically to this publication not only because it is such a scientific achievement, but also as a reminder of the timing. It is important to remember that it was only in 2004 that we were presented for the first time with an integrated synthesis of the state of the planet under anthropogenic pressures.

This integrated analysis shows a convincing trend of accelerated impacts on biophysical systems on Earth due to human pressures. It is not, as is often portrayed, only carbon dioxide emissions that have increased in an exponential way since the industrial revolution in the mid-eighteenth century. In fact, a broad range of the biophysical processes that constitute the basis for human development show the same classic 'hockey-stick' pattern of rise over the past century and millennium. Figure 1.2 shows how the trends for all the processes accelerate in the mid-1950s. This is now defined as the point of the great acceleration of human enterprise. It seems increasingly clear that 10 years after the end of World War II, there were enough people on the planet – a mere 3 billion – and a large enough proportion benefiting from the upscaling of industrialised society to create, for the first time in human history, an ecological imprint on Planet Earth. This is all the more remarkable because not only were we few compared to today, but only a minority of the inhabitants of the planet were the predominant source of the escalation in environmental problems. It was the rich minority, amounting to around 20% of the world population, who benefited from fossil-fuel-based industrialisation. There are empirical datasets covering essentially all the environment processes that generate human wealth, including emissions of greenhouse gases, interference with the global nutrient cycles of nitrogen and phosphorus, air pollution, deforestation, land degradation, overfishing and loss of biodiversity.

The social drivers generating this accelerated change, not surprisingly, show a similarly accelerated trend of change since the mid-1950s (Figure 1.2). The situation for water follows a similar pattern. The social driving forces behind pressures on finite freshwater resources are largely the same as for the Earth System as a whole, including population growth, the expansion of agriculture, consumption patterns, the globalisation of trade, and the growth of transport and energy use. The global drivers and observable impacts on systems directly related to water also show an accelerating negative trend over the past 60 years (Figure 1.3).

In sum, the operation of the global water cycle has also entered the Anthropocene era (Meybeck, 2003), and human pressures are now the dominant force determining the functioning and distribution of the global freshwater system. This includes both global-scale changes in river flows (Oki and Kanae, 2006; Shiklomanov, 1998; Shiklomanov, 2000) and shifts in vapour flows from land-use change (Gordon *et al.*, 2008). The challenge is to recognise the water implications of this human expropriation of natural capital from the Earth System. In natural capital we include both the non-living, abiotic, and the living, biotic, components of the Earth System, i.e. both finite natural resources, such as fossil energy sources, phosphorus, land and freshwater, and the living biosphere – all the living species in our landscapes and seascapes. The water implications of the human use of natural capital can be expressed under three broad headings, which are explored in-depth in this book:

- Human pressure on finite freshwater resources threatens the future ability to provision key ecosystem services, such as food, a biological gene pool, bioenergy and key ecosystem functions such as pollination and climate regulation.
- We are at risk of hitting hard-wired biophysical thresholds on a regional and planetary scale, which could induce abrupt changes in the functioning of the Earth System. Freshwater is at the heart of this concern.
- Freshwater is both a driver of change – through changes in freshwater flows and water availability affecting global change processes – and a victim of change, affected by global social–ecological change, e.g. through climate-change-induced shifts in rainfall patterns.

Freshwater determines the quality and quantity of all terrestrial and aquatic ecosystem services in human societies, and is therefore an, or even *the*, underlying determinant of social and economic growth.

Water also presents a strong social dynamic, where the impacts of human pressures on finite water resources are a reflection not only of an increasing global population, but also of a rapid increase in relative per capita water use, which in turn is a reflection of increased human wealth. This is demonstrated by the fact that freshwater withdrawals increased almost twice as fast as population growth, which increased exponentially, over the past 100 years (Lundqvist, 2000).

1.2.3 Water concern across a wider range of scales

The Anthropocene era, in which water is being impacted by a large number of global social and ecological change processes, while at the same time

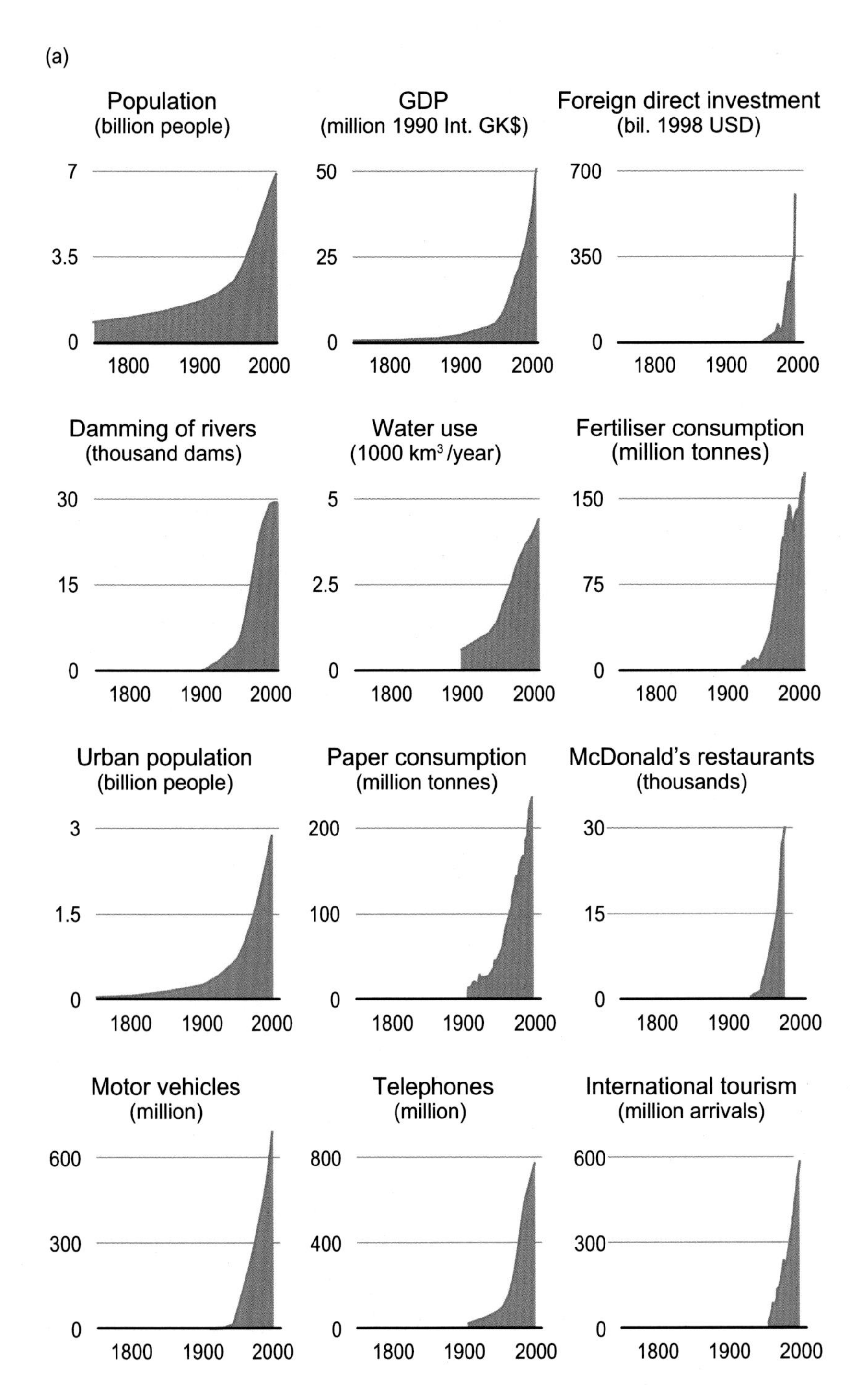

Figure 1.2 Processes affected by human change in the great acceleration of human enterprise. (a) The development of key social processes, such as population growth, water use, and consumption of various resources and goods. (b) The implications of the exponential rise in human pressures on the Earth System, for key environmental processes ranging from atmospheric concentration of greenhouse gases to loss of biodiversity (adapted from Steffen *et al.*, 2004).

(b)

Figure 1.2 *(cont.)*

CO_2 concentration (ppm)

N_2O concentration (ppb)

CH_4 concentration (ppb)

Ozone depletion (% loss of total ozone column)

N hemisphere avg temperature (°C anomaly)

Great floods (decadal flood freq)

Fisheries fully exploited (%)

Shrimp farm production (million MT)

Nitrogen flux (10^{12} mol/year)

Loss, tropical rainforest and woodland (% of 1700)

Domesticated land (% of total land area)

Biodiversity (thousand extinctions)

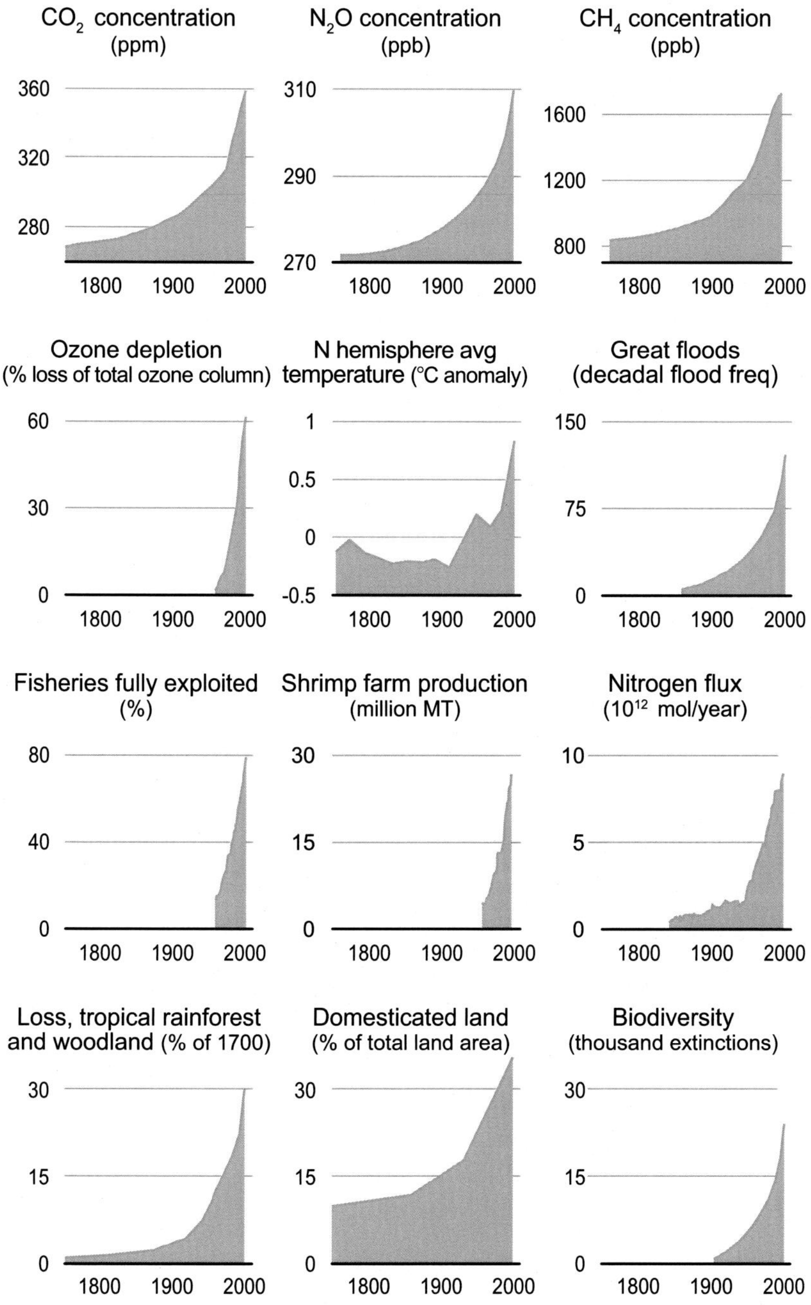

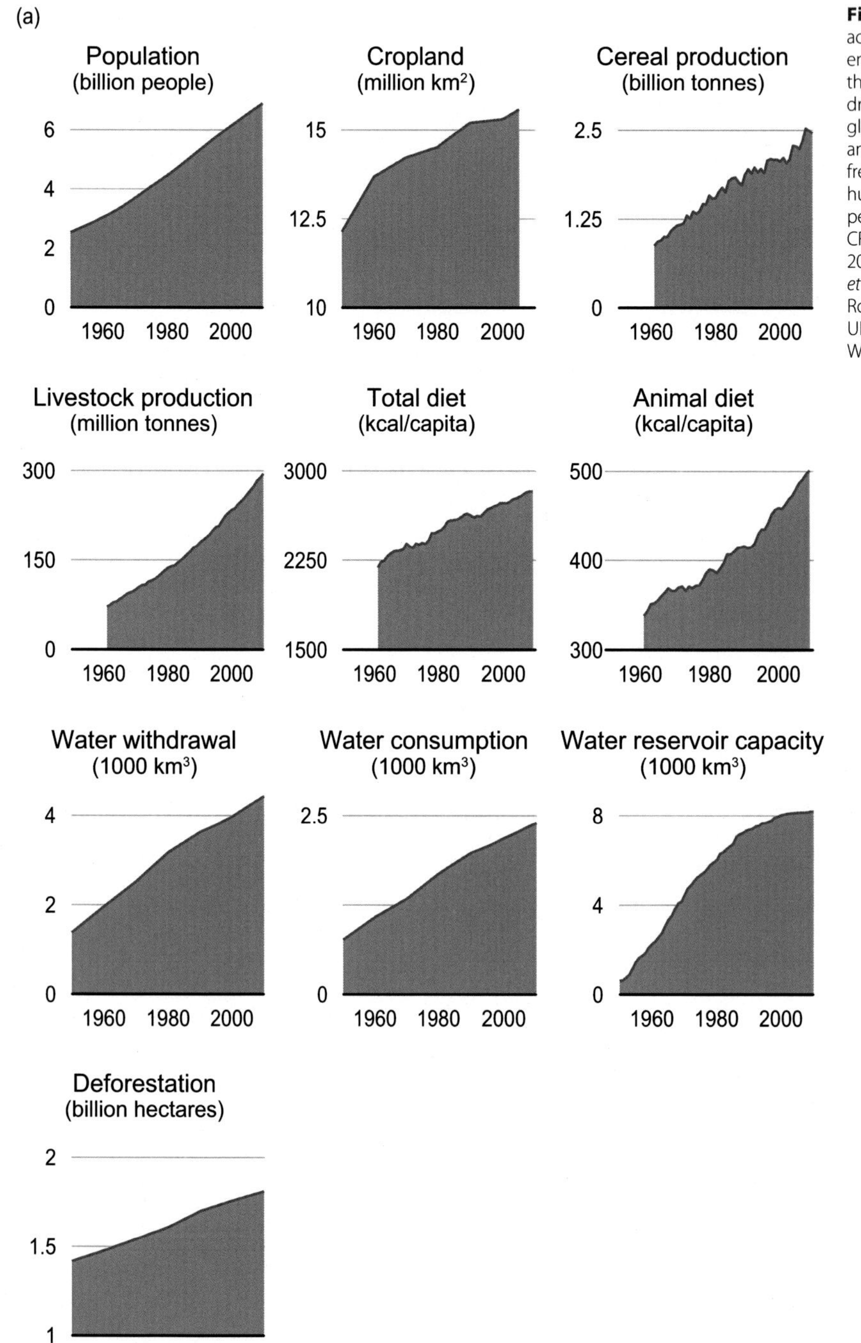

Figure 1.3 The great acceleration of the human water enterprise on Earth. Trend over the past 60 years for key global drivers with a direct impact on global freshwater resources (a), and the trend for indicators of freshwater impacts affecting human societies over the same period (b) (Chao *et al.*, 2008; CRED, 2012; Dai, 2011; FAO, 2012b, 2013; Klein Goldewijk *et al.*, 2011; Shiklomanov and Rodda, 2004; UNDESA, 2011; UNPD, 1999; Wada *et al.*, 2010; WWF, 2012).

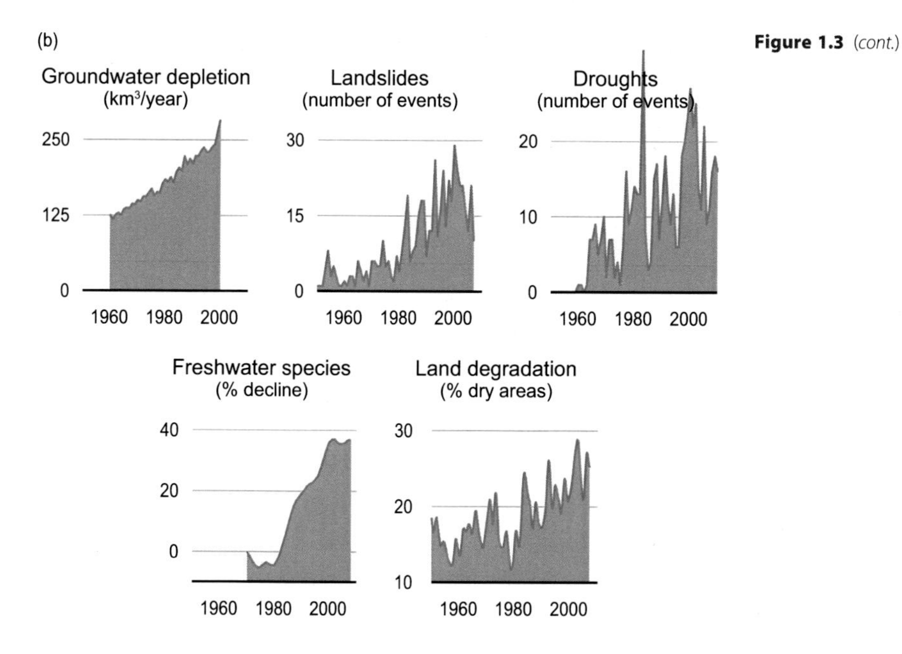

Figure 1.3 *(cont.)*

influencing biophysical change in many other systems on Earth, gives rise to a new level of water concern across a wider range of spatial scales. Our so far 'single-scale' focus on river basins will no longer suffice. We need to 'find more water' by focusing increasingly on the smaller field to catchment scale, and we need to understand the influence of larger scales, from the regional to the global scale.

Humans have altered water flows for millennia, changing the configuration of landscapes and affecting the local to regional scales. In the Anthropocene era, these are no longer the only scales of influence. Water alterations continue to occur at the local level – through changes in rainfall partitioning by management of freshwater resources in farmers' fields, to water resource alterations in small catchments and the larger river basin. The difference is that these alterations are occurring simultaneously everywhere across the planet and interacting with human-influenced systems at the global scale, such as the climate system, generating impacts beyond the local scale of water interventions. Our scale of water impacts has thus increased, where, e.g. water withdrawals, land-use change, deforestation, air pollution and human-induced climate change increasingly influence rainfall patterns at the regional scale, such as the functioning of the regional monsoon systems. In short, local water management can today be directly involved in threatening the functioning of large-scale biophysical systems, and local water resource management can no longer occur in isolation from global dynamics of change, affecting water availability at the local scale.

Water is not only involved in cross-scale interactions. It is also closely associated with changes in disturbance regimes and in generating surprise events, by inducing non-linear change, when systems abruptly and unexpectedly cross thresholds or tipping points, which can generate rapid and often irreversible change.

These four factors – how water interacts with other systems on Earth, how water interacts across spatial scales, the role of water in avoiding or inducing shocks by causing systems to cross or preventing them from crossing tipping points, and the core role of water in generating ecosystem services for human well-being – are at the heart of this book.

1.2.4 Regime shifts between multiple states

There is scientific evidence that ecosystems, ranging from local lakes (Carpenter, 2005) to regional systems such as rainfall regimes in the Sahel region

(Scheffer *et al.*, 2001), often have multiple stable states that are separated by thresholds. Lakes can shift in a non-linear fashion – often abruptly – between different stable states, e.g. from a clear, nutrient-poor oligotrophic and oxygen-rich state to a turbid, nutrient-rich eutrophied and anoxic state. Such a 'flip' occurs as a result of a long period of gradual environmental decline, which slowly erodes the ecological resilience of the system. The system, however, remains in its 'desired' state – or basin of attraction – in apparently 'good environmental health'. This gradual erosion of resilience makes it vulnerable to shocks and disturbances, and a trigger, such as a freak weather event that causes a large influx of nutrient-rich sediments or a period of abnormally low mixing of the water, pushes the system across a threshold. The system is then stuck in a new – often undesired – water state (or regime). If there is a feedback mechanism (i.e. factors that can dampen or reinforce a change process towards a new equilibrium in the function and structure of the system) involved that reinforces the new state (so-called positive feedback), this constitutes a state change, or regime shift. In the case of lakes, such positive feedbacks can be a shift in the relative importance of different phyto- and zooplankton, with an increase in species that reinforce – and maintain – the system in the new eutrophic state.

Similarly, larger systems, such as rainforests and savannah systems, appear to have distinctly separate stable states, where rainforests can flip between a wet forest state and a relatively drier savannah state (Oyama and Nobre, 2003), and where savannahs may flip from a relatively wet savannah and grass- and tree-dominated state to a much drier steppe-like bush-dominated state (Scheffer and Carpenter, 2003). The critical factor determining the existence of separate stable states and the risk of non-linear change is the existence of feedback mechanisms. For example, water withdrawals over decades from the Amu Darya and Syr Darya rivers in Central Asia gradually reduced the water level in the Aral Sea, but the regime shift from a huge inland sea to a dry, desertified state occurred when the drying process was reinforced once a dynamic feedback mechanism kicked in in the form of higher evaporative demand from the now shallower and warmer water-atmosphere layer.

Water is not always the control variable that determines whether a system will tip over a threshold, but it is often a critical factor influencing the risk of non-linear change. For the climate system, the risk of crossing tipping points, or tipping elements, is becoming a key focus of scientific attention (NAS, 2009), as it may be the defining factor guiding policy and practice on climate mitigation. If threshold behaviour in the climate system is a reality, climate mitigation must safeguard the planet from the risk of unacceptable, irreversible, flips that are catastrophic for humanity. The idea that we need to manage the unavoidable and avoid the unmanageable is given real, evidence-based teeth. If feedback mechanisms contribute to reducing or accelerating global warming, then this changes fundamentally the approach to stabilising the climate system at a 'safe' level. If water management is closely associated with our ability to avoid damaging climate thresholds, and if water presents its own set of tipping points, then this also changes our approach to water resource management.

We have, today, a pretty good picture of the subsystems on Earth that may reach potential tipping points, a critical threshold of concern which could result in abrupt, catastrophic shifts linked to continued human interference with the climate system (Figure 1.4).

Many of these are closely associated with freshwater, by being subject to climate-induced impacts and functioning as a reinforcing feedback mechanism – beyond the fact that water vapour is the dominant greenhouse gas and that gradual warming increases the intensity in the global hydrological cycle (Figure 1.5). The bistability of Sahelian vegetation, between a 'wetter' savannah state and a 'drier' steppe state, and the risk of climate-induced changes in the El Niño-Southern Oscillation (ENSO), which has major potential risks for regional rainfall patterns, are two such examples.

A major focus of this book is the direct role of water in inducing non-linear change, in order to understand where and in what ways water resource governance and management must evolve, given that such 'surprise' elements are currently not on the mainstream water resource management agenda.

It should be recognised at the outset, however, that – even without human interference – water is associated with a high degree of shocks and stresses on social–ecological systems caused by natural, often extreme, variability. Water is part of the natural disturbance regime of most terrestrial and aquatic ecosystems on Earth, influencing the structure and functioning of ecosystems through the frequency

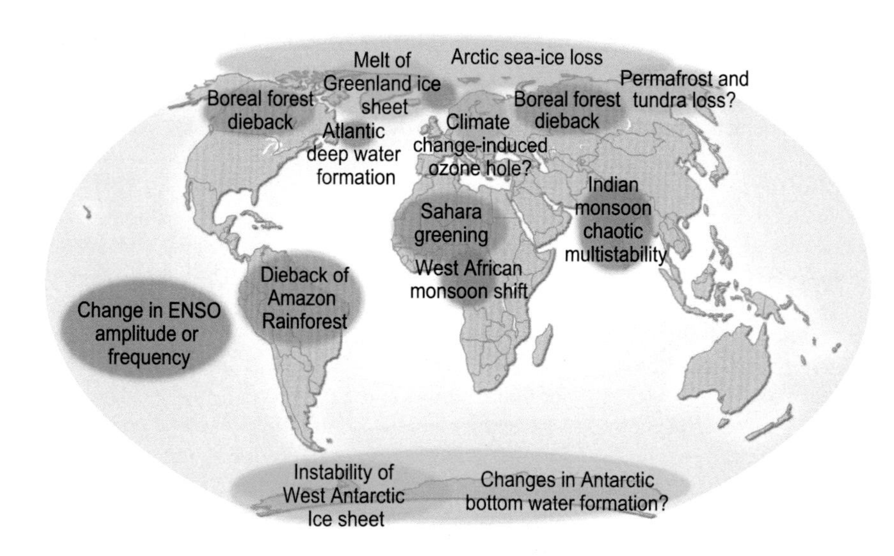

Figure 1.4 Potential tipping elements: systems at risk of abrupt, irreversible changes in the Earth System due to climate change (Lenton *et al.*, 2008).

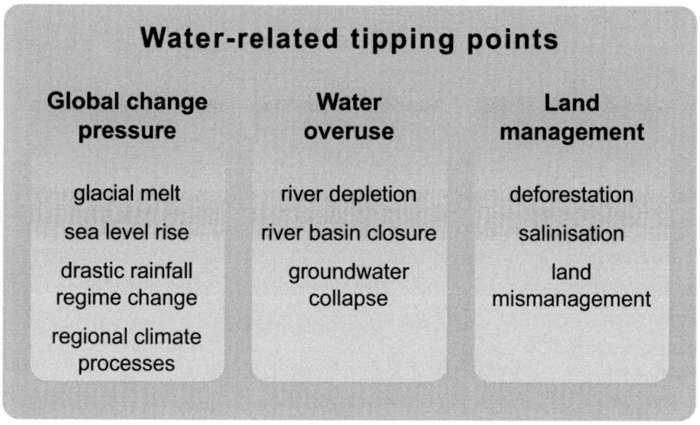

Water-related tipping points

Global change pressure	Water overuse	Land management
glacial melt	river depletion	deforestation
sea level rise	river basin closure	salinisation
drastic rainfall regime change	groundwater collapse	land mismanagement
regional climate processes		

Figure 1.5 Overview of water-related tipping points.

and amplitude of extreme flows of water, generating droughts, floods and dry spells, and shaping the characteristics of both social systems and ecosystems. The ability of an ecosystem to remain in a stable state under a natural disturbance regime is a core component of the resilience of a system, and is distinctly different from the dynamics of social–ecological systems going through a regime shift changing from one stable state to another. When farmers in the Sahel cope with dry spells using water harvesting systems, this is adaptive management for resilience at play, aimed at dealing with a natural disturbance regime – the high frequency of dry spells. This is distinctly different from a situation in which gradual desiccation linked to decades of land degradation from unsustainable land management, combined with consecutive years of low rainfall potentially reinforced by the impacts of anthropogenic climate change, cause a shift from farmland in parkland acacia savannah to desertified steppe-like conditions. Disturbance regimes change, however, and vulnerable systems are extremely sensitive to new, human-induced disturbance, for example, when overfished coastal areas, low in biodiversity and subject to sedimentation and nutrient overload from unsustainable land management, are hit by a new disease, which causes a functioning but vulnerable system to collapse.

A new approach to water resource governance and management must therefore incorporate not only cross-scale interactions (i.e. how the local community or catchment depends on and interacts with the hydrological cycle and social drivers on a regional, basin and world scale), but also the complexity of feedback mechanisms, multiple stable states and the risk of regime shifts in social–ecological systems due to the crossing of social or ecological thresholds or tipping points.

We use the terms tipping point and threshold interchangeably in this book to refer to the point, along a control variable, where the state of a social–ecological system, such as a forest or lake system, abruptly changes to a new form, e.g. a non-linear change in productivity in a farming system due to a climate change-induced shift in water availability. While tipping point and threshold denotes the position along a control variable when a state change occurs, we use the term regime shift to denote the shift in a system from one social–ecological state to another.

Apart from research on local freshwater lakes and tipping points related to changes in water quality, the risk of water-induced and water impacting regime shifts in complex social–ecological systems, such as agricultural or urban systems, remains largely unexplored. This is surprising given that water is so strongly affected by anthropogenic climate change, causing direct impacts, such as an increased frequency of droughts and floods, and feedback effects, for example, drying out rainforests and reducing carbon sequestration in soils that may trigger non-linear change.

Changes in the biophysical operations of the Earth System that affect water availability interact with social change in our societies. We live in an increasingly turbulent world, where social, financial, economic and increasingly ecological instabilities cause abrupt shocks, such as a financial crisis or a regional, social upheaval like the Arab Spring in 2010. Increasingly, we must address the challenge of interacting global dynamics, where social and ecological drivers of rapid change interact in unexpected ways with unexpected outcomes. The 2011 Arab Spring in North Africa, which led to revolutions in Tunisia, Egypt and Libya, and civil unrest in Bahrain, Syria and Yemen, was a major social eruption in response to decades of repression under dictatorial rule, enabled by the connectivity provided by the internet and mobile phones, and ignited by a tragic and symbolic event – the self-immolation of the vegetable seller, Mohamed Bouazizi, in Tunisia. It is likely, however, these social processes expanded to include the mass of the people because of rising food prices (Lagi *et al.*, 2012). The rapid spike in food prices, which more than doubled in 2008 and then again in 2010, was probably caused by a complex interaction between volatile oil prices (rising dramatically from ~90 USD/barrel in January 2008 to ~130 USD/barrel by end of 2008), rising phosphate prices (from ~100 USD/tonne in January 2008 to over 400 USD/tonne in the second half of 2008; Rimas and Fraser, 2010), the behaviour of speculators on the commodity markets, constraints on the world's grain markets due to extreme climate events (e.g. the grain export ban in Russia in 2010 owing to a fire-related reduction in yields; Rimas and Fraser, 2010), a more than decade-long drought in Australia followed by massive floods in 2010, and the biofuel policies of the United States and Europe (Lagi *et al.*, 2012). This is an illustration of socio-political factors interacting with resource constraints and ecological change, and propagating across the world in unexpected ways. The result is abrupt social change on a large scale as a result of interacting social and ecological processes that play out on a faster and larger scale. They may have a slow onset – creeping up over decades – but an abrupt impact.

1.2.5 The role of water in sustaining a desired state on Earth

The accelerating pressures on systems on Earth, and growing insights into the risk of non-linear change in local to regional systems, raises the question of whether we are at risk of threatening the stability of the Earth System. The planet has regularly moved in and out of glacial and interglacial phases in the course of the past one million years. Of particular interest for us is the past 200 000 years, which is the period of fully modern humans (Oppenheimer, 2004), with essentially the same human ability to develop civilisations as we know them. Figure 1.6 shows that for most of the past 100 000 years, during which Earth was in a glacial period, was an extremely bumpy ride for humanity in terms of the biophysical living conditions for human development on Earth. Based on ice-core data from Antarctica, it uses estimated air temperature fluctuations (in centigrade) as a proxy for the viability of living conditions on the planet. In a

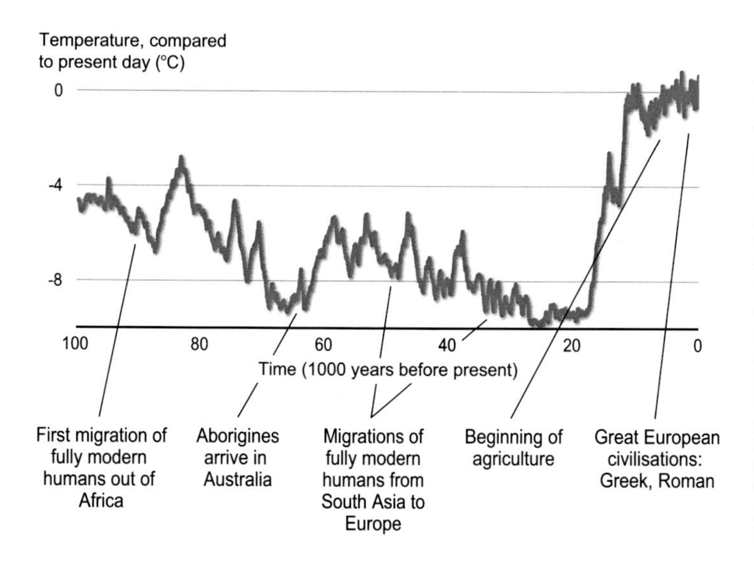

Figure 1.6 Temperature change in the past 100 000 years. For the largest part of the past 100 000 years humanity has experienced an extremely bumpy environmental journey, with extreme temperature fluctuations affecting freshwater availability and life conditions. Extreme cold periods, with large portions of the global freshwater cycle 'locked' as ice in expanding polar ice sheets, the few of us on Earth, living as hunters and gatherers, were driven to search for new sources of food, due to water scarcity and extreme hunger. A key point in the history of modern humans is when we, during an extreme cold, and thus dry, spell, about 75 000 years ago, migrated out of Africa. At that point, recent analyses indicate that we may have been down to approximately 15 000 fertile adults of modern humans on Earth. The figure shows data of the last glacial cycle of 18O (an indicator of temperature) with ice-core data from Greenland, combined with selected events in human history. The Holocene is the last 10 000 years. Adapted from Young and Steffen (2009) with temperature record from the EPICA Dome C time series (Jouzel *et al.*, 2007).

cold phase approximately 85 000 years ago, modern humans left Africa during a water crisis. Similarly, in another cold and dry period some 65 000 years ago, our ancestors reached Australia.

After the migration in waves of fully modern humans from South Asia to Europe 30–50 000 years ago, we entered the latest and current interglacial era, the Holocene. This has been a remarkably stable period, even if compared over a period of 1 million years. We were only a few thousand years into this extraordinarily stable period when we invented agriculture breaking with the hunting and gathering communities that gradually colonised the planet over 200 000 years. We domesticated animals and plants and embarked on a remarkably fast journey that led to our social history of building cultures and civilisations as we know them. From the Mesopotamian irrigation societies, to the sophisticated water reservoir management of the Maya civilisation, the Nile mastering Egyptians, the water engineers of Greek societies and the Roman empire, and the water mastering Chinese civilisations, all more or less agrarian hydraulic civilisations, we grew from a few million people to the 7 billion we are today.

It seems increasingly clear that the stable biophysical conditions provided in the Holocene era constituted and continue to constitute a precondition for the remarkable human evolution we have experienced. Water played a fundamental role in enabling this journey. All the empires and historic societies were more or less advanced irrigation societies. Without the Holocene stable state, there could be no stability in the availability of freshwater. Without stability in the supply of freshwater, there could be no agriculture. Without agriculture, there is no possibility of transforming societies from hunters and gatherers to farmers, and thus no modern societies and no exponential population growth, i.e. no civilisations as we know them today. The conclusion is that the Holocene state is our desired planetary state, the only state of the planet we know can support human societies and a world population that will soon approach nine billion people. The stability of freshwater flows in the Holocene period has played a fundamental role in enabling this human development.

Moreover, water plays a critical – but poorly understood – role in maintaining the Earth System in its current, desired Holocene state: by being a climate-regulating greenhouse gas (as vapour), by regulating energy fluxes by changing states from ice to liquid (transforming ice to water requires 334 kJ/kg ice, which is equivalent to the heat required to warm water to 80°C, which means that the current rate of ice mass loss on Greenland, ~400 Gt/year, 'consumes' ~130 PJ of energy) and by providing the basis for biomass growth and thereby regulating carbon fluxes and the flows of other greenhouse gases, such as methane and nitrous oxide.

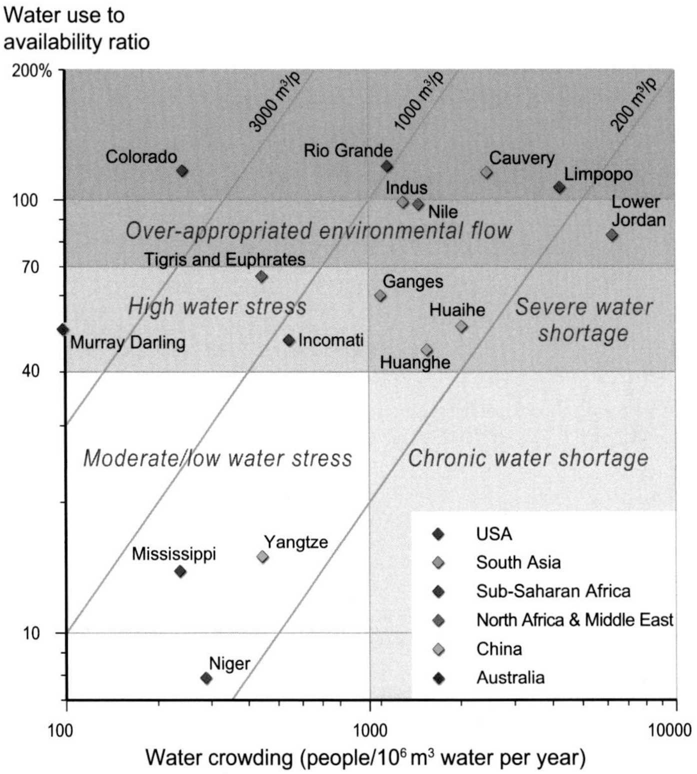

Figure 1.7 The human water scarcity quadrant for blue water use. Remaining degrees of freedom for increased consumptive use of blue water in a set of river basins. Diagram of the 'withdrawal-to-availability ratio' in per cent (vertical axis) and 'water crowding' in people per flow unit of 1 million m^3/year (horizontal axis). Diagonals show iso-lines of equivalent per capita water withdrawals. A 'rule of thumb' is that at a 70% withdrawal of blue water, basins are closed and no degree of freedom is left, due to the needs of aquatic ecosystems. Data collected by Stockholm International Water Institute from various sources.

1.2.6 Human dependence on water for life support

It goes without saying that as the bloodstream of landscapes and in the functioning of the Earth System, water is also the bloodstream of human societies. The clearest manifestation of the human dimension of water resources is our total dependence on water as our life support. This has so far focused primarily on the direct human need for blue water for irrigated agriculture and to supply water for domestic and industrial needs.

The most recognised index of blue water scarcity was developed by Malin Falkenmark in the mid-1980s (Falkenmark, 1986). The 'Falkenmark' water scarcity index shows, based on empirical evidence, that a country suffers from chronic water shortage (earlier 'absolute water scarcity') when blue water availability falls below 1000 m^3/capita per year, which corresponds to a water crowding of >1000 persons per million m^3 of available blue water. However, the degree of water scarcity or shortage experienced in a society depends on the level of water development, i.e. how large a proportion of the available water is in fact appropriated. Empirical evidence suggests that when this 'use-to-availability ratio' exceeds 40%, a society is approaching a higher level of water stress (Figure 1.7). Population growth and economic development pushes water-scarce societies toward the upper right-hand corner of this human water scarcity quadrant.

Population growth and economic development increase the pressure on water resources, particularly to meet the need for food and energy. There is a risk that by 2025, a majority of the world's population will be experiencing water stress, while 30–35% will be suffering from chronic water shortage (<1000 m^3 of blue water availability per capita per year) (Kummu *et al.*, 2010; Shiklomanov, 2003).

The growing incidence of water scarcity does not, however, capture the full complexity of the social–ecological human dependence on water. In a globalised world, virtual water is the largest (by volume and weight) traded natural commodity on Earth (approximately 2300 km^3/year of virtual water is traded, compared with < 10 km^3/year of oil). In 1998, it was estimated that 25% of world trade was driven by water scarcity (Postel, 1998) being resolved by importing food from regions with a relatively better endowment of water. International trade reduces global water use by an estimated 5% as a result of water-intensive food commodities being traded from countries with high levels of water productivity to countries with low (Hoekstra, 2010). Bioresources, including bioenergy, timber and other forest products is, together with food production, the largest conveyor of water across the world economy. Under some scenarios, the amount of freshwater required to meet global bioenergy needs by 2050 could be equivalent to current global water requirements to produce food (Berndes, 2002), which is the second biggest water-consuming economic sector in the world after forests. At the same time, the human demand for water changes with increases in human wealth. Increased meat consumption in more affluent societies could be the largest current driver of increased water use globally, given that producing a unit of meat calories consumes three to five times more freshwater than producing an equivalent unit of vegetable-based calories.

Like the biophysical dimensions of freshwater in the world, the human dimension of water use also interacts from the local to the national, regional and global scales. Obviously, the ecological and human dimensions of freshwater interact in what we define as the social–ecological dimensions of water, and play out in terms of actions and impacts on humans at the local level – the farm community, the local catchment, the wetland or the forest – that are inseparable from the rain-generating scale of the rainforest or monsoon system, or the behaviour-generating scale of the global market or regional and political stage.

This book widens the context of the human dimension of freshwater resources by integrating human dependency on water beyond direct blue water requirements to include the role of vapour fluxes, i.e. green water flows that sustain ecological functions and services beyond food and energy, and the role of water in providing social–ecological resilience.

We focus on the social and ecological dimensions of water not just to understand complex interlinkages, but primarily to develop strategies for improved governance and management. A particularly critical area of emphasis in this regard is to further concretise the urgent and fundamental part that water plays and will continue to play in human adaptation to unavoidable global change, ranging from climate and ecosystem change to shifts in lifestyles and transformations in the fabric of social globalisation.

Growing human pressures on finite freshwater resources, combined with new insights into the risk of abrupt and non-linear changes in the Earth System related to water, highlight the need to address new dimensions of water governance. In the Anthropocene era, water governance must address new social and ecological drivers, as well as cross-scale interactions, in which needs must be met locally while impacts are felt at all scales. Today, water governance must expand its agenda from its main focus on sustainable withdrawals and the allocation of water resources, to the challenges of managing shocks, uncertainty and growing risks, and helping to sustain rainfall. Governance must move away from focusing on the efficiency of water use and the optimisation of water allocation, to water governance that generates a multitude of ecosystem functions and services, and builds overall landscape resilience by maintaining an adequate degree of wetness in resilient landscapes.

1.3 Recognising the boundaries of water use

The planetary boundaries framework is a first attempt to define a safe operating space for humanity, which maintains the desirable social–ecological state of the planet provided by Holocene biophysical conditions. The stability of the global freshwater cycle is key to this analysis, and consumptive use of freshwater is proposed as one of nine key planetary boundaries for securing a safe future for humanity.

The empirical evidence of accelerated human pressures on biophysical systems on Earth, the growing understanding of the risk of non-linear change causing regime shifts when thresholds are crossed, and the realisation of the uniquely stable environmental conditions of the Holocene interglacial period – not least for water flows in the reliability of rainfall onset, runoff magnitudes and seasonality – force us to pose a new question: Are we at risk of threatening the stability of the Holocene state, the only state of the Earth that we know can support human development on a planet of soon to be 9 billion people? This question is particularly pertinent as science shows that we are currently exiting the Holocene period and entering a new 'planetary logic' of the Anthropocene era, where humanity constitutes a quasi-geological force of change at the planetary scale.

Water, as is noted above, is key to this question, both in its 'bloodstream' function, supporting and regulating functions in the biosphere and the atmosphere that determine the response of the Earth System to human-induced disturbance at the regional to global scales, and in its 'victim' function, being affected by changes in the other biophysical systems on Earth.

As we enter the Anthropocene era, freshwater becomes a social–ecological concern on a planetary scale. Evidence of this is emerging from several sources, ranging from the IPCC Fifth Assessment (2013) to the UN Millennium Ecosystem Assessment (2005), the World Water Assessment Report (2009) and the GEO 5 report (UNEP, 2012), in which water is shown to be under increasing human pressure but also key to the ability of the biosphere to buffer human disturbance and global environmental change.

1.3.1 A safe operating space for humanity in the Anthropocene era

These insights into the growing risks facing humanity in the Anthropocene era have resulted in new scientific endeavours to put forward a conceptual framework on planetary boundaries that integrates Earth System science and resilience science. This framework will enable the definition of a safe operating space for humanity on the planet in order to avoid unacceptable non-linear change in the life-support systems on Earth, and help sustain a desirable Holocene-like state on the planet.

This is a major human quest, based on the growing evidence which suggests that we are, through human actions, harming hard-wired processes at the planetary scale (Crutzen, 2002a; Steffen *et al.*, 2004, 2007). In the first application of the 'planetary boundaries' framework (Rockström *et al.*, 2009a, b), nine environmental processes were identified as being associated with Earth resilience, i.e. the stability and the ability of the Earth System and its critical biomes to provide human development in the long term.

The planetary boundaries approach, building on earlier attempts to address human development on a finite planet, ranging from Kenneth Boulding's 'Spaceship Earth' concept and the Club of Rome 'Limits to Growth' analyses in the early 1970s, to 'carrying capacity' approaches, 'ecological footprint' analyses, and 'guardrail' approaches, redefines the global agenda for human development. Instead of our current focus on predicting environmental trends in our efforts to meet human needs and reduce environmental impacts of different development trajectories, the planetary boundaries approach takes a 'back-casting' approach to human development, by first trying to identify the biophysical playing field, the safe operating space, within which human development must occur, in order to avoid, due to loss of resilience, the risk of large non-linear changes with deleterious or even catastrophic outcomes for human societies.

The framework includes the three systems associated with scientific evidence of large, regional to global thresholds: (1) the climate system, through climate change; (2) the stratospheric ozone layer, through emissions of ozone depleting substances; and (3) the oceans, through ocean acidification. The analysis also identified four 'slow' variables, large biophysical processes operating on a local to regional scale to provide the underlying resilience of the climate system, and which present local to regional-scale risks of thresholds in ecosystems and major biomes: (1) human interference with the global nitrogen and phosphorus cycles; (2) the rate of biodiversity loss; (3) land-use change, such as conversion of natural ecosystems to cropland; and (4) freshwater use.

Based on the latest science, a first attempt was made to identify control variables (key parameters directly determining whether a system may or may not cross a threshold, or proxy variables that are good indicators of when a system may go through a regime shift) for these seven planetary boundary processes, and to suggest quantifications for each. Quantifications were not proposed for the two remaining proposed planetary boundary processes – aerosol loading or air pollution by both cooling and warming agents that could influence the stability of regional weather and rainfall patterns; for example, the Asian brown cloud and chemical pollution such as the risk of affecting the human genome – due to

Planetary Boundaries

Boundary process and proposed boundary (uncertainty range)	Current position
Climate change < 350 ppm CO_2 (350–550) and < 1W/m^2	*387 ppm CO_2* *and 1.5 W/m^2*
Ocean acidification < 2.75 mean saturation state of aragonite in surface sea water (2.76–2.41)	2.90
Ozone depletion 276 Dobson units (276–261)	*283 DU*
Atmospheric aerosol loading *To be determined*	
Biogeochemical loading: Global N & P cycles < 35 Tg N/yr industrial fixation of N_2 25 % of natural fixation (35–49) < 11 Tg P/yr flowing into oceans (11–110)	*121 Tg/yr* *8.5-9.5 Tg/yr*
Global freshwater use < 4000 km^3/yr (4000–6000)	*2 600 km^3/yr*
Land system change < 15% of land under crops (15–20)	*11.7%*
Rate of biodiversity loss < 10 E/MSY (10–100)	*> 100 E/MSY*
Chemical pollution *To be determined*	

Figure 1.8 Nine proposed planetary boundary processes, including proposed control variables, boundary levels and uncertainty ranges (Rockström *et al.*, 2009a, b). A safe operating space for human development is provided 'within' the area established by the boundary levels defined by the best available science. The proposed boundary level is at the lower, most cautious, end of the current range of scientific uncertainty. Uncertainty ranges are provided in parentheses. The numbers on the right show the current estimated position for each quantified planetary boundary.

the difficulty in identifying an appropriate control variable.

Given the scientific difficulty of predicting physical thresholds for any system – not least due to interactions among boundary processes which in themselves may shift the position of thresholds – the planetary boundaries approach attempts to define an uncertainty zone around a threshold based on the latest published research. In order to create a safe operating space for each process, the planetary boundary position, i.e. the 'safe' level for each process, was placed at the lower end of the uncertainty zone. So, for example, for climate change, where the CO_2 concentration in parts per million was used as one of the control variables, the latest science indicates an uncertainty zone of 350–550 ppm (Hansen *et al.*, 2008; IPCC, 2007), within which there is a significant risk of large-scale, highly deleterious threshold effects such as the destabilisation of the Greenland ice sheet and the entire loss of the Artic summer ice. The planetary boundary is therefore set at 350 ppm in order to create a 'safe' space for human development (Figure 1.8).

1.3.2 Humanity approaching planetary boundaries

In our attempt to estimate planetary boundaries, early analysis indicated that humanity had already crossed three boundary processes – on climate change, interference with the nitrogen cycle and the rate of biodiversity loss (Rockström *et al.*, 2009a; see Figure 1.9). A more recent report on human interference with the phosphorus cycle indicates that humanity has also crossed the phosphorus boundary with respect to inducing tipping points in freshwater systems (Carpenter and Bennett, 2011). The original analysis only considered the risk of catastrophic anoxic events in the oceans caused by phosphorus overload from land. The later study shows that a planetary boundary for phosphorus is also needed for the role of phosphorus in causing regime shifts in freshwater systems on

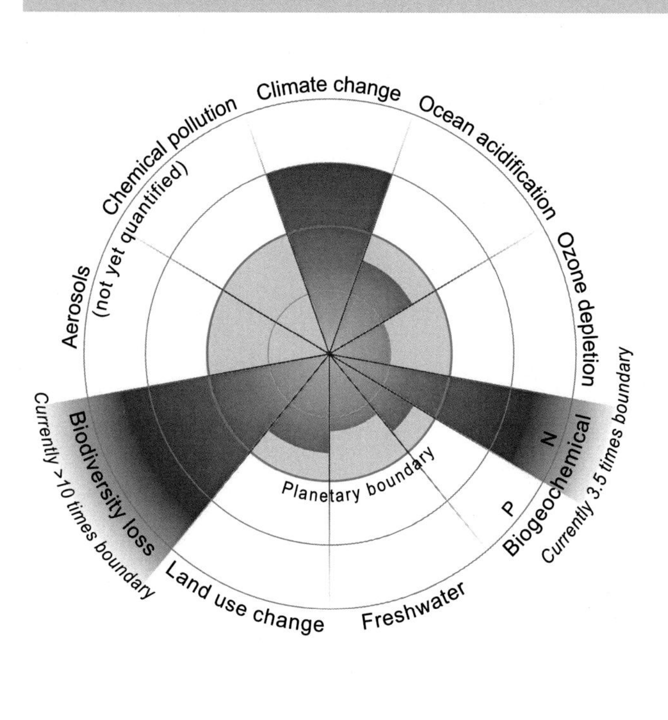

Figure 1.9 Current distance from the proposed planetary boundaries. The safe operating space, defined by the planetary boundary levels, is marked in green (adapted from Rockström *et al.*, 2009a). This estimate indicates that humanity has already reached the danger zone for the rate of biodiversity loss, climate change and interference with the global nitrogen cycle. There is still a certain degree of freedom for global consumptive use of blue water resources.

land. Before reaching the ocean, phosphorus overload causes major eutrophication of lakes and wetlands across the world, in many cases triggering abrupt shifts in ecosystem state. The boundary for phosphorus, like the boundaries for nitrogen, land, biodiversity and climate change, is thus closely intertwined with global freshwater resources.

The analysis also suggests that humanity may be rapidly approaching the upper safe level of conversion of natural ecosystems to cropland (12% of land currently under crop cultivation and a 'safe' boundary of a maximum of 15%). This 15% upper level for agricultural expansion in the biosphere was set to a large extent to protect ecosystem functions in non-agricultural ecosystems, in particular moisture feedback, ecological habitats and carbon sequestration functions in the world's remaining forests. For freshwater there remains, at least at the global level, a certain degree of freedom – with current consumptive blue water use amounting to ~2600 km^3/year, giving a remaining ~2400 km^3/year before evidence indicates that rivers will be emptied to such an extent that non-linear, irreversible changes to the functioning of ecosystems and water flows will follow. This number gives a false sense of 'sustainable' security, however, as most analyses indicate a very rapid increase in consumptive water use in the coming decades, and thereby a risk of rapidly approaching the boundary.

Furthermore, there is a need to expand the water boundary to the scale of regions and river basins by defining evidence-based thresholds for environmental water flows in order to identify the 'local' water boundaries above which there is a risk of crossing water-induced thresholds.

1.3.3 Water is closely interrelated with seven of the nine boundaries

Planetary boundary processes interact, influencing the position of thresholds and thereby the position of a safe boundary level. Water is closely interrelated with the boundaries on climate change, interference with the nitrogen and phosphorus cycles, rate of biodiversity loss, chemical pollution, aerosol loading and land-use change. Moreover, as is stated above, water both affects and is affected by other planetary boundaries. The relation between land, water and climate is particularly important in this respect. Figure 1.10 shows that ~25% of man-made emissions of CO_2 are absorbed by terrestrial ecosystems, primarily by forests, other biomass and soils. All this carbon sequestration is enabled through water consumption during the process of photosynthesis and subsequent biomass growth. Water, carbon and soils therefore interact closely, influencing the stability of the climate system on the planet. Furthermore, the

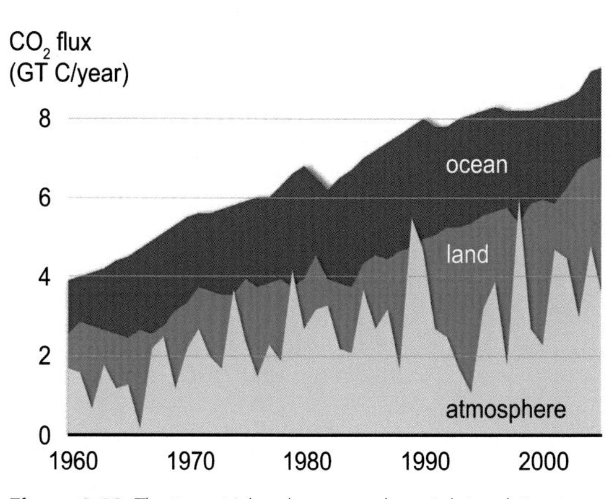

Figure 1.10 The terrestrial and ocean carbon sink in relation to the increase in CO_2 emissions, 1960–2005. Adapted from Canadell *et al.* (2007).

abrupt changes, up and down, in net annual carbon accumulation in the atmosphere, shown in Figure 1.10 (in light blue), are a result of the annual global variability in regional weather patterns, which alter the amount of freshwater flows. Major net releases of CO_2 into the atmosphere from the biosphere, shown as high spikes of CO_2 accumulation in the atmosphere, occur in years of low biomass accumulation in terrestrial ecosystems, as a result of water shocks from droughts and floods affecting biomass growth and triggering forest fires and failed vegetation growth.

Despite our understanding of the interactions among planetary boundary processes, and the close relationship between freshwater and essentially all processes except the stratospheric ozone layer (water vapour plays an important role in the stratosphere but not directly in terms of the ozone layer) (Solomon *et al.*, 2010), how these interactions operate is poorly understood. There is only limited understanding of the implications of transgressing one of the planetary boundaries for the other boundaries. There is evidence to suggest that an overshoot in one boundary process may affect the position of others, and that, for example, transgressing boundaries on climate change, nutrient loading and land use could constrain further the safe level of water consumption in river basins – as more water may have to be 'left alone' in order to provide the redundancy required to cope with a higher frequency and strength of climate-eutrophication and land-degradation-related shocks and disturbance to regimes.

We are generally on solid ground in focusing our water-related concerns on the availability and use of freshwater for human societies, and to a lesser extent for ecosystems. This central focus on water governance and management generally assumes that changes in water flows, while clearly associated with environmental impacts, are not associated with the risk of larger scale regime shifts that could have deleterious or even catastrophic, largely irreversible, impacts on societies on a regional to global scale. Now that the global hydrological cycle has entered the Anthropocene era (Meybeck, 2003), however, with unsustainable and growing human pressures on essentially all the major water systems on Earth, the question arises whether the finite global water cycle is associated with a planetary boundary for water use, beyond which we can anticipate a high risk of crossing dangerous thresholds with large-scale negative implications for the world. Evidence now indicates that the answer to this question is yes, which is one key reason why it is so important to address cross-scale interactions in water resource governance and management, be they on a catchment or river basin scale.

1.3.4 The need to consider a sustainable global level of water use

The global water cycle operates in myriad local to regional water balances, in small catchments and large river basins. This can be perceived as an argument against the existence or relevance of a global water boundary. Three arguments support the need to consider a 'sustainable' global level of water use.

First, the, now well established, knowledge that freshwater functions as the bloodstream for the entire biosphere, sustaining all life on Earth, the configuration of ecosystems and critical functions in the Earth System, such as regulating the climate system and the large cycles of, for example, carbon (as a means for carbon sequestration) and nitrogen (as a carrier of chemical compounds).

Second, the determinant role in the regional to global hydrological cycle, of the aggregate flows of runoff and vapour. A new 'agenda item' for water resource managers in the Anthropocene era is the need to manage to 'sustain rainfall' not only in local catchments, but also on a regional scale. Sustaining an adequate level of moisture feedback from landscapes, through green water flows from soil, plants and trees, is a critical criterion for resilience and the ability of all social–ecological systems on land to provide ecosystem services and human well-being.

The *third* argument is the very scale of human pressures on the Earth System in the Anthropocene era. Over millennia, and in particular over the past 150 years, humanity has gradually increased its pressure on finite freshwater resources, thereby generating pockets of major negative environmental impacts. There is evidence to show that these pressures, combined with natural variabilities in climate and hydrology, today and in the past, have triggered regime shifts on a local to regional scale. As is noted elsewhere in this volume, examples range from tipping points in lakes due to freshwater eutrophication (Carpenter, 2005), to regional-scale tipping points in lake systems due to excessive withdrawals of freshwater as, for example, in the Aral Sea system, to flips between dry and wet bistable hydro-climatic states in the Sahelian savannah (Scheffer and Carpenter, 2003). These may not in themselves generate ecological or social concern at the regional or global scale. In the past, even if major irrigation-based empires collapsed, such as the grand Mesopotamian societies along the Euphrates and Tigris rivers, due at least in part to the transgression of sustainable boundaries for water-induced salt intrusion into soils, this did not cause permanent damage at the larger scale. In the Anthropocene era, however, the situation is different. Now there is reason to be concerned about a new risk panorama, where local collapses of social–ecological water systems (from lakes to wetlands and forest systems) may occur simultaneously across the world, creating on aggregate an instability in, for example, the cumulative feedback of moisture, or the cumulative sequestration of carbon in the biosphere, which in turn could have major repercussions for, for example, regional monsoon systems or the climate system as a whole – in particular if interacting with other drivers of change, such as emissions of greenhouse gases.

The conclusion from these three strands of evidence on the role of water in global sustainability and determining the risk of crossing thresholds of particular concern for societies is that water constitutes a planetary boundary. A green and blue water resource approach was therefore adopted in an analysis of a proposed planetary water boundary definition (Falkenmark and Rockström, 2004). Two major categories were identified of water-related catastrophic thresholds that humanity needs to steer away from:

- *green water thresholds* related to a risk that too large a manipulation of global vapour flows from the land will threaten rainfall generation; and
- *blue water thresholds* related to a risk that excessive extraction from river flows will undermine freshwater ecosystems.

Based on the evidence of the genuine risk of such green and blue water thresholds (see Box 1.2), and the fact that blue water availability is a reflection of green water partitioning and use, the planetary boundary definition uses a blue water parameter, river depletion, as a proxy to reflect the full complexity of the highest risk for global water thresholds.

The proposed planetary boundary range for water is set at 4000–6000 km^3/year of blue water consumptive use. This constitutes a danger zone and a range that should not be transgressed, as it would take us too close to the risk of the blue and green water thresholds that could generate catastrophic outcomes. The actual planetary boundary position is set at the lower end of this uncertainty zone, i.e. a boundary of consumptive blue water use of 4000 km^3/year (for details see Box 1.2), as a way of applying the precautionary principle in line with the planetary boundary conceptual framework. Today, the consumptive use of blue water amounts to approximately 2600 km^3/year, which indicates a certain degree of freedom to increase consumptive water use at the global scale. A large number of river basins are already hydrologically closed, however, or closing rapidly, in terms of sustainable boundaries for water use (Molden *et al.*, 2007). Furthermore, as is shown in Box 1.2, estimates of the increase in consumptive water use in the next 40 years, just to meet the rapidly growing demand for food and the growing need for water for bioenergy and carbon sequestration, will rapidly push the world towards the planetary water boundary.

Finally, as is discussed further below, freshwater is not only a finite planetary resource affected by global environmental change. It also functions as a control variable for several other planetary boundaries, such as the regulation of water vapour feedbacks and organic carbon feedbacks in the climate system. The insight that water is a planetary boundary, i.e. at the heart not only of providing immediate human well-being through water-related ecosystem services, but also of regulating and balancing the stability of the biophysical systems on Earth, is an important additional dimension to water resource governance and management. It forces us to address the role of water resources in the Earth's resilience. Two key areas now emerge as strategic for sustainable water resource use: (1) the need to focus on sustaining rainfall on a landscape to river basin to global scale; and (2) the need to maintain a sustainable level of green and blue water wetness in landscapes, in order to maintain resilience and the ability of landscapes to continue to deliver both direct ecosystem services and wider regulating functions in the Earth System.

In operational terms, defining a global boundary for freshwater could guide international collaboration on freshwater governance and management, but will require down-scaling to regions, river basins and catchments, where actual policy and practice on water resource management occurs. This book addresses these challenges.

Box 1.2 Water: a planetary boundary

Johan Rockström, Stockholm Resilience Centre at Stockholm University

The global water challenge

The global hydrological cycle sustains life on the planet and provides humanity with freshwater for ecosystem goods (i.e. all biomass production that generates food, fibre, fuel and terrestrial biodiversity, as well as habitats for aquatic species) and services (e.g. carbon sinks and climate regulation) as well as water for domestic and industrial uses. Human pressure is now the dominating driving force determining changes in the function and distribution of the global freshwater system, threatening biological diversity and ecological functions such as carbon sequestration and climate regulation at the regional and global levels.

Recent analysis indicates that freshwater used to produce food – by far the largest freshwater consuming economic sector – will have to increase by 2000–4000 km^3/year by 2050 from the current ~7000 km^3/year. The range is due to different assumptions in water productivity improvements.

Freshwater is a finite planetary resource and functions as a control variable for several other planetary boundaries, such as the regulation of water vapour feedbacks and organic carbon feedbacks in the climate system.

The planetary boundary for water is defined by the partitioning of rainfall into green and blue water flows. The flows and stocks of freshwater in the global hydrological cycle are determined by the blue water flows – the liquid water flowing through landscapes as surface runoff and base flow in rivers, groundwater and lakes – and green water flows, i.e. vapour flow by evaporation and transpiration back to the atmosphere. Blue water flows sustain aquatic ecosystems, irrigate agriculture and supply water to humans. Green water flows sustain terrestrial ecosystem services (rainfed food, forests, grazing lands and grasslands) and regulate the terrestrial moisture feedback that sustains the bulk of the rainfall over land areas on Earth.

Green and blue water flows are intricately linked: increased green water flows reduce blue water availability, and decreased green water flows reduce rainfall and thus, in the long term, rainfall patterns and blue water generation. Furthermore, all consumptive use of water on the planet is in the form of green water flows, either directly from green water resources (infiltrated rainfall forming soil moisture in the root zone), or from evaporated blue water (from lakes or reservoirs and irrigation schemes).

Catastrophic water-related threats

The crossing of water-induced thresholds, causing abrupt changes with social–ecological implications on a regional to global scale, may occur as a result of aggregate sub-system impacts at the local (e.g. river basin) or regional (e.g. monsoon system) scale. A global water threshold might be crossed if multiple sub-system thresholds were crossed simultaneously in many places on Earth, potentially resulting in planetary effects on Earth System processes, such as a loss of carbon sequestration capacity or a triggering of climate system changes due to cumulative changes in vapour concentrations in the atmosphere. This may also result in global social impacts, such as a collapse of food markets, famine and the creation of environmental refugees, due to regional or continental reductions in agricultural yields caused by changes in rainfall patterns and/or local water availability.

There is growing evidence of the crossing of local and regional thresholds, caused by agriculture-related changes in water quality and quantity (Gordon *et al.*, 2008; see Chapter 3). It is particularly relevant to analyse water thresholds in agricultural systems, as crop and livestock production, together with forestry, are the economic sectors with the largest freshwater consumption in the world.

Two human driving forces could threaten the stability of the quantitative flows in the global

Box 1.2 (*cont.*)

freshwater system: (1) human-induced shifts in *green water* flows as a result of changes in precipitation totals and patterns, and soil moisture generation; and (2) human withdrawals of *blue water* affecting river flow dynamics.

The threats caused by changing green water flows can be related to either: (1) the drying out of landscapes, due to land degradation and desertification (changes in soil moisture generation or green water resource); or (2) moisture feedback from green water flow changes causing shifts in precipitation (large shifts in green water flow from land-use change, mainly related to deforestation). Both green threshold parameters (soil moisture availability and green water consumption) are linked to land-use change, affect the stability of the freshwater system by shifting the balance between vapour flows and runoff, and subsequently lead to river depletion.

The highest risks to crossing green water-induced thresholds that could cause regional to global impacts are concentrated in two regional hotspots: (1) the rainforests concentrated in Latin America, Central Africa and South East Asia, where changes in green water flows might alter, e.g. regional monsoon patterns; and (2) the savannah regions of the world, i.e. the dry semi-arid to dry sub-humid regions, which are host to approximately 40% of global terrestrial ecosystems and provide a high level of ecosystem services in biologically productive and diverse landscapes where water is the primary limiting biological growth factor.

Based on this reasoning, the following green water-related threats have been identified:

- collapse of biological sub-systems as a result of regional drying processes, due to the existence of alternate stable wet and dry states, such as a wet and dry Sahel, and the risk of an irreversible 'savannisation' of the Amazon rainforest, as a result of reductions in green water flows changing moisture feedback processes;
- regional desertification, where the green water resource declines below a critical threshold due to changes in rainfall patterns and land degradation;
- collapse in rainfed agricultural systems due to reductions in green water availability.

Blue water flows in rivers are determined by the amount of blue water generated from precipitation and runoff flows (as is outlined above) and the amount of blue water withdrawal and use. Beyond a certain level of blue water withdrawal and use, river depletion leads to a whole set of threats, ecological as well as social:

- collapse of *riverine ecosystems* for which stream flow is the habitat. Smakhtin *et al.* (2004) have suggested that at least 25% of the river flow should be safeguarded as 'environmental flow';
- collapse of *estuary ecosystems*, leading to an ecosystem tipping point, replacing freshwater ecosystems with brackish water ecosystems.
- collapse of *coastal ecosystems* when the river input decreases;
- collapse of *internal lakes and their ecosystems*, the Aral Sea being the classical example. As the shrinking lake ceases to influence the climate, positive feedback may further speed up evaporation. Decreasing inflow from the basin is at least a partial cause of several cases of lake water level decrease (Lake Chapala, Lake Malawi, Caspian Sea, Dead Sea, Lake Victoria and Lake Chad). Arid zone lakes are particularly vulnerable;
- collapse of *social irrigation-based systems* as demonstrated by several early irrigation-based civilisations.

Defining a global water boundary: rate of river depletion

At a global level, precipitation amounts to ~110 000–115 000 km^3/year with a variation of ±15–25% and an average runoff estimated at 42 500 km^3/year (with a range of 39 700–42 800 km^3/year) (Falkenmark and Rockström, 2004).

Green water

Total green water availability is ~70 000 km^3/year, and estimates indicate that approximately 90% of the global vapour flows from land surfaces today already contribute to sustaining terrestrial biomes, including regulating functions such as carbon sequestration in soil and forests. Global food production, which is the cause of the largest direct human manipulation of the freshwater cycle, consumes in the order of 7000 km^3/year originating from river runoff in irrigated agriculture (~2000 km^3/year) and soil moisture in rainfed agriculture (~5000 km^3/year).

To avoid savannisation, green water generation (soil moisture from infiltrated rainfall) must be at

Box 1.2 *(cont.)*

least 900 mm/year in tropical forests. Similarly, analyses show that green water generation must reach a threshold of 300–500 mm/year in order to generate blue water flows, and in hot tropical regions productive grasslands and tree savannahs only occur when green water resources reach in the order of 300 mm or above. This is the threshold, above which we start to experience sedentary rural societies practising agriculture. Therefore, below this threshold, agro-pastoral and nomadic social societies have evolved, due to water scarcity.

Blue water

Estimates indicate that 12 500 km^3/year of river runoff is potentially available for human appropriation, the rest of global runoff being constrained by its remote location or storm flow.

Global withdrawals of runoff water amount to approximately 4000 km^3/year, of which approximately 2600 km^3/year is for consumptive use. This has resulted in a severe deterioration in aquatic habitats and water shortages for downstream water-dependent social and ecological systems. An estimated 25% of the world's rivers run dry during the year due to river depletion. There is a risk that a majority of the world's population will experience water shortage by 2025, and 30–35% will suffer chronic water shortage (i.e. <1000 m^3 of blue water availability per capita and year), indicating dramatic human pressures on the global freshwater system.

Constraints on the GWS have primarily been analysed from a human water shortage perspective and more recently from an environmental water flow perspective. Experience shows that chronic water shortage is experienced when per capita availability falls below 1000 m^3/year per capita. Empirical evidence shows that when withdrawals of runoff water exceed 40% of available blue water resources, regions experience high water stress, which at a global level corresponds to withdrawals exceeding 5000 km^3/year (Ericson *et al.*, 2006).

De Fraiture and Wichelns (2010) estimate the utilisable blue water resource at ~15 000 km^3/year (a global average of 36% of the total renewable water resource of ~42 500 km^3/year), and that physical water scarcity is reached when withdrawals of water exceed 60% of the utilisable resource. This estimate indicates that water withdrawals exceeding ~6000 km^3/year is a threshold above which physical water scarcity is reached. The green water boundaries discussed above occur 'upstream' of and are interlinked with river depletion. Therefore, river depletion in the form of consumptive blue water use is chosen as a proxy for the full complexity of the highest risk for global water thresholds.

Based on the global assessment outlined above of a sustainable level of consumptive blue water use, withdrawals of 4 000–6 000 km^3/year constitute a danger zone and a range that should not be crossed, as it takes us too close to the risk of blue and green water-induced catastrophic thresholds. Furthermore, in line with the planetary boundaries conceptual framework, the boundary is set at the lower end of the uncertainty range – at 4000 km^3/year.

It may seem that this leaves a large degree of freedom for humanity, given that current consumptive water use is around 2600 km^3/year. However, this boundary is likely to be further constrained by future increases in freshwater withdrawals for irrigation and industry, and the fact that freshwater is a major prerequisite for attaining the climate boundary of 350 ppm atmospheric CO_2, and is strongly affected by climate change. Moreover, the suggested boundary for river depletion of 4000 km^3/year assumes no aggregate green water impacts on precipitation totals (i.e. moisture feedback effects) and no deterioration in precipitation levels due to climate change.

This is obviously optimistic for a number of reasons. First, projections for increases in river depletion (consumptive water use for irrigation, industry and water supply) are in the order of 1000 km^3/year until 2050. Second, estimates of carbon sequestration requirements in terrestrial ecosystems for climate mitigation indicate an increase in green water consumption of ~2000 km^3/year by 2050 in order to contribute 1–2 Gt C/year (Rockström *et al.*, 2009b). Assuming that only 50% of this increase in green water use will cause river runoff reductions because moisture feedback compensates the rest through increases in precipitation levels gives an estimated 1000 km^3/year. The 'committed' water consumption thus rises to approximately 4600 km^3/year in 2050, which suggests that we have less than 10% of available freshwater left until we reach the planetary water boundary. This leaves little freedom for additional blue water consumption to meet future food and bioenergy demands.

1.4 The role of water in social–ecological resilience

A starting point in advancing our understanding of water resilience is to clearly define what we mean by resilience and the role water plays in determining and building resilience in social–ecological systems.

We define three key features of resilience: (1) its 'persistence', i.e. the degree of disturbance (including long periods of slow change, e.g. gradual increases in water scarcity, and abrupt water shocks, such as droughts and floods) a system can be subject to without changing state or structure, or the ability to remain in a basin of attraction after a disturbance; (2) its 'adaptability', i.e. the ability of a system to adapt, self-organise and learn while remaining in the same state; and (3) its 'transformability', i.e. the ability of a system to transform into a new state after a crisis or shock that pushes the system away from its original stable state.

This broadens slightly the way resilience is generally defined (Carpenter *et al.*, 2001; Walker *et al.*, 2004; Walker and Salt, 2006), by adding the ability to transform after crossing a threshold, or being locked in a trap, to the 'core' features of resilience – the ability to persist (cope with disturbance) and adapt (navigate change, self-organise and learn). This is in line with recent discussions on the critical features of social–ecological resilience (Folke *et al.*, 2010).

In sum, resilience incorporates both stability (or robustness) and flexibility (or the capacity to adapt and transform). The definition of resilience used in this book incorporates:

- the amount of change a social–ecological system can undergo and still remain in a basin of attraction or state, i.e. still retain the same controls on function and structure;
- the degree to which the system is capable of self-organisation;
- the ability to build and increase the capacity for learning and adaptation;
- the ability to transform into a new desired state after a crisis that pushes the system out of an original state or basin of attraction.

Resilience is a critical component of, but not a substitute for, sustainability. For example, resilience can be good or bad depending on how desirable a stable state is – a degraded land in a water-scarce state in an agricultural landscape or a corrupt regime in a country can both be very resilient but are not sustainable. Resilience provides a lens to understand social–ecological interactions and feedbacks and the occurrence of thresholds and regime shifts.

The implications of resilience are profound, as has been shown in scientific and empirical evidence. Systems, from lakes and savannahs, to households and the financial system, have varying degrees of ability to remain in a desired or undesired state – resilience – but can flip into another, often several different, alternate stable states when resilience is lost or changes.

These regime shifts are associated with the crossing of one or several thresholds, and the change is non-linear, often abrupt and – depending on the timescale considered – irreversible. Importantly, non-linear shifts due to the erosion of resilience are associated with: (1) a gradual erosion of slow control variables, undermining resilience; (2) one or several shocks triggering a regime shift; and (3) feedbacks (positive or negative) which determine whether a shift turns into a state change, which changes the functioning or structure of the system. A state change can only occur if new feedbacks are set in motion. For example, when a lake flips from an oligotrophic to a eutrophic state due to a slow increase in nitrogen (N) loading (the slow control variable), the new eutrophic state is made stable by positive feedbacks (e.g. the change in phyto/animal plankton ratios in favour of N-fixating algae that reinforces the eutrophied state), which locks the system into its new stability domain. These shifts can be associated with a wide array of different non-linear process dynamics from non-linear thresholds to hysteresis effects. A threshold involves an abrupt non-linear shift in the state of a system as a result of a relatively small change in the condition of the system (i.e. pushing the system slightly further into a eutrophied or water-scarce situation, which may cause, due to triggers such as disease or an extreme weather event, an abrupt change or threshold effect). Hysteresis arises when the location – along a control variable – where a threshold occurs is path dependent. For example, loading nitrogen and phosphorus into a lake may trigger a threshold cross from a clear to a murky state of the lake at a relatively high nutrient load (the system has a good resilience to 'cope' with nutrient load). However, once it has 'flipped' and is stuck in its new anoxic and turbid state, getting it back to the original clear water state may require pushing back nutrient loads to much lower levels than the point at which the original threshold crossing occurred. This 'added cost' of return to an original state is caused by the lock-in effects that a system tends to adopt once it gets stuck in a new state (in the case of a lake, the knock-on effects related to algae growth and the consumption of oxygen in the lake).

Resilience, a sub-component of sustainable development, is of fundamental importance to the ability of social–ecological systems to provide human well-being and development. The importance of resilience is increased by the risks of an increase in turbulence and therefore the frequency and amplitude of shocks and disturbance as a result of rapid local to global change. Water shocks and disturbances, such as droughts, floods and dry spells, and slow changes, such as sea level rise, melting glaciers and changes in rainfall patterns, are among the most immediate and severe factors associated with current global environmental change.

Human development is intertwined with the biosphere, and multiple and interacting social–ecological processes determine development outcomes. An integrated approach to resilience requires a profound understanding of how human societies interact with the biosphere, and how the human world interplays with the Earth. Governance systems, economics, politics and power relations, as well as social networks, knowledge systems and human relations, play fundamental roles in influencing, together with ecological and environmental factors, the overall resilience of a household, society, nation or business.

1.4.1 The role of water resources in the resilience of social–ecological systems

Water plays a profound role in the resilience of social–ecological systems, in its function as the bloodstream of the biosphere – as the fundamental prerequisite for all life on Earth, as a cause of scarcity and shocks and as an agent of human prosperity. It is often stated, with a high degree of accuracy, that if climate mitigation is about gases, climate adaptation is about water.

Water resource research and therefore water resource governance, management and policy, have only considered this role to a limited extent. This is surprising, given that water quality has been at the core of resilience science for decades, for example in studies of regime shifts in freshwater lakes due to overfishing and eutrophication (Carpenter, 2003). Increasingly, however, science is providing disturbing evidence of regime shifts driven by changes in water resource availability, due either to climatic change (e.g. historic regime shifts in the Sahel from a wet to dry states; Scheffer *et al.*, 2001) or to land-use change (e.g. water-related regime shifts in agricultural systems; Gordon *et al.*, 2008).

These growing insights into the risk of non-linear, profound, rapid and often irreversible shifts in the state of water resources for regions or local systems – i.e. the role of resilience in water resource availability and use – have so far only partially penetrated thinking and practice on water resources. This book is an attempt to advance thinking on water resource governance and management for resilience in social–ecological systems.

Defining the role of water in resilience is not an easy task, as water is involved in all the processes that determine the resilience of a system: (1) as a control variable of social and ecological resilience; (2) as a state variable, how water is affected by other control variables; and (3) as a driving variable behind the resilience of other systems, for example, a farming system or an urban area, and of the water system at different times, i.e. future rainfall. Water is thus implicated in all facets of resilience, and is tightly coupled to both social and ecological resilience from the local to the global scales.

We distinguish between three roles of water resilience, together understood in this book as social–ecological resilience:

1. water as a direct control variable or, in other words, a 'source' of resilience;
2. water as a state variable of resilience or, in other words, how water is affected by other drivers of change;
3. water as a driving force behind critical changes that determine overall resilience or, in other words, how water influences the stability and adaptability of societies, biomes and the Earth System to deal with abrupt changes.

The predominant focus so far on water and resilience has been on water as a state variable, i.e. the impacts on water stocks and flows from changes in, for example, land use. Moreover, this focus on impacts and water as a 'victim' of change has been on blue water flows, such as changes in runoff dynamics, with only a limited focus on green water dynamics.

1.4.2 Water as a control variable of resilience

This book widens the analysis by emphasising the 'upstream' role of water as a control variable. We analyse and quantify the role of water in generating ecosystem functions and services, both in terms of the regulation of flows and the consumptive use of green water to generate, for example, biodiversity and food. Water is a key determinant of the quality and quantity of ecosystem functions and services, which in turn is fundamental in determining social–ecological resilience.

The contribution of this book, in the context of the productive role of water in ecosystems, is therefore a dual one, by advancing our understanding of two generally ignored dimensions of water for sustainability and development: (1) the role of water in generating terrestrial ecosystem functions and services, essentially producing biomass; and (2) the role of water in generating social–ecological resilience in the process of producing ecosystem functions and services.

A key question posed in this book is whether changes in water resource management – in both green and blue water flows – can generate non-linear change, or threshold effects, that in turn can cause regime shifts in the affected systems. There is little evidence of such threshold effects in terrestrial ecosystems, or in the role of water in regime shifts in social systems (e.g. changes in water use as a factor behind crises, undesired states such as traps or drastic changes in rural societies), although there is an extensive literature on the possible role of water in armed conflicts (Wolf, 2003).

Whether water resource management can generate state changes in agricultural landscapes, i.e. whether water-induced resilience or changes in resilience can trigger regime shifts, is a key question posed in this book. Because, if water matters for resilience then, given the growing evidence of turbulent environmental times ahead, we need to profoundly change the way we think about and conduct water resource governance and management. For example, if the way landscapes are configured – the types of land use – and the way water resources are managed in such landscapes can change the resilience of the landscape and even contribute to inducing abrupt and non-linear regime shifts, then the risk landscape also changes significantly, compared to the current emphasis on the allocation of water resources for different purposes and the optimisation of water use, while assuming a more or less permanently stable state in which change occurs gradually and in a predictable – and therefore controllable – way.

Water as a control variable of resilience is one of the most under-researched areas of the interaction between water and resilience. This book explores the current – albeit immature – state of knowledge on the role of water as a 'source' of resilience, with a particular focus on water for the generation of ecosystem functions and services in terrestrial agro-ecological landscapes.

1.4.3 Water as a state variable of resilience

A number of human activities and environmental processes determine the impact, or state, of water resources and water flows, from the local field or ecosystem scale, to the catchment, river basin or regional to global scales. The dominant drivers of water impacts are land-use change, chemical and nutrient pollution from industry and human societies, human extraction of water resources for agriculture, industrial and domestic purposes and global environmental change, such as the impacts on the hydrological cycle of anthropogenic climate change.

These processes, of water 'as a victim' of change and as a finite resource subject to increasing scarcity, vulnerabilities and increasing variability in space and over time, have been and are widely studied and operationalised in water resource management (e.g. in disaster risk assessments, early-warning systems and drought and flood management strategies). A key question raised in this book is: how far can the state of water be changed – under pressure from other drivers – without triggering threshold effects? Or, in other words, when do we face risks of abrupt changes in the state of water resources – from stable runoff inflows in rivers and the predictable onset of rainy seasons, to shifts in water quality in lakes and urban water supply systems?

1.4.4 Water as a driving variable of future change

Water is also a driving variable of future change. Water is a renewable resource, but it is generally – and correctly – considered to be finite in quantity (at least over decades) at the regional to global scales. However, in the Anthropocene era, the pace and scale of interacting driving forces are rapidly rising, with natural and anthropogenic changes from the local to the global scales affecting the stability of freshwater availability in its annual iterations in the hydrological cycle.

Water is a driving force regulating the stability of the Earth System through its multiple regulatory functions – as a greenhouse gas, a solvent of materials, an erosion agent and an energy carrier. The key factor of focus here, however, is on the role of water in regulating moisture feedback from land and water surfaces, which in turn determines precipitation levels, rates and distribution in future iterations of the hydrological cycle. Water is thus not only affected by change, but also affects its own future change. As is shown in Chapter 3 in particular, land and water use in terrestrial ecosystems play a very important role in determining future rainfall levels 'downwind' as a result of moisture feedback. For example, estimates indicate that over 80% of rainfall levels in central and northern China originate from moisture feedback from, in particular, forests in neighbouring Eurasian countries (Van der Ent and Savenije, 2011).

The role of water as a driving variable of future water resource availability is a highly complex, scale- and time-dependent factor. The current state of knowledge, which is touched on throughout this book, highlights the importance of considering moisture feedback from the catchment to the river basin and regional scales, in water resource governance and management, as a way of minimising the risks of changes in future water availability. This complexity is enhanced by the fact that local to regional changes in land and water management (e.g. through changes in agricultural land use and deforestation patterns) interact with global change processes, such as emissions of black carbon from burning, and warming from greenhouse-gas emissions. The risks of abrupt shifts in the global hydrological cycle are accentuated by new scientific evidence that the world is most likely already committed to a global average warming of at least 2°C compared to pre-industrial levels (Ramanathan and Feng, 2008). The significant risk of major shifts in water flows in large regional systems, such as the Asian and African monsoons, and from major biomes, such as the inland glaciers, polar regions and rainforests, forces a rapid shift in attention to the importance of land and water resource management to avoid undesirable moisture feedback changes – a core component of building resilience in intensively used landscapes.

In sum, strategies for water resilience should address the three interlinked roles that water plays in human development: as a control variable of change, as a key 'state' variable determining the sustainability of an ecosystem or urban area and as a driver of regional to global change. A first key priority area is to better understand upstream–downstream trade-offs between green consumptive use upstream and blue environmental water flows and societal water supply downstream, and how these 'control' the overall resilience of social–ecological landscapes. The second priority area is to improve our ability to address water impacts and trade-offs, i.e. the state of water, as a result of land-use change (e.g. shifts in rainfall partitioning when cutting down forest to form cropland). The third area of concern is to put forward strategies for a multi-level governance of water resources to ensure stable moisture feedback and thus a stable water supply for ecosystems and societies.

1.5 Water resilience

This book integrates three areas of scientific advances in areas at the heart of water resource governance and management: (1) an integrated approach to water resources research; (2) the integration of Earth System science with water and human development; and (3) integrating water, ecosystem services and resilience issues in order to better understand the functioning, and advance strategies for improved stewardship, of social–ecological systems.

Two key messages support the third of these areas, which runs as a thread throughout this book:

- Water plays a fundamental role in the functioning of the biosphere and the Earth System at large. Human development and well-being originate from ecosystem functions and services generated from the biosphere, which in turn contribute to the social–ecological resilience that provides human societies with the capacity to deal with change and to continue to develop.
- A resilience lens provides an opportunity to understand critical interactions, assess the risks of crossing thresholds leading to regime shifts and explore strategies to adapt to change and transform in the face of crises. Our particular interest is in advancing our understanding of the role played by freshwater resources, at different scales, in the resilience of social–ecological systems

to the risks of crossing thresholds due to the overuse of freshwater, and in strategies for building resilience.

We define resilience as the ability of a social–ecological system to deal with change while continuing to develop. As is shown in section 1.4, where we present our approach to exploring the relation between water and resilience, there are three overarching features of resilience: (1) the ability of systems to persist in a given state, or how much disturbance a system can withstand without crossing a threshold and changing state or structure; (2) the capacity to adapt, within a given state, to stress and disturbance; and (3) the ability to transform into a new (desired) state after a crisis. We distinguish between general and specific resilience in relation to water resource management, as this provides an important conceptual and operational difference in understanding and in choice of management strategy. Specific resilience defines the ability of a particular part of a system, related to a specific control variable, to cope with disturbance or shocks. It thus puts the emphasis on understanding and avoiding non-linear changes in a particular part of a social–ecological system when pushing a critical control variable too far. Examples of this are, for example, if withdrawals of river water exceed a point where environmental water flows become so low that critical habitats for fish disappear, resulting in a collapse of fish stocks, and generating a potential domino effect of feedbacks that could cause a regime shift in the entire ecosystem – as food webs change. Here, the resilience of the specific part of the system – the aquatic ecosystem – is lost, but this does not tell us what has happened to the social–ecological resilience of the system at large (e.g. the river basin and the communities hosted there). Specific resilience always answers the question 'resilience of what to what?', i.e. 'of what' – on which part of the system are we focusing our resilience attention (e.g. aquatic habitats); 'to what' – which control variable is driving change (e.g. withdrawals of water). Specific resilience is therefore our focus when we try to understand the risk of water-induced tipping points, leading to regime shifts from one stable state to another.

General resilience, on the other hand, defines the resilience of an entire system to all kinds of shocks, and is thus a measure of the general 'strength' and 'adaptability' of a social–ecological system subject to change and disturbances. Specific resilience is often possible (or at least less difficult) to quantify and analytically explore, while general resilience is more complex and often impossible to measure analytically. This distinction is important when applying a resilience framework, as specific resilience enables a more targeted assessment and operational strategy for managing water in ways that avoid pushing a part of the system too far – management for avoidance of unwanted shocks and outcomes. General resilience, on the other hand, is a useful framework in more practical terms, an additional 'lens' in the endeavour of governing and managing for sustainable development, placing the emphasis on capacities to deal with change, stress and shocks while continuing to develop in a desired direction.

The role of water in building resilience is poorly understood and at the heart of our exploration in this book. Our focus is on the role of water in the resilience of social–ecological systems in an era of rapid global change. Our shorthand for this is the term 'water resilience', which should not be interpreted as the resilience of water, as our focus is the reverse, i.e. the role water plays in the resilience of ecosystems and societies.

A starting point for this book is our conclusion that water resilience is poorly understood. The emphasis on water and resilience has so far been on the role of water quality in crossing thresholds when resilience is lost in local ecosystems, such as lakes and wetlands. Stronger emphasis is needed on cross-scale interactions, water resilience at the basin and regional scales, water-related feedbacks and the role of water in ecological resilience at the planetary scale, i.e. the ability of the Earth System to remain in the desired Holocene state.

1.5.1 Facing changes in space and time: addressing efficiency versus redundancy in water resilience

An equally fundamental starting point for this book is our working hypothesis that water plays a critical role in supporting both specific and general resilience in social–ecological systems. This, of course, includes the 'ecological core', i.e. that water plays a fundamental role in the resilience of local ecosystems, maintaining lakes, forests, wetlands, grasslands, etc., in their desired states, as well as river basins and processes at the regional and global scales (moisture feedbacks, freshwater melting dynamics in glaciers, etc.). However, our emphasis in this book is on the role of water in the resilience of connected social–ecological systems, where we focus strongly on the world's most water-dependent system – agriculture. This includes the role of water in 'social' dimensions of resilience, such as food security.

The ecological dimension of water resilience is about not only incremental changes in freshwater quality and quantity, but also maintaining the required freshwater dynamics to avoid regime shifts, in which changes in water conditions might trigger the crossing of thresholds resulting in non-linear, often irreversible, state changes in ecosystems and societies. There is strong evidence to support this hypothesis, as is demonstrated in this book, but at the same time there is growing evidence that the reverse also holds true. Global change threatens the ability to maintain water systems, such as the Himalayan glaciers which provide safe base flow of freshwater for over 1 billion people who are dependent on the current 'desired' water state of the glaciers – which is the only state we know that is conducive to human development in the southern and south-eastern regions of Asia that are fed with water from the Himalayan plateau (ranging from Pakistan and India through the Ganges, to China and the Mekong region through the Mekong river).

The social dimension of water resilience addresses the broad complex from the local management of water to the institutional dimensions of water governance and the operation of water-related sectors, such as security, health, trade and finance. A core question we repeatedly return to on the role of water in social–ecological resilience is the question of efficiency and optimisation versus redundancy in building water resilience. A critical factor in this question is to link spatial variability with temporal change, for example, when sudden social (e.g. market incentives for biofuels) and biophysical (e.g. droughts) pulses and shocks occur. Much of our water governance focuses on optimising water resource allocation and use, which is often a cost-effective way of using water. One question we raise is whether societies subject to increased social and ecological turbulence in a globalised, increasingly crowded world in the Anthropocene era, will have to invest more in social and ecological redundancy, such as insurance systems and multifunctional landscapes, which can be less efficient, as a means of generating not only human well-being but also prosperity that is resilient in the face of change.

Water-related threats to resilience play out at different scales. While changes in water quality and water quantity may erode resilience and induce regime shifts in local systems, such as lakes and catchments, an increasingly important issue is the role of water in maintaining resilience at the global, Earth System scale due to growing global pressures on freshwater resources. This issue is closely connected to the question of the timescales involved, where loss of resilience at the local, field scale or the regional, catchment scale may be associated with shifts in states at a shorter time scale of decades or even years, while loss of resilience at the regional to global levels may be associated with shifts occuring at the time scale of centuries to millennia (see Figure 3.2 in Chapter 3).

1.5.2 Water resilience and surprise feedbacks: linking development, global change and agriculture

Feedback mechanisms that help to push systems across thresholds rarely operate in a simple cause–effect manner within one social sector or environmental process. For example, it is not necessarily true that land degradation alone drives desertification. Instead, a gradual degradation of land, drying out landscapes locally, in combination with regional climate change, generates periods of increased frequency of droughts and floods, which contributes to further dry out the landscape. Together with an intensification of land use and changes in cropping systems due to demographic trends, generating higher social demands for water use, this pushes the farming landscape over the desertification threshold, triggering a social threshold, for example, a crisis, after years of adaptation efforts – generating a social migration pattern of a new magnitude with young people in the farming community leaving en masse.

It is, furthermore, increasingly clear that these social–ecological interactions, related primarily to 'management', interact also with local-to-global social drivers of change, such as political shifts in regulatory and economic policies (e.g. withdrawing subsidies for fertiliser use in extremely nutrient-poor agricultural landscapes, which accelerates the loss of biomass), shifts in international markets (e.g. shocks in world food markets) and geopolitical trends in a globalised world (e.g. investment in African land by emerging economies in Asia). Local land and water management practices interact with social drivers related to wealth, markets and lifestyles and cross-scale drivers related to regional and global climate change.

Complex and integrated social–ecological feedbacks connected to water are increasingly common on a crowded planet in a globalised world that is subject to rapid global environmental change and reduced ecological resilience at the local landscape scale. These interacting factors affect the ability to deal with change.

Our focus is on the interactions between human development, global environmental change and water resilience, with a particular emphasis on agricultural landscapes (Figure 1.11). Our emphasis on agricultural landscapes is explained by the dominant role – at the local to global scales – of agriculture in all aspects of water resource governance and management. Agriculture is the largest human activity, transforming land areas on Earth. At present, ~40% of the land surface on the planet is under some form of agriculture, including both pasture and cropland. This is by far the largest anthropogenic land use, compared, for example, with urban regions, which cover only 0.24% of the terrestrial area of the Earth (ESA, 2010). Agriculture is thus the largest land-use factor influencing rainfall

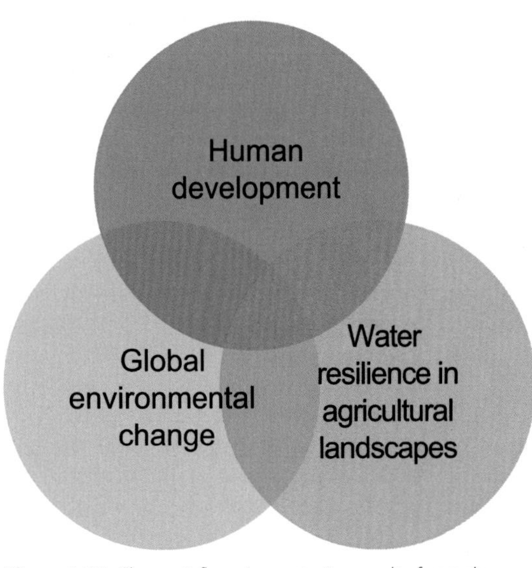

Figure 1.11 Change influencing water is a result of complex interactions between human development, global environmental change and changes in resilience in social–ecological systems. Our focus is on how these interactions play out and can be dealt with in agricultural landscapes, where social–ecological processes related to water are most intensive.

partitioning and thereby the flux of water through landscapes and return flows to the atmosphere from land. Food production is the world's largest water-consuming economic sector, even though forests consume more water. Agriculture represents the largest 'shaper' of landscapes on the planet.

Improving our understanding of how water interacts with both social and ecological feedbacks is of critical importance to future resilience, in terms of persistence, adaptive capacity and capacity to transform, in the face of global change. We define these interconnected processes as 'integrated social–ecological water feedbacks', where water interacts with other social and ecological drivers of change and thereby generates reinforcing or dampening feedbacks that affect human societies.

This book explores integrated social–ecological water feedbacks, their occurrence and relevance. Examples include changes in water use related to the rapid pace of urbanisation, and changes in consumption patterns and lifestyles in the rapidly growing economies of the world. Another example is how oil prices and food market prices interact to influence land conversions, thereby shifting the way water flows through landscapes and where water is consumed – affecting the degree of water scarcity at the local scale. A concrete example here is the recent trend for stormflow to reduce landscape features (e.g. wetlands) in Poland and Germany as a result of the European Union bioenergy directive, which provides incentives to increase the conversion of the natural ecosystems that buffer runoff flows into cropland to meet the European Union (EU) bioenergy target for 2020 (one of the '20–20–20' targets set in 2007). This may be in conflict with the European water framework directive on protecting water bodies and reducing sediment and nutrient influx to coastal areas, as well as the EU Marine Strategy Framework Directive.

A further concrete example is how climate change interacts with land-use change, affecting the regeneration of rainforests. In Bali, natural fluctuations in water regimes, with recurrent droughts associated with El Niño events, are key to the regeneration of the rainforest. The reproduction of trees in the *Dipterocarp* family, which dominate the rainforests, is tightly linked with this climate phenomenon. Up to 90% of *Dipterocarp* species synchronise their flowering with the onset of dry weather conditions, which traditionally occur on a roughly four-year cycle. Climate change together with massive land-use change, where forests have been converted to palmoil plantations, affect moisture feedback and rainfall patterns, increasing the frequency, amplitude and impact of droughts, and triggering a new, unnatural stress – forest fires. This new regime changes from being benign, with water variability as part of the inherent resilience of the system, to malign when the rainforest system moves from a desired state, with a dependence on a low frequency of droughts, to an undesirable state of too frequent and severe droughts followed by extensive forest fires. The feedback is loss of reproductive capacity, which puts the entire system at risk of long-term collapse. This loss of resilience, due to a combination of local and global processes, shifts the water dynamics and causes major social and economic problems.

The underlying question we pose about social–ecological interactions causing non-linear feedbacks is: how can such social–ecological dynamics be better understood? In addition, how can this understanding assist in building water resilience to change, in terms of persistence to maintain desired states, and the ability to adapt to change and to transform in situations of unavoidable crisis?

1.6 A new conceptual framework: water resilience for human prosperity in the Anthropocene era

As is argued above, based on an integration of new insights on the role of water in sustainable human development in the Anthropocene era, we are attempting to develop a new conceptual framework for integrated water resource governance and management. In sum, this framework is an evolution of a number of strands of major scientific advances, which are drawn on extensively in this book:

- The role of water in social–ecological resilience;
- Water and cross-scale dynamics and feedbacks;
- Water and global environmental change – water as a slow variable to sustain planetary boundaries;
- Climate change impacts on water resources (changes in disturbance regimes through increased rainfall variability, risk, droughts, dry spells, etc.);
- The green and blue water paradigm and recent advances in regional to global green and blue water resource assessments;
- The future of water and food security (diet, expansion, the virtual trade in green water);
- The use of green and blue water in agricultural development;
- Scale interactions in water resource planning and management (local- to basin-scale interactions and the growing evidence of the importance of meso-scale catchments);
- Options to build water resilience into terrestrial ecosystems, including food production;
- How to operationalise a water resilience paradigm in integrated land and water resource management in the Anthropocene era, where global change interacts with all scales and generates new disturbance regimes.

We apply a resilience lens in our quest for new insights into the sustainable governance and management of water resources. We are particularly concerned with cross-scale dynamics in this era of rapid global change and, based on our own research and research by colleagues across the world, believe that closely connecting green–blue water thinking with an integrated ecosystems and resilience approach, provides a better agenda, compared to the truncated water resources approach that has been applied so far, for resolving real-world problems, and opens up new opportunities for sustainable solutions.

1.6.1 Key insights contributing to a water resilience framework

The key features of a resilience-based framework for sustainable water resource governance and management include the integration of a number of insights on water and sustainability:

- Humanity has entered the Anthropocene epoch, an era of rapid human-induced social and ecological global change, which changes the pattern of disturbance regimes, in terms of their frequency, amplitude, scale and interdependence
- The fundamental role of water in ecosystem functions and services, which in turn determine human well-being
- That water, ecosystems and societies are interwoven in social–ecological systems, where power, equity, markets and institutions interact with water-dependent processes in the biosphere
- Social–ecological systems are complex adaptive systems with cross-scale dynamics

- A green–blue water paradigm provides a full representation of how water contributes to development and human prosperity
- The crucial role of water in resilience (here defined as water resilience), in terms of regime shifts, adaptive capacity and the ability of social–ecological systems to transform in situations of crisis.

Together, as is argued throughout this book, these insights on and features of the relations between water and human development provide the basis for a wider water resource agenda (further elaborated in Chapter 9). This new water resource agenda is articulated in: (1) new approaches to adaptive water governance and management; (2) new perspectives on the role of water in human development, where new disturbance regimes require a strong focus on water in ecosystem management and in social–ecological resilience; and (3) promoting innovations and improvements in integrated land and water resource management that provide either productivity or efficiency enhancements and resilience. This is of course expecting a lot from a new water resource agenda. As is shown in the book, however, there are many signs, from empirical examples on the ground and from current advances in science, policy and practice that we are moving towards a concretisation of these three core areas of the new agenda. As is indicated above, we suggest that the underlying 'building blocks' of this new water agenda are based on green–blue water thinking, the role of water in social–ecological resilience, the cross-scale dynamics of land and water processes, and the role of water in providing a safe operating space for humanity within planetary boundaries.

In essence, we are proposing that the current water paradigm, focused primarily on the allocation and efficient use of freshwater in ways that meet human demands while minimising environmental impacts, needs to be complemented by a focus on the role of water in sustaining ecosystem functions and services in both terrestrial and aquatic ecosystems, and the role of water in social–ecological resilience in a rapidly changing 'water world'. The implications of this widening of the water resource agenda are that the multiple roles of water in sustaining all dimensions of sustainable development are clarified, and a stronger emphasis is put on governing and managing water in ways that contribute to resilient societies being able to cope with shocks and stresses, and with being stewards of landscapes able to sustain future rainfall.

1.6.2 Integration of a green–blue water paradigm with resilience thinking

The core conceptual basis for this book is very simple. It emerges from the integration of a green–blue water paradigm into a social–ecological resilience framework and is set in the context of scale interactions and changes in disturbance regimes in the Anthropocene era (Figure 1.12).

The definitions of green and blue water flows and resources are summarised in the book glossary as well as in Box 1.2. The conceptual contribution of a green–blue water paradigm is the inclusion of all flows in the water

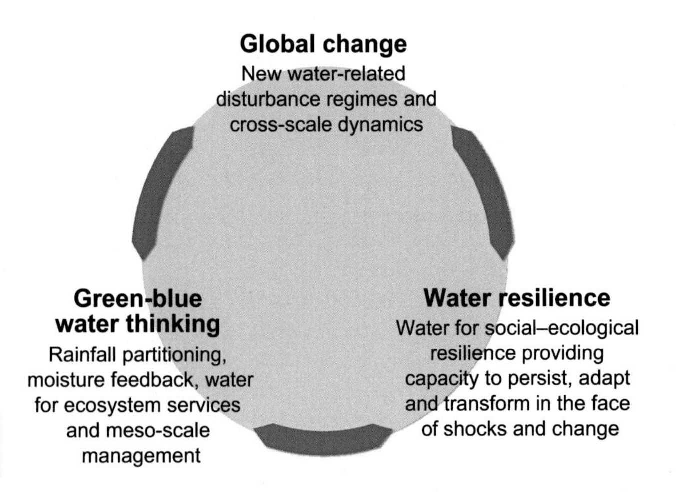

Figure 1.12 The core components of the proposed resilience-based paradigm for green and blue water governance, and management for human development in the Anthropocene era.

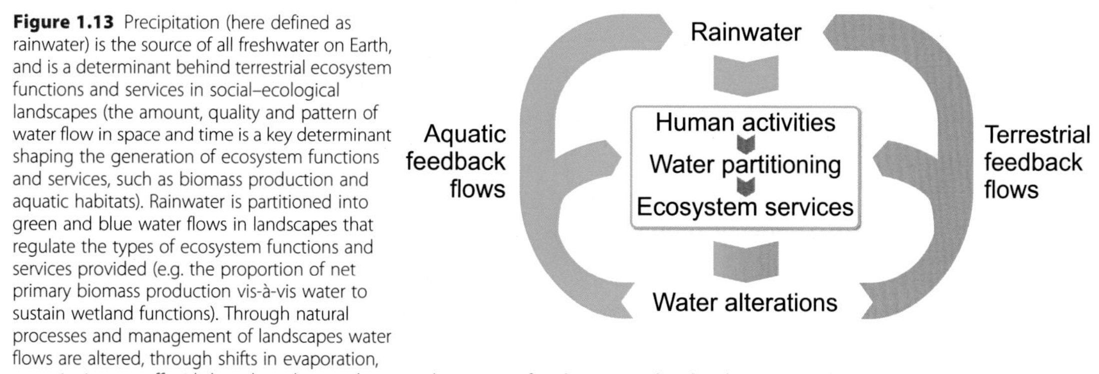

Figure 1.13 Precipitation (here defined as rainwater) is the source of all freshwater on Earth, and is a determinant behind terrestrial ecosystem functions and services in social–ecological landscapes (the amount, quality and pattern of water flow in space and time is a key determinant shaping the generation of ecosystem functions and services, such as biomass production and aquatic habitats). Rainwater is partitioned into green and blue water flows in landscapes that regulate the types of ecosystem functions and services provided (e.g. the proportion of net primary biomass production vis-à-vis water to sustain wetland functions). Through natural processes and management of landscapes water flows are altered, through shifts in evaporation, transpiration, runoff withdrawals and groundwater recharge. It is after these natural and anthropogenic alterations, related to management of land, ecosystems, agriculture, forests and water resources for industry and domestic purposes, that the amount of water feedbacks, from blue sources and green sources, are generated, which in turn influence the amount of rainfall in the next iteration of the local, regional to global hydrological cycle.

balance in the definition and quantification – and in the stewardship – of water for human development. Green water flows sustain all biomass growth on the planet, and thereby all terrestrial ecosystem functions and services.

From the broad description set out above of the social–ecological framework for water resources used in this book, it is important to emphasise that at its root, this book is advancing an ecohydrological approach to sustainable development, starting with the integration of green–blue water thinking within the framework of ecosystem functions and services (Falkenmark and Rockström, 2004). The analytical core of this integration is an emphasis on dividing rainfall into green and blue water flows across different scales, where rainwater constitutes a determining factor in the generation of ecosystem functions and services and a determinant flow behind, for example, the provision of services such as food production and regulatory functions such as carbon sequestration in soils. At the same time, the biosphere, or the landscapes on which rain falls, directly regulates the partitioning of water between different flow components in the hydrological cycle, thereby regulating green and blue water flows, which in turn alters the amount and pattern of water flows, i.e. creates water alterations. These water alterations – as a direct consequence of the governance and management of land and water – generate feedbacks for water flows in aquatic and terrestrial ecosystems, which in turn determine future rainfall amounts and patterns through impacts on the climate system and moisture feedback in the atmosphere. This simple conceptual framework constitutes the core of this book (Figure 1.13), which, in all its simplicity, contributes to thus far largely neglected features of water and human development in the way IWRM is practised, that is, the role of green and blue water resources and the relation between water and ecosystem functions and services, including feedbacks.

As is described above, the conceptual core shown is addressed in this book in the context of resilience, global change and social–ecological interactions, which provide us with a broad conceptual framework for the entire complex of integrated social–ecological water resilience.

1.6.3 An overarching framework on water resilience for global sustainability and human prosperity

Ultimately, humanity is at the mercy of the global hydrological cycle. In the past, before we entered the Anthropocene era in the mid-1950s and experienced the first signs of the impacts on rainfall and freshwater availability of human pressures on the environment and the Earth System, this was more of a rhetorical, intellectual phrase than an operational entry point for the governance and management of freshwater resources for human development. This has now changed, and the implications are not only that we need to prepare for and deal with larger risks of extreme events and gradual shifts in rainfall patterns. In parallel with the growing human pressures and impacts on freshwater resources in the Anthropocene

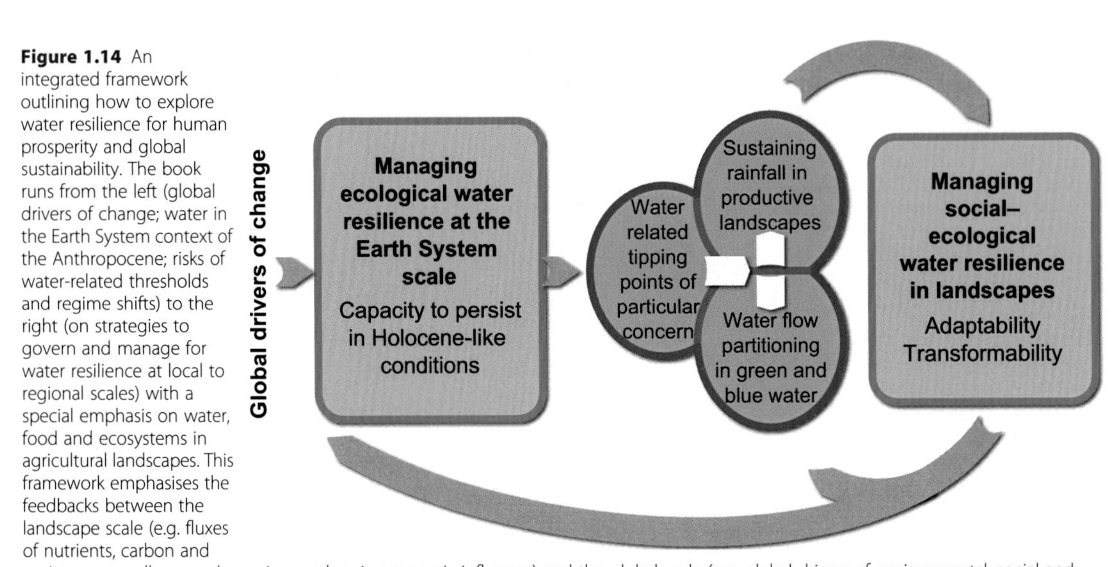

Figure 1.14 An integrated framework outlining how to explore water resilience for human prosperity and global sustainability. The book runs from the left (global drivers of change; water in the Earth System context of the Anthropocene; risks of water-related thresholds and regime shifts) to the right (on strategies to govern and manage for water resilience at local to regional scales) with a special emphasis on water, food and ecosystems in agricultural landscapes. This framework emphasises the feedbacks between the landscape scale (e.g. fluxes of nutrients, carbon and moisture, as well as goods, services and socio-economic influence) and the global scale (e.g. global drivers of environmental, social and geopolitical change). The figure emphasises the importance of cross-scale interactions in governance and management for water resilience. Of equal importance are the large, and potentially increasing, temporal pulses of water-related shocks and stresses, related to natural and anthropogenic influence on the temporal variability of water resources.

era, there is growing evidence of the importance of green and blue water flows to ecosystem services and human well-being. Moreover, there is growing evidence that ecological systems often have multiple stable states separated by thresholds. Our interest is in exploring the role of water as a control variable in determining the risk of crossing tipping points in local, to regional systems, due to alterations in water flows and changes in water feedback mechanisms. A key concern is to identify and understand how water is associated with thresholds of particular concern, such as, inspired by Biggs's concept of 'thresholds of potential concern' (Biggs *et al.*, 2003), the risk of pushing a rainforest across a drying threshold causing an irreversible shift to a drier savannah system.

The task we have set ourselves in this book is to put forward an integrated approach to water resilience for human prosperity in the Anthropocene era. To guide this journey, we have developed an integrated framework, the principle components and interactions of which are shown in Figure 1.14. This framework is further developed in Chapter 9.

The framework outlines how global drivers of change and management at the landscape scale interact to influence water resources. It emphasises a water resilience framework for sustainable water use, by exploring: (1) the role of water in threshold processes of particular concern; (2) the partitioning of water flows and the generation of ecosystem services; and (3) strategies for sustaining rainfall over landscapes. The framework advances current, dominant water resources thinking in three key ways by emphasising: (1) that in the Anthropocene era the 'supply side' of water resources – ultimately rainfall – is also changing, due to the dynamics of local to global change; (2) that water resource governance and management must occur at all scales, including planetary stewardship of water resources, and (3) resilience, in terms of understanding non-linear dynamics, interactions and thresholds, and in terms of managing water for resilience. Ultimately, the framework places a dual emphasis on water resource governance, allowing the Earth to remain in a Holocene-like state, and bringing about the basic preconditions for water management for human prosperity, both of which require an understanding of the temporal dimensions of water shocks and pulses, and of scale dimensions – that local water management can no longer occur in isolation from social (e.g. sudden shifts in trade patterns and prices) and environmental processes at the regional to global scales (e.g. climate change and deforestation).

Chapters 2, 3 and 4 focus on the left-hand side of our conceptual framework, on understanding the new global water challenges in the Anthropocene era, from global drivers of change to how water interacts with other social–ecological processes of change. They introduce the thinking behind and evidence on water-related thresholds and regime shifts in social–ecological systems. These 'global' chapters emphasise the importance of putting forward approaches to water governance that contribute to sustaining social–ecological resilience at the Earth System scale. We use the global water and food challenge (Chapter 5) as a bridge to link the global scale to productive landscapes (moving right in the conceptual framework), addressing the risks of water-related social–ecological regime shifts in agricultural systems, agricultural improvements to build resilience among rainfed farming communities (Chapter 6) and the challenges of advancing an ecosystem-based approach to water resilience in agricultural landscapes. Chapter 7 presents evidence of water-related thresholds in social–ecological systems. We explore both adaptive and transformative strategies for improved water management in landscapes. The institutional dimension of governing for water resilience (the furthest right in the conceptual framework) is addressed in Chapter 8.

1.6.4 The focus of the book: freshwater and the living systems of the biosphere

We hope that this book will form the basis for a scientific and educational dialogue on the continuing advances in integrated thinking and research on water for human development in increasingly turbulent times. Our contribution is to link our understanding of water resources and the generation of ecosystem functions and services from terrestrial and aquatic ecosystems and: (1) how these define the resilience of social–ecological systems; (ii) how human interactions with water affect ecosystems and resilience; and (iii) how improvements and innovations in water governance and management can be applied to address the new water resource agenda, originating from insights into relations between water, ecosystems, resilience and global environmental change.

This means that the book does not focus on marine ecosystems, and does not explore the important role of water for resource use (e.g. in mining) or for domestic and urban water supply and water for industrial purposes (i.e. essentially the entire domain of water pollution and water quality management, which is essential to the whole water sustainability and prosperity agenda). That said, we do take an integrated perspective on the trade-offs between water use to sustain life-support systems, resource use and other social uses. The book has a special focus on water and bioresources, particularly how to build water resilient food production in the world. This is a deliberate choice as this question has the strongest influence on how water flows in the world, because bioresources are the largest water consumer on Earth, and because what happens in the biosphere affects how water operates in the global hydrological cycle.

Our aim is that this book will stimulate a wider scientific and professional exploration of how we can be wise stewards of the fundamental core of our human prosperity – stable and high quality flows of freshwater – in a future where we increasingly define our own destiny not only in social terms, but also in terms of how the entire Earth System operates.

Summary

The role played by water in sustaining life on Earth

The current water situation is increasingly precarious. Half the world's population already faces various degrees of water scarcity, and the most dramatic trend for future demand on freshwater resources originates from the momentum of the demand for food. Water is the bloodstream of both the landscape and societies, and food production is by far the largest direct water consumer.

The future of humanity will depend on its capacity to govern and manage water in ways that build resilience in an era of rapid change and growing indications of large, undesirable risks caused by the unsustainable exploitation of ecosystems. Humans are an integrated part of the Earth System, and water availability for human needs is intimately connected to the dynamic changes in the biophysical systems on Earth. Water is a determining factor in the functioning structure and stability of biophysical systems. Water regulates the climate system, is a precondition for all biomass growth and thereby for all living species, steers the pace of the global carbon, nitrogen and phosphorus cycles, and is an agent for transport and a solvent of chemical compounds. It also delivers human well-being by generating ecosystem functions and services from terrestrial and aquatic systems.

Water is affected by a large number of social and ecological global change processes, while at the same time it influences biophysical change in many other systems. Water is closely associated with changes in disturbance regimes, and with generating surprising events by inducing non-linear change when systems abruptly and unexpectedly cross thresholds, or tipping points. This can generate rapid and often irreversible change. The stability of the global freshwater cycle is key to our analysis. Three arguments support the need to consider a sustainable global level of water use: the freshwater function as the bloodstream of the entire biosphere; the determining role of aggregate flows of runoff and vapour in the regional to global hydrological cycle; and the very scale of human pressures on the Earth System, which generates pockets of major environmental impacts.

Two key messages run as a red thread throughout the book. First, human development and wealth originate from social–ecological systems in terms of generating ecosystem functions and services, and providing social–ecological resilience. Second, over-appropriation of freshwater resources at different scales threatens social–ecological resilience, and a lack of environmental flows in terrestrial and aquatic ecosystems is an indicator of declining resilience.

The role of water for resilience

Resilience is defined as the ability of a system to cope with disturbance while continuing to develop. There are three overarching features of resilience: how much disturbance a system can withstand without crossing a threshold and changing state or structure; the capacity within a given state to adapt to stress and disturbance; and the ability to transform into a new desired state after a crisis. The ecological dimension of water resilience includes maintaining the freshwater dynamics required to avoid regime shifts. The social dimension addresses the broad complex of water management and governance. A core question is the role of efficiency and optimisation versus redundancy in building water resilience. We distinguish between three roles of 'water resilience': water as a direct control variable (water for the generation of ecosystem functions and services); water as a state variable (how water is affected by other drivers of change); and water as a driving force of future water resource availability (moisture feedback to the atmosphere).

Safe operating space

Agriculture is the largest water user and the largest 'shaper' of landscapes, influencing rainwater partitioning and thereby the flux of water through landscapes and return flows to the atmosphere from land. The empirical evidence of accelerated human pressures on biophysical systems raises the question of whether we are at risk of threatening the stability of the Holocene state – the only state on Earth we know can support human development on a planet with 7 billion people, but soon to be 9. Water is key to this question, both as the bloodstream and in its victim function, being affected by changes in other biophysical systems on Earth. Two driving forces threaten the stability of water flows: human-induced shifts in green water flows as a result of changes in precipitation and soil moisture generation, and human withdrawals of blue water affecting river flow dynamics. The tropical rainforests and the savannah regions are hotspots, most at risk of threshold effects that might cause global impacts.

The feedback mechanisms that help to push systems across thresholds rarely operate in a simple cause–effect manner. The focus is on the interactions between human development, global environmental change and water resilience, with a particular emphasis on agricultural landscapes. The book explores integrated social–ecological feedbacks, how understanding these can assist in building water resilience to change in terms of persistency to remain in desired states, and the ability to adapt to change and to transform in situations of unavoidable crisis.

This chapter provides an overview of planetary boundaries, which identify the biophysical playing field, i.e. the safe space in which to operate to avoid the risk, due to loss of resilience, of large non-linear changes with deleterious or even catastrophic outcomes for human societies. It adopts a green–blue water resources approach to the analysis of a proposed freshwater boundary: green water thresholds related to risks of too large manipulations of global vapour flows from land threatening rainfall generation; and blue water thresholds related to the risk of too large extractions of river flows undermining freshwater ecosystems.

Building blocks for the new water agenda

We attempt to develop a new conceptual framework for integrated water resource governance and management, applying a resilience lens to closely connect green–blue water thinking based on three core areas: new approaches to adaptive water governance and management; new perspectives on the role of water in human development; and the promotion of innovation and improvements.

The underlying building blocks involve: (1) green–blue water thinking; (2) the role of water for social–ecological resilience; (3) cross-scale dynamics of land and water processes; and (4) the role of water in providing a safe operating space for humanity within the planetary boundaries. Thus, the core conceptual basis for the book is set in the context of scale interactions and disturbance regimes in the Anthropocene epoch. One key concern is to identify and understand where water is associated with tipping points of particular concern, such as pushing rainforests across a drying threshold, causing an irreversible shift into drier savannah systems.

References

Alcamo, J. and Henrichs, T. (2002). Critical regions: a model-based estimation of world water resources sensitive to global changes. *Aquatic Sciences*, **64**, 352–362.

Berndes, G. (2002). Bioenergy and water: the implications of large-scale bioenergy production for water use and supply. *Global Environmental Change: Human and Policy Dimensions*, **12**, 253–271.

Biggs, H. C., Rogers, K. H., Du Toit, J., Rogers, K. and Biggs, H. (2003). An adaptive system to link science, monitoring and management in practice. In *The Kruger Experience: Ecology and Management of Savanna Heterogeneity*, ed. Du Toit, J. T., Biggs, H. C. and Rogers, K. H. Washington, DC: Island Press, pp. 59–80.

Bogardi, J. J., Dudgeon, D., Lawford, R. *et al.* (2012). Water security for a planet under pressure: interconnected challenges of a changing world call for sustainable solutions. *Current Opinion in Environmental Sustainability*, **4**, 35–43.

Brundtland, G. H. (1987). *Our Common Future.* Oxford: Oxford University Press.

Canadell, J. G., Le Quere, C., Raupach, M. R. *et al.* (2007). Contributions to accelerating atmospheric CO_2 growth from economic activity, carbon intensity, and efficiency of natural sinks. *Proceedings of the National Academy of Sciences of the United States of America*, **104**, 18866–18870.

Carpenter, S. R. (2003). *Regime Shifts in Lake Ecosystems: Pattern and Variation. Excellence in Ecology.* Oldendorf/Luhe, Germany: Ecology Institute.

Carpenter, S. R. (2005). Eutrophication of aquatic ecosystems: bistability and soil phosphorus. *Proceedings of the National Academy of Sciences of the United States of America*, **102**, 10002–10005.

Carpenter, S. R. and Bennett, E. M. (2011). Reconsideration of the planetary boundary for phosphorus. *Environmental Research Letters*, **6**, 014009.

Carpenter, S., Walker, B., Anderies, J. M. and Abel, N. (2001). From metaphor to measurement: resilience of what to what? *Ecosystems*, **4**, 765–781.

Center for Research on the Epidemiology of Disaster. (2012). *EM-DAT: The OFDA/CRED International Disaster Database Version 12.07.* Brussels: Université Catholique De Louvain. Available at: http://emdat.be (accessed 6 June 2012).

Chao, B. F., Wu, Y. H. and Li, Y. S. (2008). Impact of artificial reservoir water impoundment on global sea level. *Science*, **320**, 212–214.

Cole, D. H. and Ostrom, E. (eds) (2011). *Property in Land and Other Resources.* Cambridge, MA: Lincoln Institute of Land Policy.

Crutzen, P. J. (2002a). The 'Anthropocene'. *Journal De Physique Iv*, **12**, 1–5.

Crutzen, P. J. (2002b). Geology of mankind. *Nature*, **415**, 23.

Crutzen, P. J. and Steffen, W. (2003). How long have we been in the Anthropocene era? Comment. *Climatic Change*, **61**, 251–257.

Crutzen, P. J. and Stoermer, E. F. (2000). The 'Anthropocene'. *IGBP Newsletter*, **41**, 17–18.

Dai, A. (2011). Characteristics and trends in various forms of the Palmer Drought Severity Index during 1900–2008. *Journal of Geophysical Research: Atmospheres*, **116**, D12115.

de Fraiture, C. and Wichelns, D. (2010). Satisfying future water demands for agriculture. *Agricultural Water Management*, **97**, 502–511.

de Vries, M. and de Boer, I. J. M. (2010). Comparing environmental impacts for livestock products: a review of life cycle assessments. *Livestock Science*, **128**, 1–11.

Duda, A. M. (2003). Integrated management of land and water resources based on a collective approach to fragmented international conventions. *Philosophical Transactions of the Royal Society of London. Series B: Biological Sciences*, **358**, 2051–2062.

Enfors, E. (2013). Social-ecological traps and transformations in dryland agro-ecosystems: using water system innovations to change the trajectory of development. *Global Environmental Change*, **23**, 51–60.

Ericson, J. P., Vörösmarty, C. J., Dingman, S. L., Ward, L. G. and Meybeck, M. (2006). Effective sea level rise and deltas: causes of change and human dimension implications. *Global and Planetary Change*, **50**, 63–82.

European Space Agency (2010). *GlobCover 2009.* Paris: Globcover Consortium. Available at: http://due.esrin.esa.int/globcover/ (accessed 26 February 2013).

Ewing, B., Moore, D., Goldfinger, S. *et al.* (2010). *Ecological Footprint Atlas 2010.* Oakland, CA: Global Footprint

Network. Available at: http://www.footprintnetwork.org/images/uploads/Ecological_Footprint_Atlas_2010.pdf.

Falkenmark, M. (1986). Fresh water: time for a modified approach. *Ambio*, **15**, 192–200.

Falkenmark, M. and Chapman, T. (eds) (1989). *Comparative Hydrology: An Ecological Approach to Land and Water Resources*. Paris: UNESCO.

Falkenmark, M. and Folke, C. (2003). Freshwater and welfare fragility: syndromes, vulnerabilities and challenges. Introduction. *Philosophical Transactions of the Royal Society of London Series B-Biological Sciences*, **358**, 1917–1920.

Falkenmark, M. and Rockström, J. (2004). *Balancing Water for Humans and Nature: The New Approach in Ecohydrology*. London: Earthscan.

Folke, C., Carpenter, S. R., Walker, B. *et al.* (2010). Resilience thinking: integrating resilience, adaptability and transformability. *Ecology and Society*, **15**, 20.

Food and Agriculture Organization (2012a). *The State of Food Insecurity in the World 2012*. Rome: Food and Agriculture Organization. Available at: http://www.fao.org/docrep/016/i3027e/i3027e00.htm.

Food and Agriculture Organization (2012b). *State of the World's Forests 2012*. Rome: Food and Agriculture Organization.

Food and Agriculture Organization (2013). *FAOSTAT Online Database*. Rome: Food and Agriculture Organization. Available at: http://faostat.fao.org/.

Freydank, K. and Siebert, S. (2008). Towards mapping the extent of irrigation in the last century: time series of irrigated area per country. Frankfurt Hydrology Paper 08. Institute of Physical Geography, University of Frankfurt, Germany.

Godfray, H. C. J., Beddington, J. R., Crute, I. R. *et al.* (2010). Food security: the challenge of feeding 9 billion people. *Science*, **327**, 812–818.

Gordon, L. J., Peterson, G. D. and Bennett, E. M. (2008). Agricultural modifications of hydrological flows create ecological surprises. *Trends in Ecology & Evolution*, **23**, 211–219.

Hansen, J., Satol, M., Kharechal, P. *et al.* (2008). Target atmospheric CO_2: where should humanity aim? *The Open Atmospheric Science Journal*, **2**, 217–231.

Hoekstra, A. Y. (2010). The water footprint of animal products. In *The Meat Crisis: Developing More Sustainable Production and Consumption*., ed. D'Silva, J. and Webster, J. London: Earthscan, pp. 22–33.

Intergovernmental Panel on Climate Change (IPCC) (2007). *Climate Change 2007: Synthesis Report. Contribution of Working Groups I, II and III to the Fourth Assessment Report of the Intergovernmental Panel on Climate Change*. Geneva: Intergovernmental Panel on Climate Change.

Intergovernmental Panel on Climate Change (IPCC) (2013) *Climate Change 2013: The Physical Science Basis. Working Group I contribution to the IPCC Fifth Assessment Report*. Geneva: Intergovernmental Panel on Climate Change.

International Conference on Water and the Environment (1992). The Dublin statement on water and sustainable development. International Conference on Water and the Environment, Dublin.

Jansson, Å., Folke, C., Rockström, J., Gordon, L. and Falkenmark, M. (1999). Linking freshwater flows and ecosystem services appropriated by people: the case of the Baltic Sea drainage basin. *Ecosystems*, **2**, 351–366.

Jouzel, J., Masson-Delmotte, V., Cattani, O. *et al.* (2007). Orbital and millennial Antarctic climate variability over the past 800,000 years. *Science*, **317**, 793–796.

Kharas, H. (2010). The emerging middle class in developing countries. OECD Working papers: 285. OECD. Available at: www.oecd-ilibrary.org/development/the-emerging-middle-class-in-developing-countries_5kmmp8lncrns-en.

Klein Goldewijk, K., Beusen, A., Van Drecht, G. and De Vos, M. (2011). The HYDE 3.1 spatially explicit database of human-induced global land-use change over the past 12,000 years. *Global Ecology and Biogeography*, **20**, 73–86.

Kummu, M., Ward, P. J., de Moel, H. and Varis, O. (2010). Is physical water scarcity a new phenomenon? Global assessment of water shortage over the last two millennia. *Environmental Research Letters*, **5**, 034006.

Lagi, M., Bar-Yam, Y., Bertrand, K. Z. and Bar-Yam, Y. (2012). *Economics of Food Prices and Crises*. Cambridge, MA: New England Complex Systems Institute. Available at: http://necsi.edu/publications/food/.

Lambin, E. F. and Geist, H. J. (eds) (2006). *Land-use and Land-cover Change*. Heidelberg, Germany: Springer.

Lenton, T. M., Held, H., Kriegler, E. *et al.* (2008). Tipping elements in the Earth's climate system. *Proceedings of the National Academy of Sciences*, **105**, 1786–1793.

Liquete, C., Maes, J., La Notte, A. and Bidoglio, G. (2011). Securing water as a resource for society: an ecosystem services perspective. *Ecohydrology & Hydrobiology*, **11**, 247–259.

Loh, J., Green, R. E., Ricketts, T. *et al.* (2005). The Living Planet Index: using species population time series to track trends in biodiversity. *Philosophical Transactions of the Royal Society B: Biological Sciences*, **360**, 289–295.

Lundqvist, J. (ed.) (2000). *New Dimensions in Water Security: Water, Society and Ecosystem Services in the 21st Century.* Rome: Food and Agriculture Organization.

McIntyre, B., Herren, H., Wakhungu, J. and Watson, R. T. (eds) (2009). *Agriculture at the Crossroads: International Assessment of Agricultural Knowledge.* Washington, DC: Island Press.

Meybeck, M. (2003). Global analysis of river systems: from Earth System controls to Anthropocene syndromes. *Philosophical Transactions of the Royal Society B-Biological Sciences*, **358**, 1935–1955.

Millennium Ecosystem Assessment (2005). *Ecosystems and Human Well-being: Synthesis.* Washington, DC: Island Press.

Molden, D., Oweis, T. Y., Steduto, P. *et al.* (2007). Pathways for increasing agricultural water productivity. In *Water for Food, Water for Life: A Comprehensive Assessment of Water Management in Agriculture*, ed. Molden, D. London: Earthscan, pp. 219–310.

Molle, F. and Wester, P. (eds) (2009). *River Basin Trajectories: Societies, Environments and Development.* Wallingford, UK: CAB International.

National Academy of Sciences of United States of America (2009). PNAS tipping elements in Earth systems special feature. *Proceedings of the National Academy of Sciences*, **106**, 1068–1072.

Odum, E. (1969). The strategy of ecosystem development. *Science*, **164**, 262.

Oki, T. and Kanae, S. (2006). Global hydrological cycles and world water resources. *Science*, **313**, 20561–20563.

Oppenheimer, S. (2004). *Out of Eden: The Peopling of the World.* London: Constable & Robinson.

Oyama, M. D. and Nobre, C. A. (2003). A new climate-vegetation equilibrium state for tropical South America. *Geophysical Research Letters*, **30**, 2199.

Pahl-Wostl, C., Lebel, L., Knieper, C. and Nikitina, E. (2012). From applying panaceas to mastering complexity: toward adaptive water governance in river basins. *Environmental Science & Policy*, **23**, 24–34.

Poff, N. L., Allan, J. D., Bain, M. B. *et al.* (1997). The natural flow regime. *BioScience*, **47**, 769–784.

Poff, N. L. and Zimmerman, J. K. H. (2009). Ecological responses to altered flow regimes: a literature review to inform the science and management of environmental flows. *Freshwater Biology*, **55**, 194–205.

Postel, S. L. (1998). Water for food production: will there be enough in 2025? *Bioscience*, **48**, 629–637.

Ramanathan, V. and Feng, Y. (2008). On avoiding dangerous anthropogenic interference with the climate system: formidable challenges ahead. *Proceedings of the National Academy of Sciences*, **105**, 14245–14250.

Richter, B., Baumgartner, J., Wigington, R. and Braun, D. (1997). How much water does a river need? *Freshwater Biology*, **37**, 231–249.

Rimas, A. and Fraser, E. D. (2010). *Empires of Food: Feast, Famine, and the Rise and Fall of Civilizations.* New York: Free Press.

Ripl, W. and Hildmann, C. (2000). Dissolved load transported by rivers as an indicator of landscape sustainability. *Ecological Engineering*, **14**, 373–387.

Rockström, J., Gordon, L., Folke, C., Falkenmark, M. and Engwall, M. (1999). Linkages among water vapor flows, food production, and terrestrial ecosystem services. *Conservation Ecology*, **3**, 5.

Rockström, J., Steffen, W., Noone, K. *et al.* (2009a). A safe operating space for humanity. *Nature*, **461**, 472–475.

Rockström, J., Steffen, W., Noone, K. *et al.* (2009b). Planetary boundaries: exploring the safe operating space for humanity. *Ecology and Society*, **14**, 32.

Scheffer, M., Carpenter, S., Foley, J. A., Folke, C. and Walker, B. (2001). Catastrophic shifts in ecosystems. *Nature*, **413**, 591–596.

Scheffer, M. and Carpenter, S. R. (2003). Catastrophic regime shifts in ecosystems: linking theory to observation. *Trends in Ecology & Evolution*, **18**, 648–656.

Shiklomanov, I. A. (1998). *World Water Resources: A New Appraisal and Assessment for the 21st Century.* Paris: United Nations Educational, Scientific and Cultural Organization, pp. 369–384.

Shiklomanov, I. A. (2000). Appraisal and assessment of world water resources. *Water International*, **25**, 11–32.

Shiklomanov, I. A. (2003). World water use and water availability. In *World Water Resources in the Beginning of the 21st Century*, ed. Shiklomanov, I. A. & Rodda, J. C. Cambridge: Cambridge University Press.

Shiklomanov, I. A. and Rodda, J. C. (2004). *World Water Resources at the Beginning of the 21st Century. International Hydrology Series.* Cambridge: Cambridge University Press.

Smakhtin, V., Revenga, C. and Döll, P. (2004). A pilot global assessment of environmental water requirements and scarcity. *Water International*, **29**, 307–317.

Solomon, S., Rosenlof, K. H., Portmann, R. W. *et al.* (2010). Contributions of stratospheric water vapor to decadal changes in the rate of global warming. *Science*, **327**, 1219–1223.

Steffen, W., Crutzen, P. J. and McNeill, J. R. (2007). The Anthropocene: are humans now overwhelming the great forces of nature. *Ambio*, **36**, 614–621.

Steffen, W. L., Sanderson, A., Tyson, P. D. *et al.* (2004). *Global Change and the Earth System: A Planet Under Pressure. Global Change: The IGBP series, 1619–2435.* Berlin: Springer.

Tilman, D., Balzer, C., Hill, J. and Befort, B. L. (2011). Global food demand and the sustainable intensification of agriculture. *Proceedings of the National Academy of Sciences*, **108**, 20260–20264.

United Nations Department of Economic and Social Affairs (2011). *World Population Prospects: The 2010 Revision.* Rome: United Nations Department of Economic and Social Affairs. Available at: http://esa.un.org/unpd/wpp/Excel-Data/population.htm (accessed 21 November 2012).

United Nations Environment Programme (2012). *Global Environment Outlook GEO 5.* Nairobi: United Nations Environment Programme.

United Nations Population Division (1999). *The World at Six Billion ESA/P/WP.154.* Rome: United Nations Population Division. Available at: http://www.un.org/esa/population/publications/sixbillion/sixbillion.htm.

Van der Ent, R. and Savenije, H. (2011). Length and time scales of atmospheric moisture recycling. *Atmospheric Chemistry and Physics*, **11**, 1853–1863.

Wada, Y., van Beek, L. P. H., van Kempen, C. M. *et al.* (2010). Global depletion of groundwater resources. *Geophysical Research Letters*, **37**, L20402.

Walker, B., Holling, C. S., Carpenter, S. R. and Kinzig, A. (2004). Resilience, adaptability and transformability in social-ecological systems. *Ecology and Society*, **9**, 5.

Walker, B. and Salt, D. (2006). *Resilience Thinking: Sustaining Ecosystems and People in a Changing World.* Washington, DC: Island Press.

Wolf, A. (2003). 'Water Wars' and other tales of hydromythology. In *Whose Water is it? The Unquenchable Thirst of a Water-hungry World*, ed. Jehl, D. & Mcdonald, B. Washington, DC: National Geographic, pp. 109–124.

World Trade Organization. (2012). *International Trade Statistics 2012.* Geneva: WTO Available at: http://www.wto.org/english/res_e/statis_e/its2012_e/its12_charts_e.htm.

World Water Assessment Programme. (2009). *World Water Development Report 3: Water in a changing world.* Paris: United Nations Educational, Scientific and Cultural Organization.

World Wildlife Fund for Nature. (2012). *Living Planet Report 2012: Biodiversity, Biocapacity and Better Choices.* Gland, Switzerland: World Wildlife Fund for Nature.

Young, O. R. and Steffen, W. (2009). The Earth System: sustaining planetary life-support systems. In *Principles of Ecosystem Stewardship*, ed. Chapin, F. S., Kofinas, G. P. & Folke, C. Heidelberg, Germany: Springer, pp. 295–315.

Living in a human-dominated world

Street scene from Vietnam, where vendors come to the cities to sell their produce. Urban areas, where trends show population increase, are dependent on agricultural products from rural areas.

Human modification of the Earth System

The focus in this chapter is the human forcings in the Earth System at large, and their origin in multiple drivers of change: demographic, economic development, urbanisation, technological development, international trade and foreign direct investment, climate change, and national and international policies. The chapter shows in what ways the Earth System has responded by different kind of impacts and feedbacks, and discusses the risk of approaching crucial thresholds and tipping points in the Earth System. Special attention is paid to the world's most water-dependent system: agricultural production and human food security.

2.1 Humans have altered the Earth System through multiple drivers of change

Anthropogenic pressures are multiple, complex and equal in magnitude to some of the great forces of nature – and they are accelerating. They interact with each other and can trigger abrupt, non-linear changes if they cross critical thresholds. It has been recognised for some time that key environmental parameters have moved well beyond the range of natural variability, and that the magnitude and rate of change are unprecedented (from the Amsterdam Declaration on Global Change; Moore *et al.*, 2001).

The Earth System is seen, in this book, as the highest level unit containing connected sub-systems and components at all scales, with levels of organisation beyond individual building blocks. Given the complexity of this system, it will not be possible to anticipate all the pressures, interactions and feedbacks between the different components. We should expect further surprises, including non-linear responses and sudden regime shifts or abrupt, often unexpected, changes resulting from a disturbance or shock – usually resulting in an alternate stable state – as well as large and persistent changes in structure and function.

A key driver of change is the **need to increase food production** and associated biomass appropriation for a growing, more affluent and increasingly urbanised population (resulting in increasing competition for water, land and other natural resources). The impacts of this driving force vary between different regions.

2.1.1 Demographic trends

The greatest acceleration in the human population took place during the second half of the twentieth century, from 2.5 to 6.1 billion, or close to 150%, in 50 years. Despite decreasing fertility rates, demographic trends indicate that the world population will increase to about 9 billion people by 2050 before stabilising, according to the UN's medium projection (UNPD, 2004). Figure 2.1 gives high, medium and low scenarios for projected population change by 2100: an increase to 14 billion or a decline to close to 5 billion.

Continuing high fertility rates in much of Africa and western Asia mean that almost all the increase in population by 2050 will take place in the developing world. Population growth rates are usually significantly higher in urban areas than in rural areas. This is currently true for sub-Saharan Africa, where current net annual population growth is 4% in urban areas compared to 2% in rural areas. However, the rural population in the 'least developed' countries, many in sub-Saharan Africa, will continue to increase (Figure 2.2).

Growing demand for water is currently a much more important driver of water scarcity than reductions in water availability due to climate change (Kummu *et al.*, 2010) and in several regions economic development is becoming an even stronger driver of water demand than population growth.

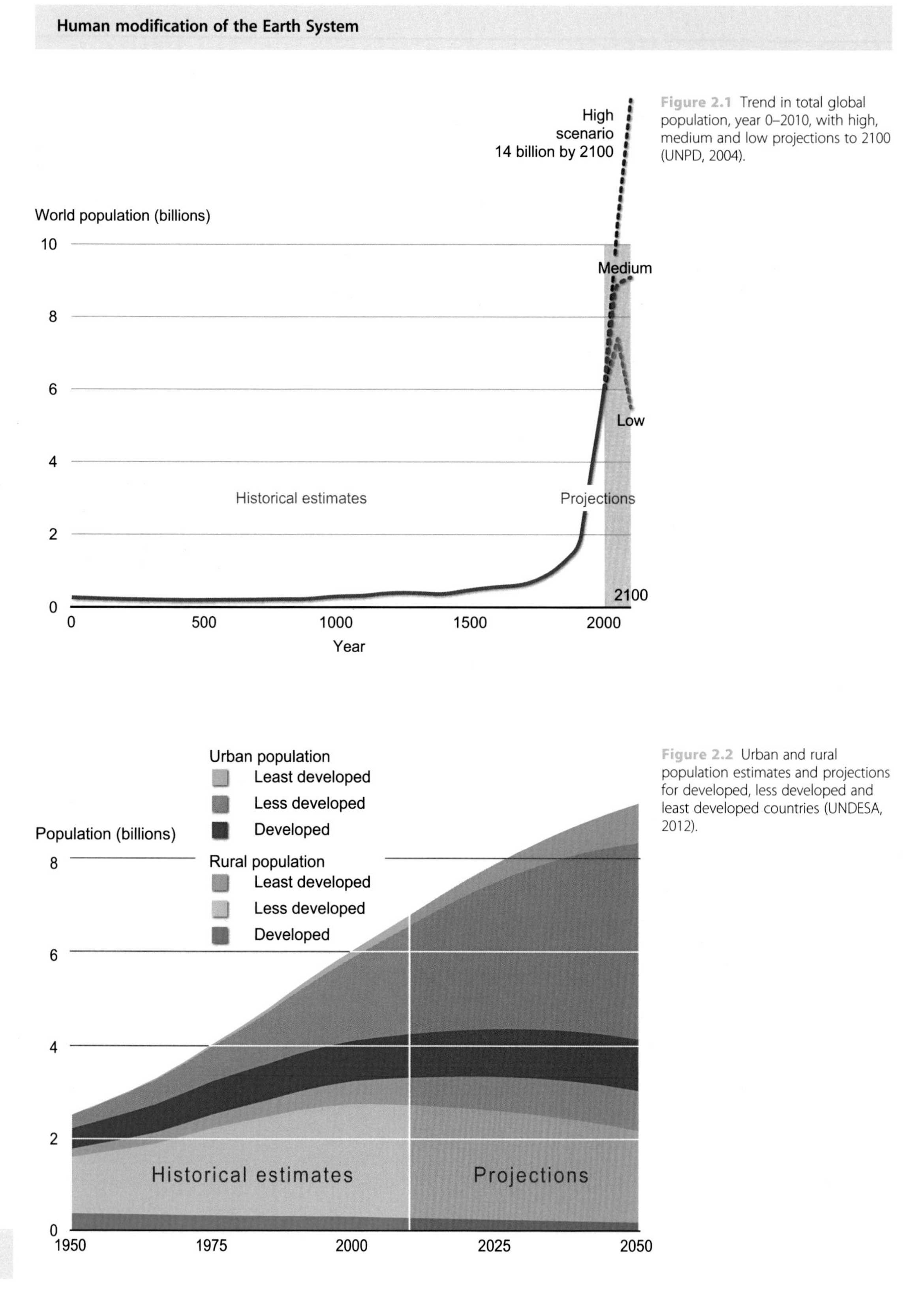

Figure 2.1 Trend in total global population, year 0–2010, with high, medium and low projections to 2100 (UNPD, 2004).

Figure 2.2 Urban and rural population estimates and projections for developed, less developed and least developed countries (UNDESA, 2012).

2.1.2 Economic development

Economic development generally increases resource use per capita, up to a certain level of development. Human **diets** respond to economic development and personal income, but also to urbanisation, globalisation, increases in trade and the advertising and other strategies of food corporations.

Globally, per capita calorie intake has increased over time (FAO, 2013), as shown in Figure 2.3. The increase is both in total calories and the amount and proportion of animal-source foods (from meat, milk and eggs). The OECD (Organisation for Economic Cooperation and Development) group clearly stands out, with an average per capita daily food intake close to 3500 kcal, close to 1000 kcal of which is from animal-source foods (28%). By contrast, sub-Saharan Africa and South Asia still show low levels, both in absolute terms of less than 2200 kcal/capita per day and with less than 10% from meat, milk and eggs. Of particular note is the extremely rapid increase in northeast Asia, the region that includes China, Japan and South Korea. The formerly centrally planned countries of Eastern Europe, Central Asia and Russia, are the only group to show a decrease over time, with a particularly severe fall after 1991 but with a rapid pick-up in the past decade. Most worrying is the fact that both regions that are projected to have the highest population increase in the coming decades, i.e. sub-Saharan Africa and South Asia, show an almost stagnant per capita kcal supply, indicating persistent hunger and malnutrition.

The increasing per capita demand for food and water following the rapid changes in diet in China since 1980, after the introduction of the economic reforms initiated by Deng Xiaoping in 1978, are clearly illustrated in Figure 2.4 (Liu and Savenije, 2008). Like many rapidly developing countries, diets have changed from a high proportion of staple grains to increasing consumption of more water-intensive vegetables, fruit, dairy products and meat.

Livestock in particular require large amounts of crops for feed as well as large grazing areas. About one-third of global cereal production is fed to animals. Given the low conversion efficiency of feed to livestock products, meat requires a lot more water and energy per kcal compared to vegetable-based diets (Mekonnen and Hoekstra, 2010). Compared to cereals, fruit and vegetables require about twice as much water per calorie produced, dairy products 3–5 times more and meat 3–10 times as much – or more (Lundqvist *et al.*, 2008). The virtual water content of meat is about 10 times higher per calorie than that of cereals.

One side effect of economic development and increasing wealth is the overconsumption and wastage of food, which highlights the consumption side of the food supply chain (Kummu *et al.*, 2012). Less developed countries suffer more from losses on the production side, linked to food storage and transportation issues (Godfray *et al.*, 2010) – more details in

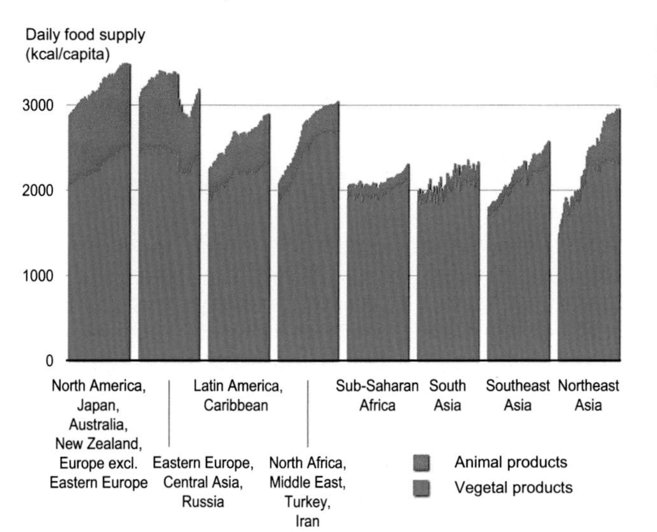

Figure 2.3 Trend in daily global calorie food supply, 1961–2007: vegetal and animal-source foods for different regions and the global average (FAO, 2013).

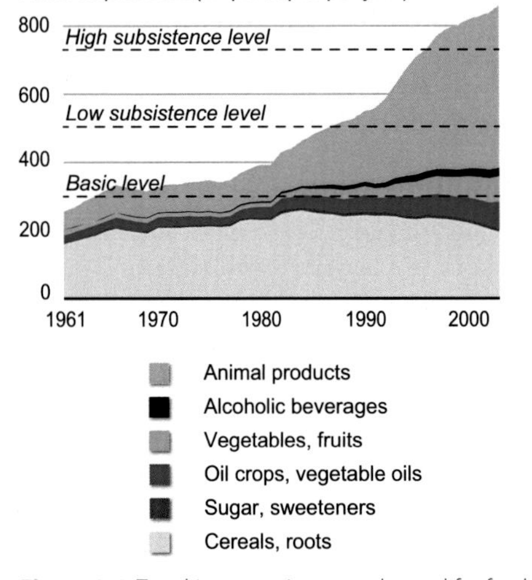

Figure 2.4 Trend in per capita water demand for food in China and relationship with economic development (after Liu and Savenije, 2008).

Chapter 5. A decoupling of economic development and improved diets from resource use will be required in order to avoid transgressing critical boundaries.

2.1.3 Urbanisation

Urbanisation is another driver of change, itself driven by globalisation, improved infrastructure, economic opportunities and degrading livelihoods in rural areas. Cities account for only about 2% of the Earth's land surface (Bicheron *et al.*, 2008) but house about 50% of the world's population, a proportion which is projected to increase to 70% by 2050 (United Nations Department of Economic and Social Affairs, 2012) as is shown in Figure 2.2. Cities today consume three-quarters of all natural resources and account for 71% of the world's energy-related CO_2 emissions (IEA, 2008).

Cities concentrate demand for energy, food, water and other resources, so that only part of these demands can be met locally from within the city itself, while most will have to be imported from surrounding areas – sometimes across large distances (Hoff *et al.*, 2013). Accordingly, production and consumption of, for example, biomass are decoupled, and cities accumulate waste products, including wastewater, that are re-exported to the surrounding areas, where ideally they are recycled. However, some resources, for example, some nutrients, do not flow in closed cycles. Non-sustainable practices mean that there are significant losses, for example, of phosphorus, which is transported by rivers to the sea from where recovery is almost impossible, leading to phosphorus scarcity and negative impacts on global food security (Cordell *et al.*, 2009).

The combination of high population numbers and high densities in cities, with generally stronger economies and more resource-intensive lifestyles and diets, and more consumption of externally produced food results in larger ecological, carbon and water footprints compared to rural areas. The major part of cities' water footprints for food production occurs outside the city as external footprints. It should be noted that the water footprint of a city, country or other unit as typically calculated (e.g. Hoekstra *et al.*, 2011) only depicts the volume of freshwater that is embodied in the goods and services traded. This does not describe the genuine footprint, i.e. the impact in the production region. It is merely a measure of 'shoe size' (Pfister and Hellweg, 2009).

Another side-effect of urbanisation is the loss of agricultural land and associated green water, giving way to expanding cities. Cities are often located in highly productive, water-endowed regions of the country.

Given the rapidly increasing urban population and its role in the global economy and environmental footprints, but also their concentrated knowledge, and financial, social and institutional resources, cities must become nuclei of sustainable consumption and resource efficiency.

2.1.4 Technological development

Technological development and the diffusion of technologies across the globe have various effects on spatio-temporal patterns and trends in resource use and environmental change. Richer countries generally have higher resource productivities, but they also use more resources per capita. Technological development, which improves resource productivity, tends to be at least partially offset by higher consumption levels, the so-called rebound effect (Hertwich, 2005). Increasing productivity in the agricultural and processing sectors (including plant breeding and biotechnology) can reduce resource demand per unit of food or other goods and services produced. Indeed, over the past five decades, grain production has more than doubled while the area of cropland has increased by only about 9% globally (Pretty, 2008). Technological development also enables ever larger infrastructure, however, such as dams or water transfers, more rapid exploitation of natural resources, including fossil fuels, and cheaper transportation of goods around the world - and, with that, accelerating globalisation. In addition, technological development on the supply side can motivate and slow down improvements in eco-efficiency and demand-side management.

2.1.5 Trade: a decoupling of production and consumption

Another effect of globalisation is a spatial decoupling of biomass growth and resource extraction from the manufacture, consumption and disposal of products. Trade in agricultural products responds to growing - including out of season - demand, local scarcity and comparative advantages in certain production factors as well as changing consumption patterns, technological developments in the transport sector, trade liberalisation and other drivers. Trade in agricultural products has grown by 3.5% per year on average over the past 60 years, and by about 4% per year over the past 20 years (WTO, 2008). The trade in food globally has been growing faster than production. However, there are clear regional differences (Anderson, 2010) as is shown in Figure 2.5. The trade balances of developing countries are turning increasingly negative (Bruinsma, 2003).

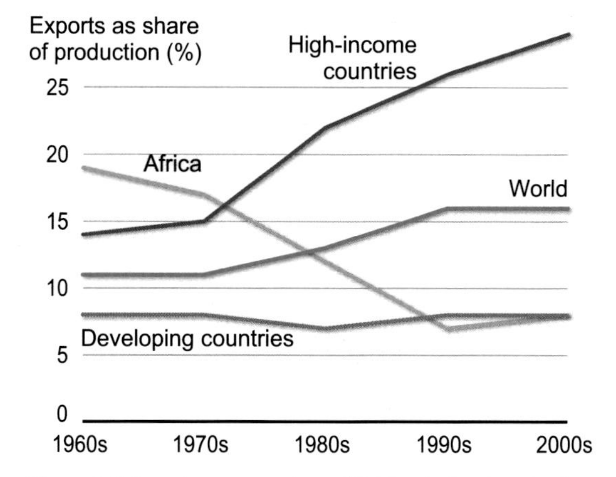

Figure 2.5 Exports as a percentage of total agricultural production. Numbers indicate the first half of the respective decade (Anderson, 2010).

Increasing trade along with increasing transportation capacity allow the global distribution and consumption of crops which are only cultivated in a few places. A prominent example of this decoupling is the cultivation and consumption of soy beans, primarily used as soy cakes for animal feed. While soy is used as livestock feed in many parts of the world, it is only grown in a few locations. The United States and Brazil are the two largest exporters (Monfreda *et al.*, 2008), as is shown in Figure 2.6.

Such a concentration of crops - often, like soy production in Brazil, in the form of large monocultures - can cause significant environmental and social externalities, such as a reduction in local soil carbon storage, the displacement of other land uses into protected areas, and the marginalisation of rural farmers (see Fearnside, 2007; Fearnside *et al.*, 2009, and Weinhold *et al.*, 2011), all of which can further accelerate change and negatively affect the resilience of the respective social–ecological system.

Trade also allows countries to decouple their economic growth from local resource exploitation or overexploitation, by sourcing from other regions and externalising negative side-effects and costs - often across large distances. Thus, trade weakens the message of environmental Kuznets curves - which describe how resource use intensity decreases with increasing GDP (gross domestic product) in developed countries. While importing countries reduce pressure on their own resources and externalise large parts of their resource footprints, sometimes increasing their total footprint beyond national resource availability, in the exporting countries this trade acts as an external

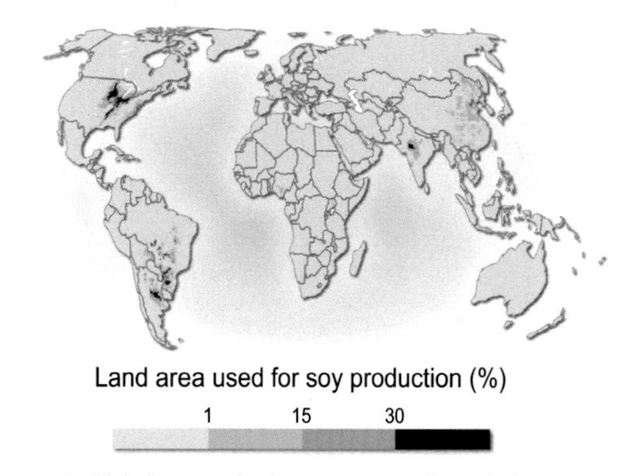

Figure 2.6 Global soy production: concentrated to only four major areas (Monfreda *et al.*, 2008).

driver of change, for example on local or regional water resources (Hoff *et al.*, 2013).

There is an ongoing debate about the possibility of incorporating sustainability criteria into trade regimes or levying cross-border taxes on commodities produced from non-sustainable resource extraction (e.g. Hoekstra and Chapagain, 2007). Initial provisions for a more equitable trading system between developing and least developed countries already exist, for example, in the form of preferential trade agreements between the developed and least developed countries (WTO, 2011).

2.1.6 Decoupling of production and consumption through foreign direct investment

Like trade, foreign direct investment (FDI), in particular in land, redirects growing demand to other regions (Figure 2.7) and becomes an external driver of change in exploiting local water resources. FDI has been a response to growing scarcities and price volatilities in the international commodity markets.

The World Bank (Deininger *et al.*, 2011) estimates that two-thirds of recent land acquisition (by area) has taken place in sub-Saharan Africa. Friis and Reenberg (2010) estimate that 50–60 million ha of land in Africa (2–20% of the agricultural land in the respective countries) has been sold or is under negotiation. In particular, China, India and the Arab nations are acquiring land and associated water resources in sub-Saharan Africa on a large scale, including in countries with water and food insecurity.

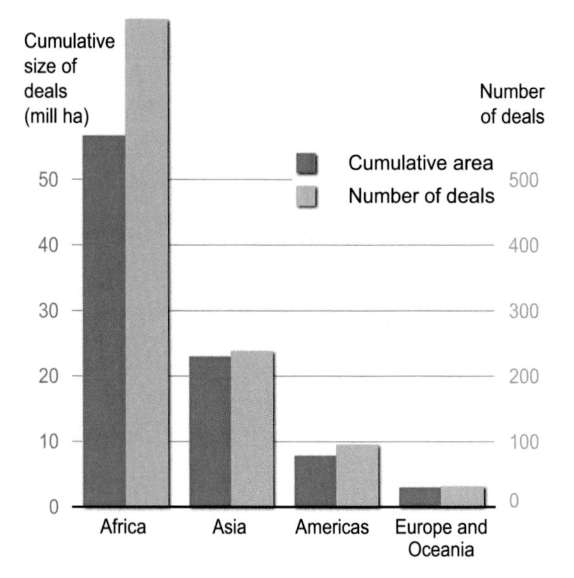

Figure 2.7 Foreign land acquisitions by target region, area and number of projects (Anseeuw *et al.*, 2012)

FDI in land, in combination with growing production for export, can lead to a reduction in resource availability and access for the local population. In such cases, terms like 'land grabbing' or 'water grabbing' are frequently used (e.g. Allan, 2012; Bues, 2011). In principle, FDI could have positive effects in the target country, for example increasing water and land productivity by sharing knowledge and technology, especially in Africa, where agricultural yields are well below the world average (Hoff *et al.*, 2012). To reconcile the economic goals of FDI with social and ecological goals, it will be important to regulate investment within countries by establishing environmental and socio-economic minimum criteria to enable benefits to be shared with the local population (e.g. FAO, 2010b; von Braun and Meinzen-Dick, 2009).

2.1.7 Climate change

Climate change is an additional driver of change, on top of, interacting with and feeding back into the drivers described above, with regionally very different impacts on the water system. Anthropogenic climate change is driven by greenhouse gas (GHG) emissions and land-use change. Land-use change is responsible for about 10% of total CO_2 emissions (Le Quéré *et al.*, 2009). Food production is a key driver of climate change through GHG emissions – in particular the powerful GHGs methane (CH_4) and nitrous oxide (N_2O) – and land-use change. According to Sachs *et al.* (2010), one-third of total GHG emissions originate directly or indirectly from the food sector. At the same time, food production is projected to be seriously affected by climate change (e.g. Knox *et al.*, 2012; Müller *et al.*, 2011).

The primary effects of climate change are warming and increasing evapotranspiration, triggering further effects in the hydrological cycle (intensification) and in water systems. Strzepek *et al.* (2010) suggest that a 4°C increase in temperature would require at least a 10% increase in precipitation to balance higher evaporative losses. Potentially the most critical effects of climate change on water resources are associated with higher variability, more common, more intense and longer extreme events, lower water availability in many drylands, and higher levels of uncertainty, for example, in terms of the onset, length and intensity of the rainy season (Bates, 2008; Milly *et al.*, 2008). Climate change impacts on water resources are projected to increase in the coming decades, including relative to other drivers of change such as population growth.

The impacts of climate change can be mitigated somewhat by increasing atmospheric CO_2 concentration, primarily through higher photosynthetic water-use efficiency, which reduces transpiration requirements and hence also mitigates water scarcity, and may possibly alter crop yields, the competitive strengths of different plant species, and vegetation composition and density under water limitation.

Although a number of uncertainties remain about the direction and magnitude of climate change in particular regions, some robust statements about the impacts on water resources can be made:

1. Already water stressed regions, such as the Middle East and North Africa and Southern Africa, will experience further aridification due to increasing temperatures (higher evaporative demands) and in some cases also decreasing precipitation, with possibly severe effects on agriculture and other ecosystems.
2. Variability and extremes (disturbances and shocks) are generally expected to increase in regions with already highly variable climate, i.e. so-called drylands, interacting with aridification trends and other slow changes.
3. Glacier melting due to global warming will reduce water availability in the dry season in the long run, for example in the Himalayan or Andean basins.
4. Sea level rise will cause saltwater intrusion into coastal aquifers.

Many of these impacts are expected to hit poor people in developing countries hardest, the group that has contributed least to the cumulative CO_2 emissions that cause climate change. Climate change interacts with the other drivers listed above. Together these drivers severely affect the Earth System and its components. While we do not know the intensity and frequency of future climate shocks and their impacts, we can use recent extreme events, such as droughts and floods, as an indicator on what to expect.

Observations of trends and simulations of future hydro-climatology indicate that past statistics on variability and extremes can no longer guide water management and planning or, as Milly *et al.* (2008) put it, 'stationary is dead'. The Intergovernmental Panel on Climate Change (IPCC, 2011) somewhat more carefully states with *medium confidence* that some regions of the world have experienced more intense and longer droughts, in particular in southern Europe and West Africa, and also with *medium confidence* that droughts will intensify in the twenty-first century in some seasons and areas. These climate change impacts on water resources and water-dependent social–ecological systems will potentially increase vulnerability to the other pressures described above – and vice versa.

2.1.8 Political interactions: national and international policies

Other drivers of change include national and international policies with increasingly far-reaching consequences, such as, for example, the EU Biofuel Directive aimed at substituting renewables for fossil fuels, which causes land-use change in the form of agricultural expansion and intensification, and subsequent changes in water demand, landscape diversity and resilience. These effects are not limited to the country or region in which a particular policy has been established. Globalisation and increasing interconnectedness mean the impacts can be felt in distant regions (see teleconnections in Chapter 4). Some of Europe's demand for biofuels – or for oil crops to substitute for European oil crops now going into biofuels – is met by production in Latin America, South East Asia and sub-Saharan Africa.

Internationally, there is a growing number of multilateral environmental agreements (Figure 2.8) and water initiatives, which also affect local water resource management, resource availability and productivity.

Examples of the systemic and potentially far-reaching effects of international agreements are provided by the Clean Development Mechanism (CDM) and Reducing Emissions from Deforestation and Forest Degradation in Developing countries (REDD +) under the UNFCCC. They facilitate emission reduction and emission removal projects in developing countries, including afforestation and reforestations. If implemented on a larger scale, these mechanisms could have implications for

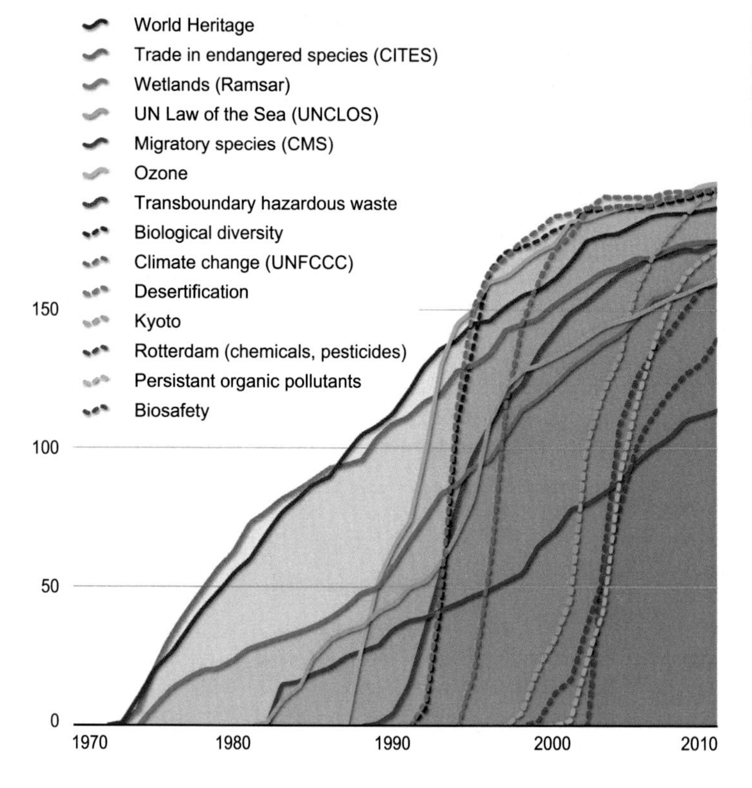

Figure 2.8 Number of countries that have ratified major multilateral environmental agreements over time (UNEP, 2011). The shading indicates the cumulative number of parties to the treaties and protocols.

land–atmosphere interactions, in particular due to the higher level of transpiration from trees than from other types of vegetation. While the conservation of primary forests generally stabilises the existing hydrology, new large-scale tree planting activities can increase vegetation water demand and hence significantly reduce landscape water availability (Calder, 2005).

Two types of change can be identified as resulting from the above drivers, which act at different spatial scales and which require different management and governance responses:

1. Truly global systemic changes, in particular atmospheric CO_2 and climate change;
2. Pandemic local changes that cumulatively have global effects, such as land- and water-use changes or environmental pollution.

It is projected that global, systemic changes will become relatively more important in the future (Steffen *et al.*, 2004).

Most of the above drivers are interconnected with **globalisation**. The **increasing connectivity** of regions, in particular through trade or FDI, not only propagates economic disturbances around the world, but also causes social and environmental impacts such as changed livelihoods and resource overexploitation in regions far away from the original driving force. If institutions in the affected regions are weak or fragmented, and international activities such as FDI not well regulated, undesirable environmental and social externalities are likely. Some international investments may even track levels of social or environmental standards, in search of the lowest.

Globalisation will remain a major driver of change, and it will be important to regulate and shape it – including through economic incentives – towards sustainable development. For example, trade can help to overcome national food self-sufficiency problems and at the same time may increase overall resource use efficiency (see e.g. Chapagain and Hoekstra, 2008).

2.2 Interacting anthropogenic drivers, impacts and feedbacks in the Earth System

Responses to different human pressures interact with each other, sometimes damping the effects of the original force and sometimes amplifying them. Impacts and responses become feedbacks, which in turn can lead to further pressures (Steffen *et al.*, 2004). Feedbacks occur when a change in system A triggers changes in system B, which eventually influence system A again. Feedbacks may be self-regulating (negative) or self-amplifying (positive). Positive feedbacks are particularly important in socio-ecological systems as they can push a system over a critical threshold and into a new regime.

Some of the impacts of and responses to the drivers and feedbacks listed above include:

- Land conversion: humans have directly transformed about 50% of the ice-free land surface (Steffen *et al.*, 2004). Anthropogenic biomes, i.e. terrestrial biomes with sustained direct human interaction, now cover more than 75% of the ice-free land surface; less than 25% remains as wild land (Ellis and Ramankutty, 2008).
- Agricultural intensification, in particular shifts to input-intensive mechanised agriculture and irrigation, but also the expansion of agricultural land, have resulted in almost a tripling of agricultural production over the past 50 years (FAO, 2011b), which has a number of environmental side-effects.
- Alterations of ecosystems, ecosystem services and primary productivity: 70% of temperate forest and Mediterranean biomes, 60% of tropical and subtropical dry forest, and 50% of tropical and subtropical savannah have already been converted (Millennium Ecosystem Assessment, 2005). Global net primary productivity (NPP) has been reduced by about 10% (Haberl *et al.*, 2007).
- There are alterations of biogeochemical cycles, for example constantly growing CO_2 emissions, and anthropogenic nitrogen fixation from the atmosphere exceeds all the Earth's natural terrestrial processes combined (Rockström *et al.*, 2009b), and there is overexploitation of water resources.

These trends in anthropogenic drivers, impacts and feedbacks in the Earth System are discussed further below.

2.2.1 Land conversion and the implications for carbon stocks

Humans are now a major force in shaping the land surface (Figure 2.9). The fact that almost the entire terrestrial land surface has been transformed to some extent by human action is captured in the term 'anthromes', i.e. anthropogenically modified biomes. Land conversions have been largely determined by population development. For example, temperate woodland biomes have been developed and transformed more intensively over longer periods of time, compared to savannahs, scrubland and grassland which have been subject to rapid changes more recently (Ellis, 2011). However, globalisation increasingly externalises the demands resulting from population increases and economic development to other – often less densely populated – regions.

Globally, agricultural areas expanded by 14% between 1961 and 2005, and they are projected to expand by another 9% by 2050. These past rates of and future projections for the expansion of agricultural areas are 23% and 21% in developing countries, and 40% and 30% in Latin America (Bruinsma, 2009). However the largest increase in agricultural production has been and will be achieved from the intensification of production on existing agricultural land.

Primary forests now account for just 36% of the total area of forest. The highest rate of forest loss is currently observed in Brazil and Indonesia, but many South East Asian and African forests are also coming under increasing pressure. In contrast, forest cover in Europe and China is increasing (FAO, 2010a) due to agricultural land abandonment and active reforestation programmes, respectively.

Any changes in land cover, including forests, directly affect the stocks of carbon stored in biomass and the soil. Increasing forest cover means that temperate regions are increasingly turning from carbon sources into sinks (FAO, 2011a). By contrast, in the tropical regions ongoing land conversions, in particular deforestation, result in large carbon losses (Figure 2.10).

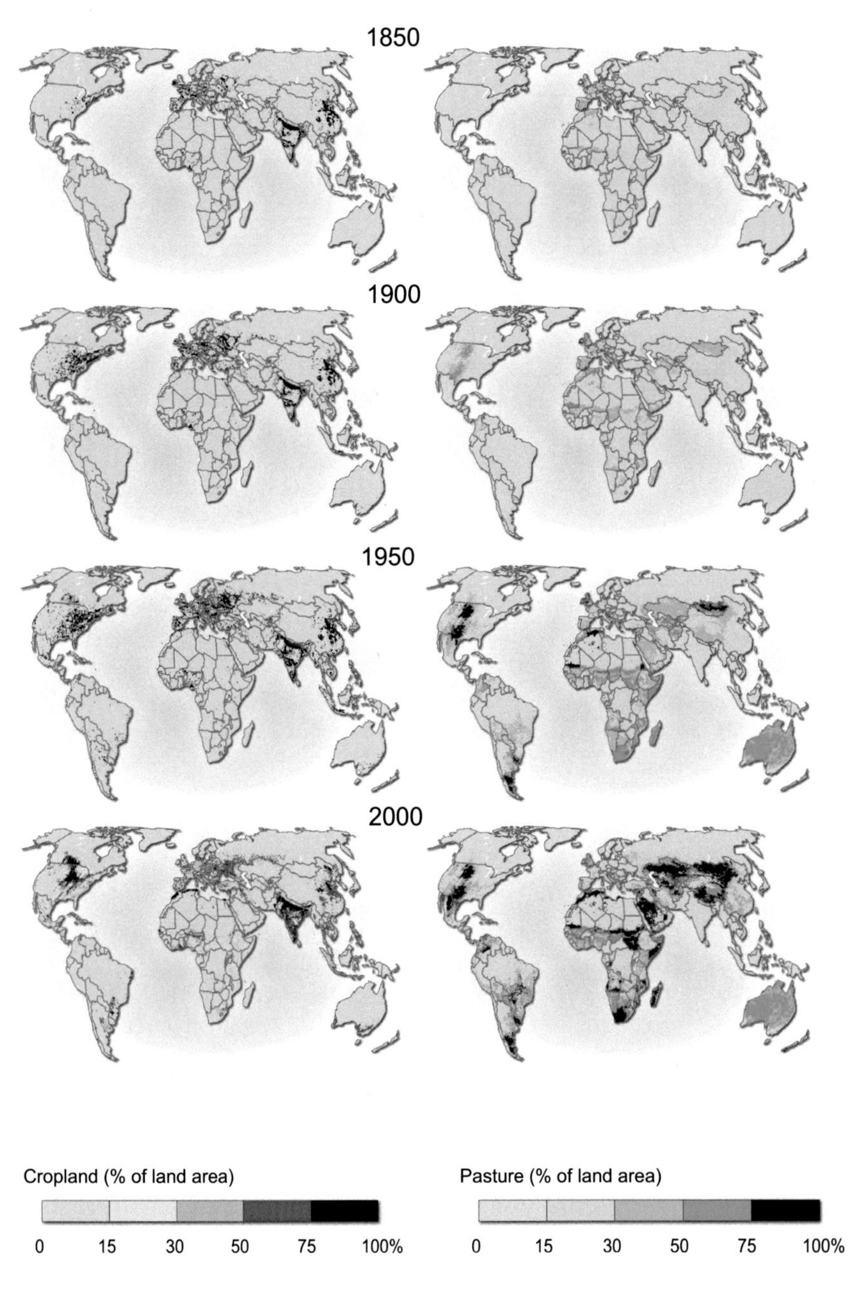

Figure 2.9 Expansion of arable land and pastures, 1850–2000 (Klein Goldewijk *et al.*, 2011).

2.2.2 Agricultural expansion and intensification

The world's agricultural production has grown 250–300% in the past 50 years, in response to the growing demand for food. More recently, the demand for biofuel has also been increasing rapidly (Berndes, 2008) in response to the need for alternatives to fossil fuels for climate protection and energy security reasons. Box 2.1 explains the trends for water demand and consumptive water use for biofuel growth and production.

(a)

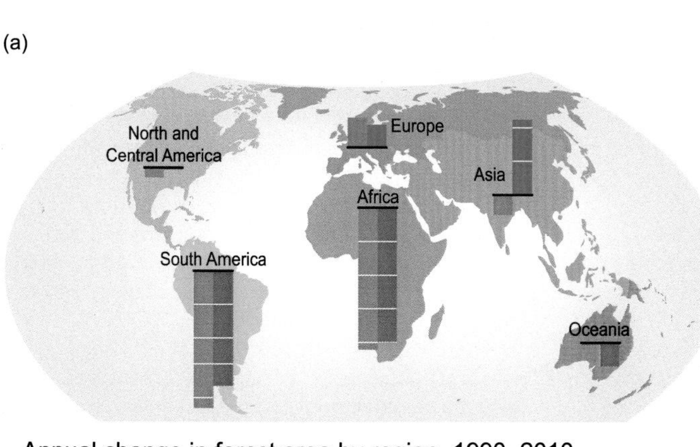

(b)

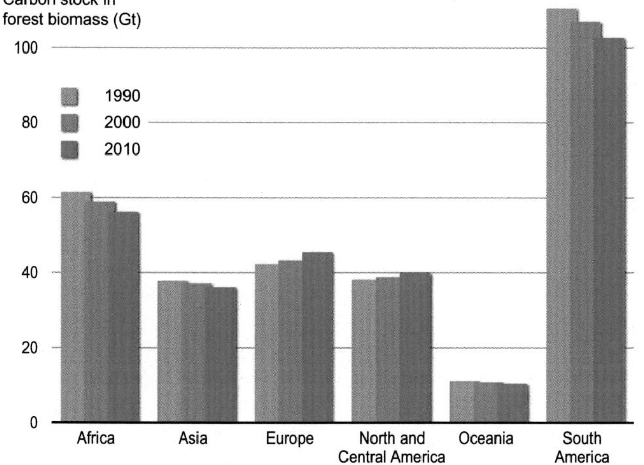

Figure 2.10 Changes in forests per continent 1990–2010 indicating a loss of carbon sink capacity. (a) Change in area from deforestation and reforestation. (b) Change in carbon stocks (FAO, 2010a).

Box 2.1 Bioenergy and water

Göran Berndes, Chalmers University of Technology

Biomass meets about 10% (50 EJ) of the annual global demand for primary energy. A major part of this biomass is used for cooking, space heating and lighting, generally by poorer populations in developing countries. It is commonly called traditional bioenergy. Modern bioenergy use for industry, heating, power generation and transportation fuels corresponds to about 10 EJ (primary energy) but this share is growing rapidly. For instance, government policies in various countries led to a five-fold increase in global biofuels production in the period 2000–2008.

Longer-term energy scenarios (to 2050) commonly foresee a substantial increase in bioenergy use – up to 700 EJ/year – especially in cases where such scenarios include climate targets, such as the 2° target. This bioenergy use can be compared with today's global industrial wood production of 15–20 EJ/year, and the global harvest of major crops (cereals, oil crops, sugar crops, roots and tubers and pulses), which corresponds to about 60 EJ/year (FAO, 2013).

A wide range of biomass feedstocks are used for bioenergy – including forest and agricultural residues, organic municipal solid waste and other organic waste streams, tree plantations, conventional

Box 2.1 *(cont.)*

food/feed crops as well as dedicated energy crops – and a variety of processes are employed or developed to convert these feedstocks into gaseous, liquid and solid fuels to be used for transport or for producing electricity and heat. The scale of production can vary from farm level biomass conversion plants to centralised plants requiring biomass supply infrastructures similar to those of large pulp and paper plants. Bioenergy options have varying levels of technical maturity and economic competitiveness. Some are already deployed as competitive options on the energy markets (e.g. sugar cane ethanol, and heat and power generation from waste and residues), while other options still require significant government support in the form of subsidies and/or funding for research and development, and demonstration purposes.

Water requirements

Like many other industrial activities, the process of biomass conversion can require a substantial volume of water. Most of this water is returned to rivers and other water bodies and is therefore available for future use. Unless suitable equipment is installed, however, negative impacts can occur due to chemical and thermal pollution loading the aquatic systems. This problem is not restricted to the biomass-based industry. It is a general problem in countries with less strict environmental regulations or insufficient law enforcement capacity.

The water quality consequences of feedstock production depend on feedstock type. Strategies that focus on biofuels for transport and mainly lead to increased cultivation of conventional agricultural food/feed crops, such as cereals, oil crops and sugar crops, will have consequences for water quality that resemble those associated with the food sector. If, however, mainly lignocellulosic feedstocks are produced, the consequences for water quality can be both negative and positive. For instance, unsustainably high extraction of harvest residues may result in higher soil erosion rates and sediment loads on aquatic ecosystems, and also increased eutrophication, as fertiliser demand may increase when residue recirculation in the fields goes down. Conversely, if farmers cultivating sloping and otherwise sensitive soils shift from conventional annual food crops to producing perennial herbaceous and woody plants as bioenergy feedstocks, water quality would improve. The cultivation of different types of lignocellulosic plants represents an opportunity for farmers to make productive use of land that is otherwise excluded from cultivation due to associated water quality impacts. Furthermore, feedstock production systems can be shaped to help mitigate existing water quality problems. For instance, some plants can be cultivated as vegetation filters for treatment of nutrient-bearing water (e.g. pre-treated wastewater from households and runoff from farmland), reducing eutrophication. Soil-covering plants and vegetation strips can also be located to limit water erosion, reduce evaporating surface runoff and enhance infiltration to soils.

From a water quantity perspective, bioenergy is unique among energy options in that feedstock production can require much more water than the subsequent processing into various fuels. The total water requirement can therefore be several hundred times larger than for the fossil fuels it is commonly intended to replace. Water use in biomass feedstock production is also different in that much of the water is evapotranspired into the atmosphere, and is consequently not immediately available for further use by humans and the ecosystem in the watershed from which it was originally withdrawn. Rainfed feedstock production does not require water extraction from groundwater, lakes and rivers, but it can still reduce downstream water availability by redirecting precipitation from runoff and groundwater recharge to crop evapotranspiration.

There are certainly places where an expanding bioenergy industry is unlikely to be constrained by lack of water. There are also places where increased demand for water for bioenergy may place considerable additional stress on available water resources. The water quantity consequences will be very different depending on how bioenergy expansion takes place. Some plants that are suitable as bioenergy feedstock are also drought-tolerant and relatively water efficient. These crops can be grown in areas not suitable for conventional food and feed crops. Plants that are cultivated in multi-year rotations can also make better use of rain that falls outside the growing season of more conventional crops.

Thus, opportunities exist to improve water productivity in agriculture and alleviate competition for water as well as the pressure on other land-use systems. However, these opportunities need to be carefully assessed from a water balance perspective. For instance, the use of marginal areas with sparse vegetation to establish high-yield bioenergy plantations may lead to substantial reductions in downstream water availability. This may be an unwelcome effect that requires the management of a trade-off between upstream benefits and downstream costs.

The recent growth in harvested areas has been concentrated in a few key commodities, in particular soybean, maize, oil palm and rapeseed (Deininger *et al.*, 2011). Oil crops for food and feed have seen the fastest expansion of any crop type over the past decade (Bruinsma, 2003). The area covered by oil palm plantations in Indonesia has doubled over the past 10 years. Indonesia is the largest producer, followed by Malaysia (see Chapter 4 for the environmental and resilience effects of the rapid expansion of oil palm in Borneo). Soybean expansion is concentrated in the Argentinean Pampas and the Brazilian Cerrado. In Brazil, this expansion is driven by exports growing at a rate of 15% per year, resulting in one of the world's fastest land transformations and the consequent expansion of the agricultural frontier into Amazonia (Lapola *et al.*, 2010; see also Chapter 4 on the environmental and resilience effects). This expansion may further accelerate as foreign investors begin to purchase land in Brazil, in particular for soy production. Furthermore, increasing domestic and international demand for biofuel, primarily from sugar cane, competes with soybean for land and water, and pushes other land uses such as grazing towards Amazonia (Martinelli and Filoso, 2008).

The expansion of agricultural land at the expense of other land-cover types, in particular natural and semi-natural ecosystems, is hitting sustainability limits in several world regions, impinging on other ecosystem services, protected areas and biodiversity as well as on carbon storage. An increased vulnerability of social–ecological systems to shocks may result from such land conversions to intensively used, high-input – often monoculture – agricultural systems.

An alternative to agricultural expansion is intensification on existing agricultural land, for example through multiple cropping, specialisation, mechanisation and additional inputs such as fertiliser or irrigation. Irrigation provides higher yields in particular in arid and strongly seasonal climates. More than 40% of the total increase in food production in the past 50 years has been achieved in irrigated areas, which have doubled in area during that period – compared to a growth in total agricultural area by 12% during that time (FAO, 2011b). This growth in irrigation has been dominated by a handful of countries (see Figure 2.11).

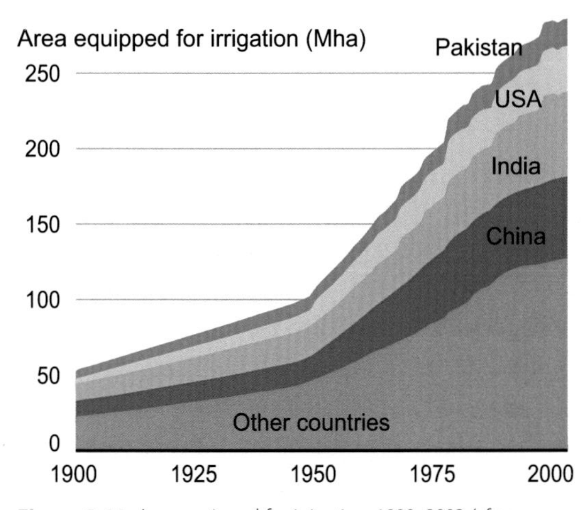

Figure 2.11 Area equipped for irrigation, 1900–2003 (after Freydank and Siebert, 2008).

However the growth rate of irrigated areas has slowed. It peaked in the 1970s at 3% per year, dropping to 1.1% in the 1980s and was only 0.2% in the 2000s. It has also been recognised that increasing reliance on irrigation, which was initially a measure to increase buffer capacity against periodic water scarcity, may in some cases weaken the resilience of social–ecological systems, for instance if groundwater resources are overexploited (see India's irrigation economy in Chapter 4) or groundwater recharge decreases, for example due to climate change. A historical example of the breakdown of an irrigation-based culture can be found in Mesopotamia *c.* 2100–2000 BC, where the third dynasty of Ur relied on unsustainable large-scale irrigation systems (Tainter, 2006).

Both the expansion of cropland and an intensification of the use of existing cropland have environmental side-effects and could contribute to the transgression of planetary boundaries (Rockström *et al.*, 2009a). While expansion causes a reduction of other ecosystems and their services, intensification can negatively affect the surrounding and downstream environment as well as *in-situ* agrobiodiversity (see Keys *et al.*, 2012). Hence, new development pathways need to be based on sustainable intensification as well as demand management, for example through reducing overconsumption and wastage (Kummu *et al.*, 2012).

High levels of land degradation – an effect of non-sustainable agricultural expansion or intensification – are often found in the regions where poverty levels are highest (FAO, 2011b), indicating the need to address agricultural productivity and development in tandem.

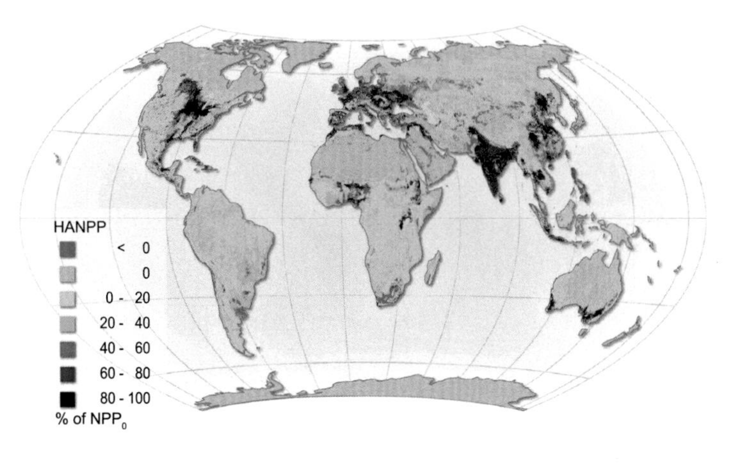

Figure 2.12 Human appropriation of net primary productivity (NPP) as a percentage of the NPP of potential natural vegetation (NPP_0). Green or negative values indicate that excess biomass above NPP_0 remains in the system after harvest (Haberl *et al.*, 2007).

2.2.3 Alterations of ecosystems, ecosystem services and primary productivity

Natural ecosystems have been reduced by the expansion of agricultural systems:

- Cropland – mostly converted from forest land – now covers 12% of the global land area, ranging from 7.5% in the Middle East and North Africa region to 40% in South Asia.
- Pasture – mostly converted from grassland – now covers 22% of the global land area, ranging from 2% in South East Asia to 30% in tropical Africa (Ellis, 2011; Houghton, 1999).

The conversion of natural land to agriculture and agricultural intensification – often towards monocultures – has a number of side-effects:

- A reduction in the diversity of crop and livestock species and agrobiodiversity.
- A reduction in the diversity of soil biota.
- A decrease in soil carbon stocks (Lal, 2007).
- An increase in nitrate, other nutrient and pesticide concentrations in downstream water bodies, resulting in eutrophication and low oxygen conditions all the way to the receiving marine systems (Matson *et al.*, 1997).

Land degradation due to non-sustainable intensification globally has affected 561 million ha of cropland, 685 million ha of pasture and 719 million ha of forest and woodland (Lal, 2007).

Land-use change also redirects green and blue water flows to other ecosystems and their services – in particular to provisioning services such as food, feed, biofuel and timber production – often at the expense of other supporting and regulating services, such as carbon sequestration and climate regulation (see also moisture recycling in Chapter 4). The Millennium Ecosystem Assessment (2005) investigated the extent and scales of these crucial ecosystem services. Since water is central to all ecosystem services, any changes in the hydrological cycle can directly or indirectly affect several of the planetary boundaries (see Chapters 4, 5 and 7).

Human land use has changed the composition, structure and functioning of ecosystems globally. One aggregate measure of ecosystem state and functioning is **net primary production (NPP)**. This can be defined as the amount of vegetable matter produced per unit of time, the net flux of carbon from the atmosphere into vegetation, or gross primary production (photosynthesis) minus plant respiration. Human land use, primarily the conversion of natural ecosystems to agricultural systems, has caused a global NPP reduction of about 10%, with distinct spatial patterns and regional differences (Haberl *et al.*, 2007). The highest anthropogenic NPP reductions have been observed for parts of Africa as well as western, South and South East Asia. Significant increases in NPP have occurred in intensively cultivated parts of Europe as well as irrigated drylands, for example in North Africa and South Asia.

According to Steffen *et al.* (2004), humans now use or co-opt about 50% of global NPP. Figure 2.12 shows the spatial distribution of this human appropriation of growing biomass – indicated by NPP, which follows population density but is also a function of human consumption levels and natural NPP.

2.2.4 Alterations of biogeochemical cycles, including water

Human activities such as agriculture, industry and trade, among others, have altered **biogeochemical cycles**, including those of carbon, nitrogen and water.

By burning fossil fuels, humans have increased the concentration of atmospheric CO_2 from about 280 ppm in pre-industrial times to 390 ppm. It is currently rising faster than almost all the previous projections of the IPCC (IPCC, 2000). This is likely to lead to significant changes in the Earth's climate. Land-use change, in particular from natural ecosystems to agro-ecosystems, often reduces soil and vegetation carbon storage, leading to subsequent additional GHG emissions.

Human-related nitrogen fixation from the atmosphere is greater than that by all other natural terrestrial sources combined (Vitousek *et al.*, 1997). There are large regional variations in mineral nitrogen fertiliser use, ranging from above 100 kg/ha per year in parts of Europe, the United States and China to less than 10 kg/ha per year in many African countries. This results in nitrogen leaching and atmospheric emissions from the former and nitrogen depletion in the latter (Liu *et al.*, 2010).

The leaching of nitrogen into water bodies and emissions into the atmosphere, with subsequent downstream accumulation or downwind deposition that in many regions is an order of magnitude greater than natural depositions, cause environmental damage, drive biodiversity losses and contribute to or interfere with climate change through the production of N_2O, among other things (Galloway *et al.*, 2008). Another effect of increasing nutrient inputs and concentrations in agricultural systems is a potential increase in pathogens and pest populations (Matson *et al.*, 1997).

Water-related impacts of, responses to and feedbacks from human-driven change include:

- Increased global runoff of almost 2000 km^3/year (Rost *et al.*, 2008) caused by the conversion of natural ecosystems, mostly to agricultural land which generally lowers evapotranspiration. These hydrological changes are often accompanied by additional erosion and the transportation of sediments and nutrients.
- A redirection of evaporative fluxes from natural to agro-ecosystems. Consumptive water use from rainfed and irrigated cropland amounts to more than 7000 km^3/year, or about twice that amount if all grazing land is included (Hoff *et al.*, 2010).
- Human appropriation of about 50% of the accessible runoff (Postel *et al.*, 1996).
- Temporary or permanent 'closure' of a growing number of river basins, such as the Nile, Yellow, Colorado and Ganges rivers due to large water withdrawals. River basins are closed when all renewable blue water is used and additional commitments can only be met at the cost of existing ones, and no more water reaches the sea (Falkenmark and Molden, 2008).
- A reduction in the extent of inland water bodies, such as the Aral Sea and Lake Chad.
- Groundwater overdraft above recharge rates in critical regions such as India, China and the Middle East and North Africa region in the order of 160 km^3/year (Kemper, 2007).
- The damming of rivers has increased the time water resides in river courses globally by 1–2 months and has decreased total sediment flow by 25–30%, with consequences for water availability, water quality, river morphology and nutrient availability in floodplains and aquatic ecosystems (Vörösmarty *et al.*, 2003).
- Reductions in environmental flows, with impacts on aquatic ecosystems and their services, in particular in North Africa, western and South Asia and the south-western United States (Smakhtin *et al.*, 2004).
- A loss of wetland ecosystems (in some regions more than 50% of wetland was converted in the twentieth century), causing a decline in associated ecosystem services, for example water purification and provisioning, fish supply and climate regulation (Millennium Ecosystem Assessment, 2005).
- Significant but as yet unquantified loss of soil water storage due to land degradation.

Human influence now exceeds natural forces in continental aquatic systems in many parts of the world (Meybeck and Vörösmarty, 2004).

Human activity has led to an 'aging and drying of landscapes, breakdown of vegetation, opening up and degradation of water and matter cycles, reduced resource efficiency, and interference with the bloodstream of the biosphere'. Ripl (2003) defines the sustainable development of landscapes as 'the most reduced rate of ageing' (see Box 4.1, Chapter 4).

These interferences with water, as the 'bloodstream of the biosphere', and associated biomass production can also affect carbon fluxes between land and the atmosphere, and hence contribute to observed and projected switches of land carbon sinks to sources of carbon, thereby contributing to climate change (e.g. Le Quéré *et al.*, 2009; Morales *et al.*, 2006).

Whether any of these changes will lead to abrupt regime shifts or even a collapse of social–ecological systems will depend – among other factors – on the rate and magnitude of change and the vulnerability, adaptability and transformability of the respective system.

2.3 We now live in the Anthropocene era, and are approaching various water-related tipping points

Humans are now changing their environment in all parts of the globe. In many regions they have become the major driver of change, through processes entrenched for some time before their importance was recognised (Vitousek *et al.*, 1997). This increasing human dominance of the Earth System over two centuries is recognised in the term Anthropocene, which indicates that we have entered a new geological era, the onset of which was marked by the beginning of the Industrial Revolution at the end of the eighteenth century (Steffen *et al.*, 2007; Steffen *et al.*, 2011). In addition to human dominance of landscapes, the Anthropocene era is also characterised by a warmer climate, changed hydrological and other biogeochemical cycles and an impoverished biosphere.

From the analysis of early land-use change (e.g. Pongratz *et al.*, 2008, see above) it appears that significant human influence on the Earth System began about 1000 years ago. Ruddiman (2003) dates the beginning of significant human impacts on the Earth System, which he also terms 'Anthropocene', even further back to 5000–8000 years ago – the start of significant forest clearances and conversion to agriculture, as well as irrigated rice cultivation and associated greenhouse-gas emissions in Eurasia, parts of China and India. However, the rate of land clearances and cultivation was constrained at that time by the available energy supply, i.e. human and animal power.

That energy constraint was overcome when first water power and then fossil fuels became available as new energy sources. Within 100 years, some 50% of land ecosystems had been anthropogenically transformed (Ellis, 2011). A second stage of the Anthropocene era began in the middle years of the twentieth century, notable by the various hockey-stick curves that depict various global change indicators (see Chapter 1). The speeding up of many anthropogenic processes after World War II began the 'great acceleration' (Hibbard *et al.*, 2007) – human dominance over natural forces, resulting in much more rapid rates of change than in any previous era of the Earth System.

Social–ecological systems at all scales may hit a number of critical thresholds, beyond which regime shifts occur and future human security could be threatened. These critical thresholds have been called tipping points – a critical point at which any further change triggers a transition to a different state, i.e. a regime shift. A number of 'tipping elements', or large-scale components in the Earth System which may pass a tipping point due to climate change impacts, have been identified by Lenton *et al.* (2008) (see Chapter 1). Chapter 3 further covers water-related tipping points.

Summary

Human pressures on the Earth System

Anthropogenic pressures are multiple, complex and equal in magnitude to some of the great forces of nature – and they are accelerating. They interact and, if they cross critical thresholds, can trigger abrupt, non-linear changes. It has been recognised for some time that key environmental parameters have moved well beyond the range of natural variability, and that the magnitude and rate of change are unprecedented. Two types of change can be generated: global systemic changes and pandemic local changes with cumulatively global effects.

The four main drivers are:

- Demographic: the human population continues to expand despite decreasing fertility rates. Higher fertility rates in much of Africa and West Asia mean that nearly all the increase takes place in the developing world, often in regions with almost stagnant development. The total world population is projected to stabilise after 2050.
- Economic development has a general tendency to increases per capita resource use. Human diets respond to income levels: some 3500 kcal/capita per day is consumed in the OECD countries,

including close to 30% from animal sources; in sub-Saharan Africa and South Asia, the figure is less than 2200 kcal/capita per day, with less than 10% from animal sources. Increased wealth often leads to overconsumption and food wastage. To avoid transgressing critical boundaries, it will be necessary to decouple economic development from resource use.

- Urbanisation is a major driver of change and is driven by push and pull factors: degrading livelihoods in rural areas and economic opportunities in cities.
- Technological development, finally, and the diffusion of technologies across the globe have various effects in terms of resource use and are contributing to environmental change. Often, development that improves resource productivity tends to be partially offset by higher consumption levels in the rebound effect.

Climate change is a more recent additional driver of change that interacts with and feeds back into other drivers with regionally very different impacts on water systems. Furthermore, national and international policies can have increasingly far-reaching cross-regional consequences. A growing number of multilateral environmental agreements also affect water resources management. Finally, both trade and FDI have been rapidly expanding over time and have caused a decoupling of production from consumption. They have allowed countries to decouple their economic growth from their local natural resource situation. Trade in agricultural products, although growing more slowly than other commodity trade, is still growing faster than agricultural production. This makes developing countries' agricultural trade balances increasingly negative.

Earth System responses

The Earth System is responding in complex ways to all these human activities and interacting drivers, impacts and feedbacks, in particular in terms of land conversion, ecosystem change and degradation, altered biogeochemical cycles and changing water systems.

Responses to different human pressures interact with each other, sometimes damping the effects and sometimes amplifying them. Responses become feedbacks, which in turn can lead to further pressures. Self-amplifying feedbacks in social–ecological systems are particularly important as they can push a system over a threshold into a new regime.

Patterns of human-dominated land – anthropogenic biomes or so-called anthromes – are largely determined by population density. The remaining primary forest accounts for only 36% of total forest area. In response to the growing demand for food, the world's agricultural production has grown by 250–300% over the past 50 years. More recently, demand for biofuels has been increasing in response to the need for alternative fuels. While the growth in irrigated area has slowed in recent years, the expansion of agricultural land at the expense of other land-cover types is approaching its limits in the natural and semi-natural ecosystems in several regions of the world. Human land use has changed the composition, structure and functioning of ecosystems. Highly distinct regional patterns of anthropogenic land-cover change mean that there has been a global reduction in NPP. Overall, human activity has altered biogeochemical cycles, including those of carbon, nitrogen and water.

The water-related impacts of and feedbacks from the human modification of the Earth System have not been quantified, reflecting manifold alterations in the global water circulation system: precipitation, moisture feedback, groundwater recharge, river flow and seasonality, frequency of floods and droughts, consumptive water use, environmental flow, and so on. Interferences with associated biomass production contribute to observed switches of land carbon sinks to carbon sources, further contributing to climate change.

An era of significant human influence

Humans are now changing their environment in all parts of the globe. Significant human influence on the Earth System began about 1000 years ago, accelerated 200 years ago, when energy constraints were overcome by the use of fossil fuels, and entered a new era of great acceleration some 50 years ago, when human forces began to dominate natural ones. With time, anthropogenic modifications to the Earth System may approach several tipping points – critical thresholds beyond which regime shifts may occur.

The magnitude and rate of change are unprecedented. Key environmental parameters have already moved well beyond the range of natural variability. Although societal awareness of the magnitude of change has existed for more than a decade, efforts to elicit serious, large-scale governance responses have been a failure.

References

Allan, J. A. (2012). *Handbook of Land and Water Grabs in Africa: Foreign Direct Investment and Food and Water Security*. London: Routledge.

Anderson, K. (2010). Globalization's effects on world agricultural trade, 1960–2050. *Philosophical Transactions of the Royal Society B: Biological Sciences*, **365**, 3007–3021.

Anseeuw, W., Boche, M., Breu, T. *et al.* (2012). Transnational land deals for agriculture in the global south: analytical report based on the land matrix database. The Land Matrix Partnership. Available at: http://landportal.info/landmatrix/media/img/analyticalreport.pdf.

Bates, B. (2008). Climate Change and Water: Technical Paper of the Intergovernmental Panel on Climate Change, IPCC Secretariat, Geneva.

Berndes, G. (2008). Water demand for global bioenergy production: trends, risks and opportunities. *Journal of Cleaner Production*, **15**, 1778–1786.

Bicheron, P., Defourny, P., Brockmann, C. *et al.* (2008). *GLOBCOVER Products Description and Validation Report*. Paris: European Space Agency.

Bruinsma, J. (2003). *World Agriculture: Towards 2015/2030: An FAO Perspective*. Rome: Food and Agriculture Organization.

Bruinsma, J. (2009). *The Resource Outlook to 2050: By How Much do Land, Water and Crop Yields Need to Increase by 2050?* Rome: Food and Agriculture Organization.

Bues, A. (2011). Agricultural foreign direct investment, water rights and conflict: an institutional analysis from Ethiopia. MSc thesis, Humboldt University Berlin and PIK Potsdam.

Calder, I. R. (2005). *Blue Revolution: Integrated Land and Water Resource Management*. London: Earthscan.

Chapagain, A. K. and Hoekstra, A. Y. (2008). The global component of freshwater demand and supply: an assessment of virtual water flows between nations as a result of trade in agricultural and industrial products. *Water International*, **33**, 19–32.

Cordell, D., Drangert, J. O. and White, S. (2009). The story of phosphorus: global food security and food for thought. *Global Environmental Change*, **19**, 292–305.

Deininger, K., Byerlee, D., Lindsay, J. *et al.* (2011). *Rising Global Interest in Farmland: Can it Yield Sustainable and Equitable Benefits?* Washington, DC: World Bank.

Ellis, E. C. (2011). Anthropogenic transformation of the terrestrial biosphere. *Philosophical Transactions of the Royal Society A: Mathematical, Physical and Engineering Sciences*, **369**, 1010–1035.

Ellis, E. C. and Ramankutty, N. (2008). Putting people in the map: anthropogenic biomes of the world. *Frontiers in Ecology and the Environment*, **6**, 439–447.

Falkenmark, M. and Molden, D. (2008). Wake up to realities of river basin closure. *International Journal of Water Resources Development*, **24**, 201–215.

Fearnside, P. M. (2007). Brazil's Cuiabá-Santarém (BR-163) highway: the environmental cost of paving a soybean corridor through the Amazon. *Environmental Management*, **39**, 601–614.

Fearnside, P. M., Righi, C. A., Graça, P. M. L. A. *et al.* (2009). Biomass and greenhouse-gas emissions from land-use change in Brazil's Amazonian 'arc of deforestation': the states of Mato Grosso and Rondônia. *Forest Ecology and Management*, **258**, 1968–1978.

Food and Agriculture Organization (2010a). *Global Forest Resources Assessment 2010*. Rome: Food and Agriculture Organization.

Food and Agriculture Organization (2010b). Principles for responsible agricultural investment that respects rights, livelihoods and resources. Discussion Note prepared by FAO, IFAD, UNCTAD and the World Bank Group. Rome: Food and Agriculture Organization.

Food and Agriculture Organization (2011a). *State of the World's Forests 2011*. Rome: Food and Agriculture Organization.

Food and Agriculture Organization (2011b). *The State of the World's Land and Water Resources for Food and Agriculture (SOLAW): Managing Systems at Risk*. Rome and London: Food and Agriculture Organization and Earthscan.

Food and Agriculture Organization. (2013). *FAOSTAT Online Database*. Rome: Food and Agriculture Organization. Available at: http://faostat.fao.org/ (accessed multiple dates).

Freydank, K. and Siebert, S. (2008). Towards mapping the extent of irrigation in the last century: time series of irrigated area per country. Frankfurt Hydrology Paper 08. Institute of Physical Geography, University of Frankfurt, Germany.

Friis, C. and Reenberg, A. (2010). Land grab in Africa: emerging land system drivers in a teleconnected world. GLP Report: 1. Global Land Project International Project Office.

Galloway, J. N., Townsend, A. R., Erisman, J. W. *et al.* (2008). Transformation of the nitrogen cycle: recent trends, questions, and potential solutions. *Science*, **320**, 889–892.

Godfray, H. C. J., Beddington, J. R., Crute, I. R. *et al.* (2010). Food security: the challenge of feeding 9 billion people. *Science*, **327**, 812–818.

Haberl, H., Erb, K. H., Krausmann, F. *et al.* (2007). Quantifying and mapping the human appropriation of net primary production in Earth's terrestrial ecosystems. *Proceedings of the National Academy of Sciences of the United States of America*, **104**, 12942–12945.

Hertwich, E. G. (2005). Consumption and the rebound effect: an industrial ecology perspective. *Journal of Industrial Ecology*, **9**, 85–98.

Hibbard, K., Crutzen, P., Lambin, E. *et al.* (2007). Decadal-scale interactions of humans and the environment. In *Integrated History and Future of People on Earth*, Dahlem Workshop Report, ed. Costanza, R., Graumlich, L. and Steffen, W. Santa Barbara, CA: National Center for Ecological Analysis and Synthesis, pp. 341–378.

Hoekstra, A. Y. and Chapagain, A. K. (2007). Water footprints of nations: water use by people as a function of their consumption pattern. *Water Resources Management*, **21**, 35–48.

Hoekstra, A. Y., Chapagain, A. K., Aldaya, M. M. and Mekonnen, M. M. (2011). *The Water Footprint Assessment Manual: Setting the Global Standard.* London: Earthscan.

Hoff, H., Döll, P., Fader, M. *et al.* (2013). Water footprints of cities: indicators for sustainable consumption and production. *Hydrolology and Earth System Sciences*, **10**, 2601–2639.

Hoff, H., Falkenmark, M., Gerten, D. *et al.* (2010). Greening the global water system. *Journal of Hydrology*, **384**, 177–186.

Hoff, H., Gerten, D. and Waha, K. (2012). Green and blue water in Africa: how foreign direct investment can support sustainable intensification. In *Handbook of Land and Water Grabs in Africa: Foreign Direct Investment and Food and Water Security*, ed. Allan, J. A., Keulertz, M., Sojamo, S. and Warner, J. London: Routledge, pp. 359–375.

Houghton, R. A. (1999). The annual net flux of carbon to the atmosphere from changes in land use 1850–1990. *Tellus B*, **51**, 298–313.

Intergovernmental Panel on Climate Change (2000). Special report on emissions scenarios: a special report of Working Group III. Geneva: Intergovernmental Panel on Climate Change.

Intergovernmental Panel on Climate Change (2011). Special report managing the risks of extreme events and disasters to advance climate change adaptation (SREX), summary for policy makers. Geneva: Intergovernmental Panel on Climate Change.

International Energy Agency (2008). *World Energy Outlook 2008.* Paris: International Energy Agency.

Kemper, K. E. (2007). Instruments and institutions for groundwater management. In *The Agricultural Groundwater Revolution: Opportunities and Threats to Development*, ed. Giordano, M. and Villholth, K. G. Wallingford: CABI, pp. 153–172.

Keys, P. W., Barrón, J. and Lannerstad, M. (2012). *Releasing the Pressure: Water Resource Efficiencies and Gains for Ecosystem Services.* Stockholm: United Nations Environment Programme and Stockholm Environment Institute.

Klein Goldewijk, K., Beusen, A., Van Drecht, G. and De Vos, M. (2011). The HYDE 3.1 spatially explicit database of human-induced global land-use change over the past 12 000 years. *Global Ecology and Biogeography*, **20**, 73–86.

Knox, J., Hess, T., Daccache, A. and Wheeler, T. (2012). Climate change impacts on crop productivity in Africa and South Asia. *Environmental Research Letters*, **7**, 034032.

Kummu, M., de Moel, H., Porkka, M. *et al.* (2012). Lost food, wasted resources: global food supply chain losses and their impacts on freshwater, cropland, and fertiliser use. *Science of The Total Environment*, **438**, 477–489.

Kummu, M., Ward, P. J., de Moel, H. and Varis, O. (2010). Is physical water scarcity a new phenomenon? Global assessment of water shortage over the last two millennia. *Environmental Research Letters*, **5**, 034006.

Lal, R. (2007). Anthropogenic influences on world soils and implications to global food security. *Advances in Agronomy*, **93**, 69–93.

Lapola, D. M., Schaldach, R., Alcamo, J. *et al.* (2010). Indirect land-use changes can overcome carbon savings from biofuels in Brazil. *Proceedings of the National Academy of Sciences*, **107**, 3388–3393.

Le Quéré, C., Raupach, M. R., Canadell, J. G. and Marland, G. (2009). Trends in the sources and sinks of carbon dioxide. *Nature Geoscience*, **2**, 831–836.

Lenton, T. M., Held, H., Kriegler, E. *et al.* (2008). Tipping elements in the Earth's climate system. *Proceedings of the National Academy of Sciences*, **105**, 1786–1793.

Liu, J. and Savenije, H. H. G. (2008). Food consumption patterns and their effect on water requirement in China. *Hydrology and Earth System Sciences*, **12**, 887–898.

Liu, J., You, L., Amini, M. *et al.* (2010). A high-resolution assessment on global nitrogen flows in cropland. *Proceedings of the National Academy of Sciences*, **107**, 8035–8040.

Lundqvist, J., de Fraiture, C. and Molden, D. (2008). *Saving Water: From Field to Fork: Curbing Losses and Wastage in the Food Chain.* Stockholm: Stockholm International Water Institute (SIWI).

Martinelli, L. A. and Filoso, S. (2008). Expansion of sugarcane ethanol production in Brazil: environmental and social challenges. *Ecological Applications*, **18**, 885–898.

Matson, P. A., Parton, W. J., Power, A. and Swift, M. (1997). Agricultural intensification and ecosystem properties. *Science*, **277**, 504–509.

Mekonnen, M. M. and Hoekstra, A. Y. (2010). The green, blue and grey water footprint of farm animals and animal products. Value of Water Research Report Series: 48. Delft, the Netherlands: UNESCO-IHE Institute for Water Education. Available at: http://www.unesco-ihe.org/Value-of-Water-Research-Report-Series/Research-Papers.

Meybeck, M. and Vörösmarty, C. (2004). The integrity of river and drainage systems. In *Vegetation, Water, Humans and the Climate*, ed. Kabat, P., Claussen, M., Dirmeyer, P. A. *et al.* Heidelberg, Germany: Springer Verlag, pp. 297–397.

Millennium Ecosystem Assessment (2005). *Ecosystems and Human Well-being: Synthesis*. Washington, DC: Island Press.

Milly, P. C. D., Betancourt, J., Falkenmark, M. *et al.* (2008). Climate change – stationarity is dead: whither water management? *Science*, **319**, 573–574.

Monfreda, C., Ramankutty, N. and Foley, J. A. (2008). Farming the planet: 2. Geographic distribution of crop areas, yields, physiological types, and net primary production in the year 2000. *Global Biogeochemical Cycles*, **22**, GB1022.

Moore, B., Underdal, A., Lemke, P. and Loreau, M. (2001). Amsterdam Declaration on Global Change. In *Challenges of a Changing Earth*, ed. Steffen, W., Jäger, J., Carson, D. J. and Bradshaw, C. Heidelberg, Germany: Springer Verlag, pp. 207–208.

Morales, P., Hickler, T., Rowell, D. P., Smith, B. and Sykes, M. T. (2006). Changes in European ecosystem productivity and carbon balance driven by regional climate model output. *Global Change Biology*, **13**, 108–122.

Müller, C., Cramer, W., Hare, W. L. and Lotze-Campen, H. (2011). Climate change risks for African agriculture. *Proceedings of the National Academy of Sciences*, **108**, 4313–4315.

Pfister, S. and Hellweg, S. (2009). The water 'shoesize' vs. footprint of bioenergy. *Proceedings of the National Academy of Sciences*, **106**, E93-E94.

Pongratz, J., Reick, C., Raddatz, T. and Claussen, M. (2008). A reconstruction of global agricultural areas and land cover for the last millennium. *Global Biogeochemical Cycles*, **22**, GB3018.

Postel, S. L., Daily, G. C. and Ehrlich, P. R. (1996). Human appropriation of renewable fresh water. *Science*, **271**, 785–788.

Pretty, J. (2008). Agricultural sustainability: concepts, principles and evidence. *Philosophical Transactions of the Royal Society B: Biological Sciences*, **363**, 447–465.

Ripl, W. (2003). Water: the bloodstream of the biosphere. *Philosophical Transactions of the Royal Society of London. Series B: Biological Sciences*, **358**, 1921–1934.

Rockström, J., Falkenmark, M., Karlberg, L. *et al.* (2009a). Future water availability for global food production: the potential of green water for increasing resilience to global change. *Water Resources Research*, **45**.

Rockström, J., Steffen, W., Noone, K. *et al.* (2009b). A safe operating space for humanity. *Nature*, **461**, 472–475.

Rost, S., Gerten, D., Bondeau, A. *et al.* (2008). Agricultural green and blue water consumption and its influence on the global water system. *Water Resources Research*, **44**.

Ruddiman, W. F. (2003). The anthropogenic greenhouse era began thousands of years ago. *Climatic Change*, **61**, 261–293.

Sachs, J., Remans, R., Smukler, S. *et al.* (2010). Monitoring the world's agriculture. *Nature*, **466**, 558–560.

Smakhtin, V., Revenga, C. and Döll, P. (2004). A pilot global assessment of environmental water requirements and scarcity. *Water International*, **29**, 307–317.

Steffen, W., Crutzen, P. J. and McNeill, J. R. (2007). The Anthropocene: are humans now overwhelming the great forces of nature. *Ambio*, **36**, 614–621.

Steffen, W., Grinevald, J., Crutzen, P. and McNeill, J. (2011). The Anthropocene: conceptual and historical perspectives. *Philosophical Transactions of the Royal Society A: Mathematical, Physical and Engineering Sciences*, **369**, 842–867.

Steffen, W. L., Sanderson, A., Tyson, P. D., *et al.* (2004). *Global Change and the Earth System: A Planet Under Pressure. Global Change. The IGBP Series, 1619–2435.* Berlin: Springer.

Strzepek, K. and Boehlert, B. (2010). Competition for water for the food system. *Philosophical Transactions of the Royal Society B: Biological Sciences*, **365**, 2927–2940.

Tainter, J. A. (2006). Archaeology of overshoot and collapse. *Annual Review of Anthropology*, **35**, 59–74.

United Nations Department of Economic and Social Affairs (UNDESA) (2012). *World Urbanization Prospects 2011. On-line Data: Urban and Rural Population*. New York: United Nations Department of Economic and Social Affairs. Available at: http://esa.un.org/unpd/wup/unup/index_panel1.html (accessed 23 May 2012).

United Nations Environment Programme (2011). *UNEP/GEO Core Indicators: Number of Parties to Multilateral Environmental Agreements, 1971–2009*. Available at: http://geodata.grid.unep.ch/extras/indicators.php (accessed 28 March 2012).

United Nations Population Division (2004). *World Population to 2300*. New York: United Nations Population Division. Available at: http://www.un.org/esa/population/publications/publications.htm.

Vitousek, P. M., Mooney, H. A., Lubchenco, J. and Melillo, J. M. (1997). Human domination of Earth's ecosystems. *Science*, **277**, 494–499.

von Braun, J. and Meinzen-Dick, R. S. (2009). *'Land Grabbing' by Foreign Investors in Developing Countries: Risks and Opportunities*. Washington, DC: International Food Policy Research Institute.

Vörösmarty, C. J., Meybeck, M., Fekete, B. *et al.* (2003). Anthropogenic sediment retention: major global impact from registered river impoundments. *Global and Planetary Change*, **39**, 169–190.

Weinhold, D., Killick, E. and Reis, E. J. (2011). *Soybeans, Poverty and Inequality in the Brazilian Amazon*. Available at: http://mpra.ub.uni-muenchen.de/29647/.

World Trade Organization (2008). *International Trade Statistics 2008*. Geneva: World Trade Organization. Available at: http://www.wto.org/english/res_e/statis_e/its2008_e/its08_toc_e.htm.

World Trade Organization (2011). *World Trade Report 2011: The WTO and Preferential Trade Agreements: From Co-existence to Coherence*. Geneva: World Trade Organization.

Curtains of rain fall on a forest in Tanzania. With global change, dry lands of the world are increasingly vulnerable to regime shifts in precipitation patterns.

Chapter 3

Balancing on a threshold of alternate development paths: regime shift, traps and transformations

This chapter addresses the involvement of water in abrupt and unexpected changes in social–ecological systems. It looks more closely at water's many roles in reinforcing processes and stabilising feedbacks, and how systems suffer when feedback processes weaken or break. It discusses water-related disturbances and feedbacks, different categories of regime shift and ways of enhancing resilience. The issue of traps is discussed, both green water-related poverty traps, involving self-enforcing mechanisms, and blue water-related rigidity traps, when people and institutions resist change. The chapter explains the use of transformation management to get out of traps.

3.1 Rapid, accelerating and surprising changes in the Anthropocene era

Many of the changes in the GWS are rapid and can be surprising. Some are accelerating, but change can also be smooth and gradual. Most importantly, some change can be turbulent and abrupt, and result in non-linear responses from human, economic or ecological systems, with effects that are difficult, if not impossible, to reverse. Documented cases include the collapse of marine cod fisheries in 1992 (Hutchings, 1996), the collapse of Saharan vegetation 6–8000 years ago (Foley *et al.*, 2003) and the financial crises of 2008, all of which caused rapid and unprecedented reductions in resource availabilities with abrupt, unforeseen and negative effects on human well-being.

The Anthropocene era has seen an increase in the links between humans and the environment, at multiple scales. Freshwater plays a significant role in this context, through its capacity to link different systems and places. Intricate cross-scale interactions play out in novel ways, which connect distant peoples and places (Adger *et al.*, 2008; Galaz *et al.*, 2010) reshaping the capacity of the biosphere to sustain human well-being (Folke *et al.*, 2011) and increasing the risk of abrupt change.

An example of how unexpected connections can lead to rapid shifts is the outcome of wildfires in Russia in 2010. Fuelled by record temperatures and a summer drought, they destroyed much of Russia's wheat harvest, leading to a ban on grain exports, which contributed to rising global food prices. Rising food prices were identified as one of the triggers of Tunisia's unrest in December that year, and of the uprisings that spread across much of North Africa, including Egypt where wheat is a subsidised staple food item for millions of poor people, becoming known as the 'Arab Spring' (Fraser and Rimas, 2011).

Climate change exacerbates this new situation. The interplay between the water cycle and water-related disasters, such as floods and droughts, has major effects on food supplies, health and the economy, sometimes even triggering social upheavals. For example, economic losses from natural disasters, including floods and droughts, increased three-fold between the 1960s and the 1980s (Kabat *et al.*, 2012). Riverine floods are a good example of the damage that can arise, and on average 100 million people are affected by such floods every year. From 1990 to 1996 there were six major floods throughout the world in which the number of fatalities exceeded 1000, and 22 floods with losses exceeding USD 1 billion each. Floods in the period 1971–95 affected more than 1.5 billion people worldwide, making over 81 million people homeless (CRED, 2012).

Sometimes, these abrupt changes are easily reversed, but at other times they cause major changes in ecosystem processes, water flows or society, that make them hard, or even impossible, to reverse. These changes can be said to push the system into a new state, which is often substantially different to how it behaved before, with a new 'identity, function, feedbacks and structure' (Folke *et al.*, 2004). This change of state is called a **regime shift**.

3.2 What are regime shifts?

In short, regime shifts are large, abrupt and persistent changes in the structure and function of systems, such as a forest, lake, household or city. Sometimes these changes are relatively easy to reverse if the pressures on the systems are reduced. At other times they are very difficult to reverse or even irreversible. This would be the case if the change were to cause a shift between alternate regimes, each sustained by a different set of mutually reinforcing feedbacks (Holling, 1973; Scheffer *et al.*, 2001).

A regime shift can be explained using the metaphor of sailing a small boat in heavy winds. As the wind increases, the boat might lean sideways more and more, while at the same time the vessel probably picks up speed satisfactorily and the journey continues as planned. Suddenly, the unexpected happens. A slight increase in wind, perhaps from a different direction, grips the sail. You lose control and unexpectedly the boat capsizes. Once this has happened, the situation becomes new and substantially different. A significant decline in wind speed does not help your position and you have to navigate from a totally new set of circumstances. You must decide whether to try to reverse the situation as quickly as possible in order to continue the journey as planned, head to shore to find another way to continue the voyage, or perhaps return home.

The existence of alternate regimes in social–ecological systems is being increasingly documented (Folke *et al.*, 2004; Gordon *et al.*, 2008; Walker and Meyers, 2004). The situation is similar to that of the sailing boat. Just as substantial investment and changes in practice are needed to get the boat turned the right way up, substantial investment and changes in management practice or social organisation are needed to get the system back on its previous development trajectory. If you cannot get it back, you have lost many of the services the sailing boat – or the social–ecological system – gave you.

Figure 3.1 illustrates the differences between gradual ecological change and three different types of regime shift exemplified by precipitation and vegetation interactions. The feedbacks that maintain a system in a given regime are represented as the shape of stability landscapes, and the configuration of the system is represented by a ball. Multiple dips in the landscape represent the potential existence of alternative regimes. It illustrates that small pushes on the system will still keep it in the same dip, but a larger push, or a change in the depth of the dip, can cause the ball, i.e. the system, to end up in a new dip – or regime.

System dynamics are shaped by internal dynamics, such as vegetation growth or a boat swaying back and forth as you move around in it, and external forces, such as changes in precipitation, the size of waves and wind speed. Regime shifts occur when external forces or gradual internal changes alter a system so that its organisation shifts from being organised around one set of mutually reinforcing processes. This can be vegetation enhancing precipitation or the wind moving the boat forward. It can also be reduced vegetation with less influence on precipitation leading to overall drier conditions or an upside down boat

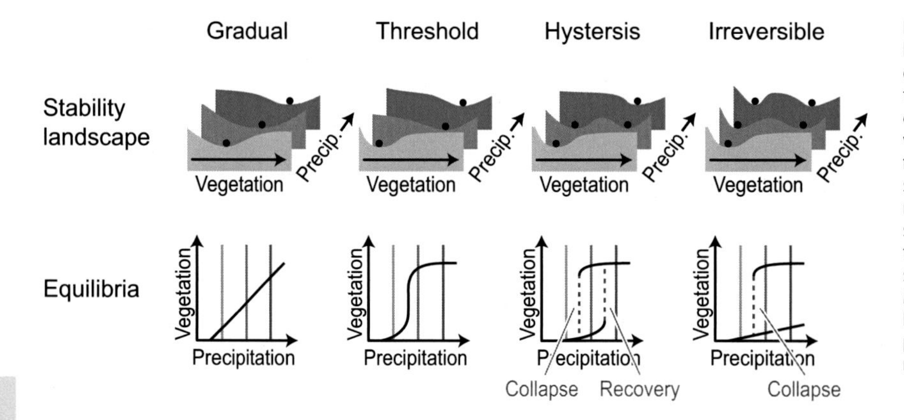

Figure 3.1 Differences between gradual ecological change and three different types of regime shifts, exemplified by precipitation–vegetation interactions. The feedbacks that maintain a system in a given regime are represented as the shape of stable landscapes, with the configuration of the system represented by a ball. Multiple dips in the landscape represent the potential for alternative regimes (Gordon *et al.*, 2008).

that has to be towed to move forward (Scheffer *et al.*, 2005). External shocks, such as a drought or an intense rainfall event, can change the system configuration, moving the ball across the landscape. The stability landscape (Figure 3.1) can also change as the forces defining the feedback processes in the system change. Maybe, for instance, all the passengers on the boat move to one side. If a dip in our stability landscape completely vanishes or an external shock pushes the system from one dip to another, the system undergoes a regime shift.

Changes in the internal variables that alter the stability landscape that defines a regime are often slower than exogenous disturbances. They are also often less monitored. For example, people tend to monitor yield levels as an indicator of how well agriculture is doing, but to ignore slower changes in soil attributes, such as organic matter or nutrient status, that enable sustained levels of crop production. These internal variables are often referred to as 'slow variables'.

Figure 3.1 shows the difference between gradual change, thresholds and hysteretic change. Gradual change occurs in a case when precipitation increases from drier to wetter conditions in a landscape that contains species with diverse growth responses that are relatively evenly distributed, so that vegetation cover gradually increases with precipitation. Threshold change occurs when vegetation cover rapidly increases at a specific amount of precipitation, in a landscape where most species' responses are similar. Declines in precipitation, which is the external driver, can push the system from dense to sparse vegetation cover, but in the case of threshold effects it recovers quickly once precipitation increases again.

A hysteresis effect can occur if there is a strong feedback in the form of moisture recycling from vegetation that stimulates precipitation. This implies that the precipitation thresholds at which vegetation will quickly increase – or 'recover' – are higher than the precipitation levels at which vegetation will collapse. Once vegetation has collapsed, it therefore requires larger increases in external precipitation to recover. Irreversible change is a stronger form of hysteresis, where vegetation is unable to recover to its pre-collapse levels even if rainfall increases substantially. This type of dynamic occurs when the ability of vegetation to recover is lost from the system during a collapse.

3.3 Regime shifts can happen across the whole hydrological cycle

A rapidly growing body of evidence suggests that modifications to the quality and quantity of hydrological flows can increase the risk of ecological regime shifts (Gordon *et al.*, 2008). However, previous large-scale assessments of water (Molden, 2007) and ecosystem services (Millennium Ecosystem Assessment, 2005) have revealed a lack of an integrated understanding of how human modifications to the hydrological cycle, through e.g. land use and climate change, regulate the prevalence and severity of surprising non-linear change in ecosystems. Table 3.1 synthesises the current understanding of how regime shifts are produced from growing demand for water, agricultural products such as food and biofuels, and/or climate changes. The table exemplifies regime shifts that happen across the full hydrologic cycle – in soil, groundwater, lakes and coastal zones, in the atmosphere and on land in water phase-transitions, where such shifts are linked to runoff, groundwater and evapotranspiration.

The table is a simplification of known regime shifts that are clearly linked to water. What drives these shifts is often a complex mix of social and ecological interlinkages, as can be seen from an example on Borneo (Folke *et al.*, 2011), where rainforest that has been acting as a carbon sink has shifted to fragmented oil palm plantations, becoming a carbon source while at the same time losing its capacity for self-renewal. It is a telling example of the interactions between disturbance events, regeneration, resilience and vulnerability.

In Borneo's rainforest, droughts following El Niño naturally serve as a trigger for regenerating the rainforest and its biodiversity by inducing synchronised reproduction, among rainforest trees. In recent decades, there has been a widespread expansion of palm oil plantations to produce cheap biofuel, food and feed, as well as concession-based timber extraction. Largely driven by global market demand for palm oil and tropical timber, and enabled by weak institutions this has resulted in highly fragmented and degraded forests (Curran *et al.*, 2004). Wildfires have become much more common as a result, and it is estimated that widespread El Niño-related wildfires in Borneo in 1997 released between 0.81 and 2.57 Gt of carbon into the atmosphere, equivalent to 13–40% of the mean annual global carbon emissions from fossil

Table 3.1 Comparison of regime shifts related to water, consequences of regime shift, key endogenous variables, agricultural drivers of change, other drivers, and assessment of the evidence for each shift, with references.

Hydrological place	Regime shift	Regime A	Regime B	Impacts of shift from A to B	Human intervention	Endogenous slow variable	Water as a state or control variable	Other drivers	Evidence	References or known cases
Groundwater	Groundwater depletion	Available groundwater	Depleted groundwater	No water availability impacts on households, increased health security risk	Overpumping	n/a	Control (water availability determines regime)	Reduced precipitation, altered vegetation	Strong	Gujarat, North-eastern China Plain and Ogallala
Runoff to lakes and seas	River depletion	Natural river flow	Little or no flow during parts of or entire year	Loss of habitat, impact on fisheries, impact on downstream livelihoods, increased risk of conflicts	Irrigation, or other upstream consumptive water use	n/a	State (water availability in river), control (river flow pulses create and maintain habitats)	Reduced precipitation, altered vegetation	Strong	Nile, Colorado
Runoff to lakes and seas	Lake depletion	Inflow and evaporation in equilibrium	No/Decreasing water body	Increased salinity, shrinking water body, dust storms from lake sediment, loss of habitat for and of terrestrial species of birds, animals, and aquatic species of fish, amphibians, turtles, etc.	Irrigation, or other upstream consumptive water use	n/a	State (water volume as habitat)	Reduced precipitation, altered vegetation	Strong	Aral Sea
Runoff to lakes and seas	Freshwater eutrophication	Eutrophic	Non-eutrophic	Reduced access to recreation, drinking water, and risk of fish loss	Fertiliser use	Sediment and watershed soil phosphorus	State (water quality as habitat/drinking)	Flooding, landslides	Strong	Carpenter (2003); Carpenter (2005); Genkai-Kato and Carpenter (2005); Scheffer and van Nes (2007)
Runoff to lakes and seas	Coastal hypoxic zones	Hypoxic	Not hypoxic	Fishery decline, loss of marine biodiversity	Nutrient use	Aquatic biodiversity	State (water quality as a habitat)	Flooding	Strong	Boyer *et al.* (2006); Diaz (2001); Galloway *et al.* (2004)
River fragmentation	Controlled flow and migration barriers	Natural river flow and natural migration path	River flow human controlled, barriers like reservoirs and weirs	Changed seasonality and rhythm, loss of habitats and species migration along 'river continuum' leading to loss of species and invasion of new species	Control of river flow including dams	n/a	State (water quantity as a habitat)	Land-use changes	Strong	

Soil	Organic matter	Sustained higher productivity	Sustained low productivity	Yield decline, reduced drought tolerance	Too high withdrawal of biomass/ nutrients	Soil organic matter	State (water holding capacity impacts water)	Droughts, dry spells	Weak	Bossio (2007); Enfors and Gordon (2007)
Soil	Salinisation	Sub-surface	Near surface	Contamination of drinking water, yield declines, salt damage to infrastructure and ecosystems	Reduced woody vegetation, irrigation	Water table	Control (water leakage in the soil, irrigation, increase water table)	Wetter climate	Strong	Anderies *et al.* (2006); Clarke *et al.* (2002); Cramer and Hobbs (2005); Gordon *et al.* (2003)
Soil	Landscape patchiness	Spatial pattern	No spatial pattern	Productivity declines and erosion	Grazing pressure, land-use change	Vegetation pattern	Control/state (water determines runoff, transport/infiltration and thus vegetation)	Fires, droughts	Medium	Bestelmeyer *et al.* (2006); Ludwig *et al.* (2005); Peters and Havstad (2006); Rietkerk *et al.* (2004)
Atmosphere	Cloud forest	Cloud forest	Woodland	Loss of productivity, reduced runoff, biodiversity loss	Clearing of forest	Vegetation surface area	Control (amount of evaporation) and state (amount of precipitation)	Fog frequency	Medium	Dawson (1998); del-Val *et al.* (2006); Holmgren *et al.* (2006)
Atmosphere	Forest Savannah	Forest	Savannah	Loss of biodiversity, altered agriculture suitability (+/−)	Clearing of trees	Moisture recycling, energy balance	Control (amount of evaporation) and state (amount of precipitation)	Fires	Medium	Cowling *et al.* (2004); Feddema *et al.* (2005); Hutyra *et al.* (2005); Oyama and Nobre (2003); Sankaran *et al.* (2005); Sternberg (2001)
Atmosphere	Savannah Desert	Savannah	Dry savannah/ desert	Loss of productivity, yield declines, droughts/dry spells	Clearing of trees/land degradation	Moisture recycling, energy balance	Control (amount of evaporation) and state (amount of precipitation)	Droughts, fires	Medium	Dekker *et al.* (2007); Foley *et al.* (2003); Los *et al.* (2006)
Atmosphere	Monsoon shifts	Monsoon	Weak/no monsoon	Risk of crop failures and changed climate variability	Irrigation, or reduced irrigation	Energy balance, advective moisture flows	Control (amount of evaporation) and state (amount of precipitation)	Change in sea surface temperatures	Weak	Douglas *et al.* (2006); Fu (2003); Miller *et al.* (2005); Zheng and Eltahir (1997); Zickfeld *et al.* (2005)

fuels (Page *et al.*, 2002). As a consequence, the massive land-use change in recent decades has changed the role of El Niño events from creating opportunities for the renewal of the forests to hindering renewal, since they now trigger wildfires, disrupt the fruiting of the rainforest trees, interrupt wildlife reproductive cycles and erode the basis for rural livelihoods (Curran *et al.*, 2004).

The Borneo case clearly illustrates how global markets can affect social–ecological systems in locations with weak and fragmented institutions and turn disturbance events, like El Niño, from a regenerative into a destructive force. In the Borneo landscape, a biodiversity-rich multifunctional tropical rainforest acting as a carbon sink has become a simplified palm oil landscape and a carbon source. The example illustrates how land-use changes driven by emergent markets directly affect local biodiversity and livelihoods, and eventually contribute to climate change.

3.3.1 Land use and climate are important drivers of water-related regime shifts at different scales

Many of the shifts shown in Table 3.1 are sensitive to the impacts of climate change. Climate change is expected to generate unprecedented changes in precipitation, soil moisture and runoff (see Chapter 4), which will make some regions more prone to regime shifts – while a small number might become less vulnerable. For example, the melting of the permafrost in the Arctic is likely to increase with a warmer climate, which could increase the risk of regime shifts (Karlsson *et al.*, 2011). On the other hand, some projections point to a drier climate in Western Australia (Hughes, 2003), which could reduce the effects of dryland salinisation since this is triggered by increased water volumes in the soil. However, drier conditions in Australia could also create extra vulnerability. Box 3.1 discusses how 10 relatively dry years in Australia caused a belief among farmers that salinity was 'yesterday's problem', which made them less prepared for the potential for wetter conditions to rapidly increase levels of salinisation.

Land-use change is another important driver of regime shifts, where increasing demand for food, fibre and biofuels causes particular pressures. Land-cover changes from forest or wet savannah to cropland, for example, are the main driver behind 'savannisation' and 'dry savannah–wet savannah' regime shifts (Table 3.1). Increased irrigation is a major driver of the depletion of aquifers, rivers and lakes. Unsustainable soil practices are a major factor behind soil-related regime shifts.

There is also a wide variation in the scale of these regime shifts, in terms of both the spatial distribution of the shift and the time it would take to reverse its effects (Figure 3.2). Agriculture–aquatic system regime shifts occur at catchment to river basin scales, but vary from years to millennia in their reversibility. For example, freshwater eutrophication is often irreversible or only reversible after massive reductions in phosphorus inputs over decades or longer. Internal cycling of phosphorus within the lake system and continued inflows to the lake of phosphorus leaking from historically accumulated deposits in the catchment soils keeps the phosphor concentration in the lake high until nutrients become bound in the lake sediment or are removed from the lake (Carpenter, 2005).

Agriculture–soil regime shifts tend to operate at the field to landscape scale, with varying degrees of reversibility. Although soil structure regime shifts occur at small spatial scales, their impact can cascade across the landscape, as exemplified by the development of the Dust Bowl in the United States in the 1930s. The Dust Bowl started at the scale of individual fields and expanded nonlinearly to affect the main agricultural regions of the United States. Wide-scale weather patterns caused individual fields to become highly connected, creating massive dust storms that exponentially aggravated the situation (Peters *et al.*, 2004).

Finally, agriculture–atmosphere regime shifts tend to operate at relatively large spatial and temporal scales, although uncertainty remains about the scale of land–atmosphere feedbacks. For example, evapotranspiration from forests is the main source of water for precipitation in the Amazon, but patchy regional deforestation that increases landscape heterogeneity can contribute to an increase in rainfall through the establishment of anomalous convective circulations, while large-scale deforestation would substantially decrease precipitation at distant locations (D'Almeida *et al.*, 2007).

3.3.2 Water can be both a control and a state variable in these shifts

Water is involved in different regime shifts as both a state and a control variable. Sometimes it is water

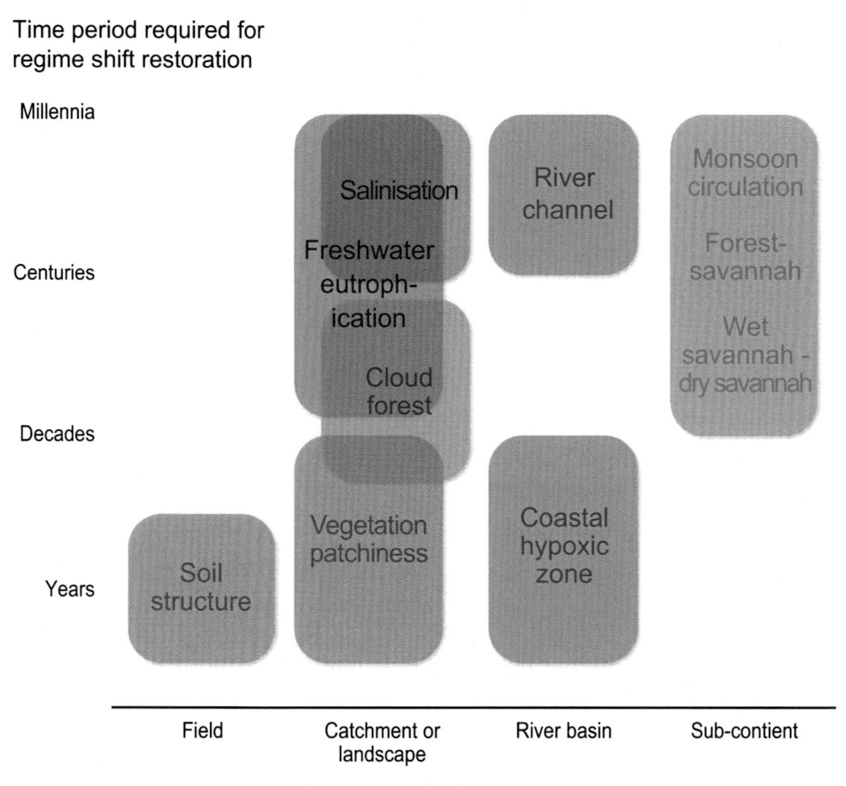

Figure 3.2 Estimates of the spatial and temporal scales at which regime shifts operate. Blue indicates agriculture and aquatic systems, brown indicates agriculture and soil, and green indicates agriculture and atmospheric regime shifts (Gordon *et al.*, 2008).

itself that is affected, as in groundwater regime shifts or eutrophication shifts where it is the water quality that changes. At other times, water is the control variable in that it is changes to water flows that control the shift, such as in soil salinisation caused by irrigation. However, it becomes abundantly clear from the synthesis table that to avoid water regime shifts, it is not enough to manage water by itself. Water management needs to be combined with stewardship of the whole landscape, including ecological processes such as vegetation change and organic matter in soil. Water is involved in all these shifts but it is the feedbacks between water and other processes that create the vulnerability or can help build resilience.

3.3.3 Different regime shifts are more common in some parts of the world

Figure 3.3 presents tentative global maps of regions vulnerable to specific types of water-related regime shifts. The first map (Figure 3.3a) shows regime shifts triggered by changes in water flows as a result of land-use change, by three main drivers (1) deforestation, (2) land mismanagement, and (3) salinisation related to irrigation practices. Deforestation in the Amazon rainforest can reduce evaporation and open up the system so it becomes more prone to rising air turbulence, potentially pushing it across the rainforest-to-savannah threshold. In the Sahel, land mismanagement can reduce evaporation, reducing the moisture available for precipitation, potentially also influencing monsoon dynamics (Table 3.1; wet-to-dry savannah). In Australia, intensive irrigation practices combined with inherently saline soils has resulted in a high risk of a water-induced salinisation threshold, similar to the salinity thresholds in arid irrigation systems in the Middle East, the Aral Sea and in Northern China.

Figure 3.3b shows a tentative global map of hotspot regions/systems for thresholds triggered by water overuse, including two categories, unsustainable withdrawals of river flow (causing basins to close) and overuse of groundwater (causing abrupt collapse of

(a)

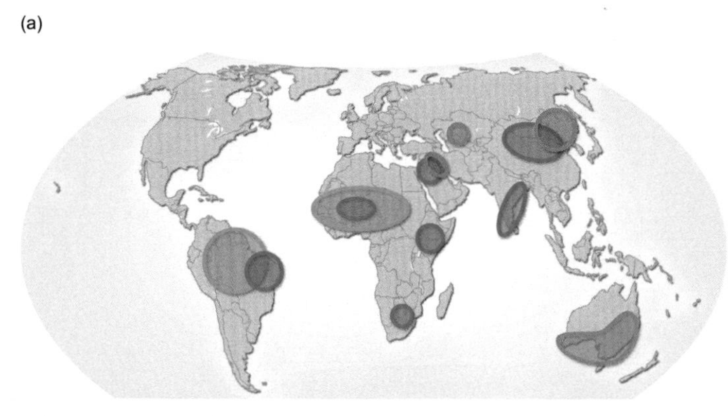

Tipping points, regional risks due to land management issues

Deforestation moisture feedback
Land mismanagement (e.g. soil loss, land degradation)
Salinisation

(b)

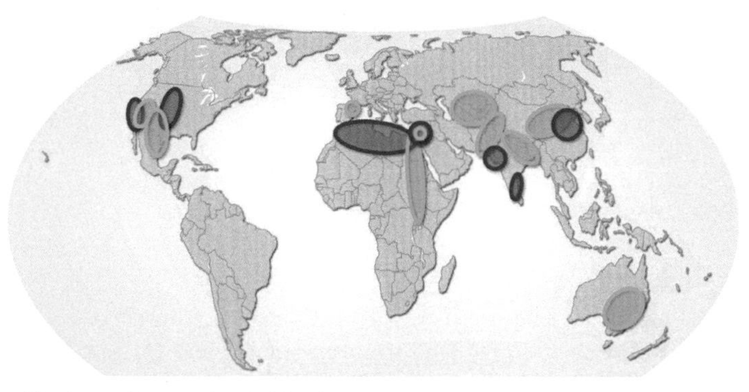

Tipping points, regional risks due to water overuse

Groundwater collapse
River basin closure/river depletion

Figure 3.3 First attempt, based on expert judgement, to map global risks of water-induced thresholds for bottom-up processes related to land use (a) and water use (b), and top-down processes related to global change (c). To put the tipping points into context, the map (d) displays the adjusted human water stress index (Vörösmarty *et al.*, 2010).

available water quality and/or permanent state of scarcity). The regime shifts in Figure 3.3a and b, operate 'bottom-up', i.e. are triggered by the scale and pace of local to regional management and use of land and water resources in landscapes. A third critical driver behind water-related thresholds, operates 'top-down' through global change. The water-related hotspots for global change derived regime shifts are tentatively presented in Figure 3.3c. These large shifts, e.g. the risk of destabilising the Asian monsoon or the ENSO triggering of the tropical climate regime, are primarily related to global changes in the climate system. However, even these large systems interact with (1) regional pollutants such as black carbon from forest fires and biomass burning, which is strongly linked to periods of drought and (2) to land and water management practices (e.g. large-scale deforestation in Indonesia influencing the regional rainfall patterns).

Even though poorly understood, it is increasingly clear that the three water-related regime shift processes described in Figure 3.3a, b and c, interact in ways that generate feedbacks that can dampen (e.g. carbon sequestration in soil when CO_2 concentration

(c)

Figure 3.3 (cont.)

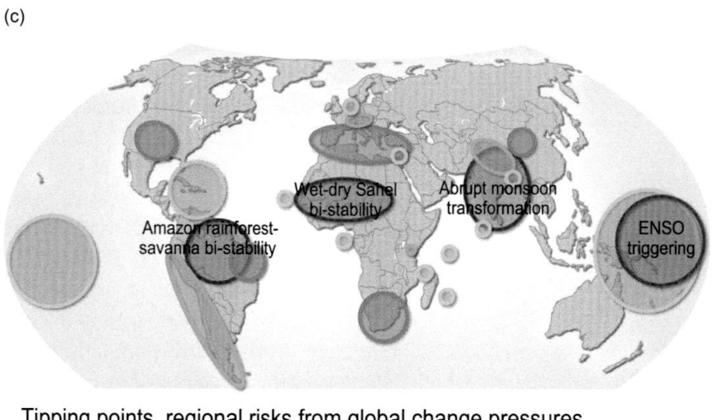

Tipping points, regional risks from global change pressures

Regional processes
Sea-level rise and salt-water intrusion
Drastic rainfall regime change
Glacier melt

(d)

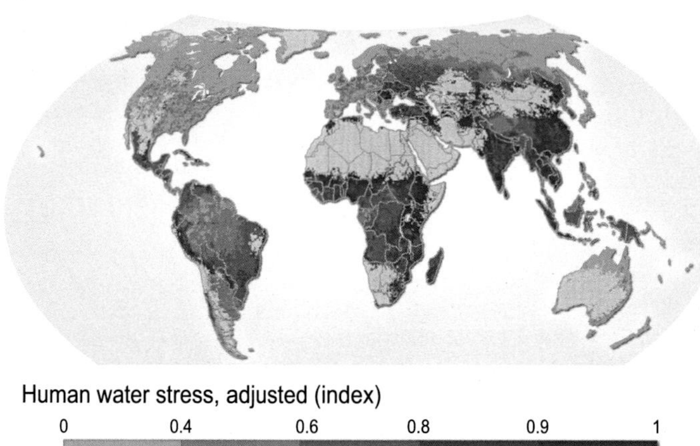

rises in atmosphere) or reinforce change (shifts in moisture feedback abruptly shifting forests to savannahs). Furthermore, critical biomes are interconnected through regional climate and hydrological systems, generating hydrometeorological teleconnections between regions in the world. Large-scale deforestation of rainforests in Amazonia and Central Africa has been shown to potentially reduce rainfall in the US Midwest, and deforestation in South East Asia influences rainfall in China (Snyder, 2010).

3.4 Regime shifts involve both social and ecological processes

The above sections only examine biophysical regime shifts, but all those discussed were triggered by changes in management. There is growing evidence that regime shifts happen in interlinked social–ecological systems (see e.g. the special issue of *Ecology and Society*, especially the introduction by Walker *et al.*, 2006). Linked social–ecological regime shifts have been conceptualised and analysed in three slightly different ways:

- regime shifts that happen in either the social or the ecological domain, but which are driven by gradual or sudden changes in the other;
- regime shifts in biophysical, social and economic domains that happen individually, but interact and trigger each other;
- truly linked social–ecological system regime shifts.

3.4.1 Gradual change in the biophysical domain causes a regime shift in the social domain

The examples of biophysical regime shifts in Table 3.1 are caused by changes in the social domain. One example of a gradual change in the biophysical domain that has been described in relation to a regime shift in the social domain in some places is land degradation in some parts of the world. This degradation exists despite large investments in projects to encourage the adoption of soil conservation technology. Antle *et al.* (2006) found stable states in these areas, with low and high levels of soil degradation, respectively. These were separated by a threshold level of soil degradation beyond which conservation investment did not yield a positive return. The study showed that when a parcel of land crosses a productivity threshold, soil degradation becomes **economically irreversible**, that is, it is not profitable to invest in soil conservation, even though the degradation may be **technically reversible**. We might then talk about economically irreversible soil degradation, which implies that once economically irreversible degradation has occurred, subsidies for the adoption of soil conservation practices are unlikely to lead to permanent adoption by farmers.

3.4.2 A regime shift in one domain can cascade to other domains

At other times, regime shifts in the biophysical system interact with regime shifts in the social and economic system. Box 3.1 shows these interacting thresholds in the ecological, social and economic domains in relation to wheat production in Goulburn–Broken catchment, South East Australia. The success and spread of wheat cultivation at the expense of native woody vegetation has resulted in rising water tables in many areas of Australia. In Western Australia, for example (Kinzig *et al.*, 2006), it has resulted in a soil salinity regime shift that now covers 16% of the region and affects agricultural productivity and infrastructure such as roads and buildings. This has coincided with a decline in the human population since the 1960s. It has been suggested that there was a crossing of a critical threshold in the socio-cultural domain in the 1970s, with more amalgamation of farms. The number of farming enterprises peaked in the late 1960s but has fallen sharply due to an annual decline of about 7%. There are thresholds of population size, below which towns become unviable when local services such as health care and schools are withdrawn – further reducing the population.

Box 3.1 Water management and resilience in the Goulburn–Broken Catchment, Australia

Brian Walker, Commonwealth Scientific and Industrial Research Organisation (CSIRO)

Located in the Murray–Darling Basin in South East Australia, the Goulburn–Broken (GB) catchment is one of Australia's main food bowls. Its upper, mountainous parts are mostly forested, the mid-catchment is about half cleared and used for dryland cropping and grazing, and the lower riverine plains have just 2% native vegetation left and are intensively used for irrigated dairy and fruit production. Clearing of native vegetation began over 100 years ago, and has caused saline water tables to rise. Modelling the catchment's water dynamics (Anderies *et al.*, 2006) has shown that once more than 15–20% of the deep rooted native vegetation had been cleared, the stable state of the water table was at the surface, although it took many decades for it to rise and for this to become evident. Groundwater pumping has become necessary to prevent salinisation but it leads to the discharging of salt into the Murray River, which is unacceptable to downstream users.

Two related threshold effects are involved in the water dynamics. The first, at the catchment and landscape scales, is the threshold amount of native vegetation cover that is needed for the water table to remain at depth (80–85%). Once this threshold has been exceeded, the second threshold effect, which becomes evident at a local scale, is the level of the water table above which water (and the salt dissolved in it) is drawn to the surface of the soil by capillary action. Depending on soil texture, this is somewhere around 15 cm. Over 300 pumps are in

Box 3.1 *(cont.)*

operation in the irrigation area to keep the water table below this threshold.

The water table in this system is a controlling slow variable with threshold levels that cause shifts into alternate system regimes (productive vs. salinised). Managing this slow water table variable has to be integrated with managing other slow variables that also have threshold effects. A resilience assessment of the GB social–ecological system (Walker *et al.*, 2009) revealed 10 interacting thresholds that occur at different scales and in different domains of the system (see Figure 3.b1.1). Depending on the kind of shock that the system has to deal with, one or other of these thresholds may be crossed. If crossed, the subsequent changes in dynamics can lead to cascading effects on the other slow variables, either increasing or decreasing the likelihood that a threshold for one of them may be crossed. The details of the different thresholds are given in Walker *et al.* (2009).

Thresholds (1) and (2) are the two related to the water table. The controlling (slow) variable for (1) has already been crossed, with a delayed action in the change in the response variable (the water table), and this now involves considerable expense in artificially keeping the state of the system away from its dynamically stable state (water at the surface). In the past 10 years, however, this region has experienced one of the worst droughts in memory, and the very significant change in the amount of water entering the system slowed down the rise of water tables, and in fact caused it to drop in some areas, calling for much reduced pumping. The mood of farmers became 'salinity is yesterday's problem: water shortage is now our problem'. However, the results of modelling showed that any above-average rainfall could cause a rapid spike in water table rise, and this happened in early 2011 after one of the wettest seasons on record.

Allocation of water to irrigators from reservoirs adds to the rate of water table rise (acting like increased rainfall) and any significant change in the proportional allocation of water from storage and rivers to environmental flows, vs. agriculture, would significantly influence the likelihood of crossing thresholds (2), (4), (5), (6) and (7) – some for the better, some for the worse. Such a change in allocation is considered by the region's stakeholders itself to be a threshold effect (10). As the slow variable of societal pressure for environmental conservation increases, farmers fear – and conservationists hope – that a tipping point will occur in water regulation, favouring the environment over agriculture. Such a decision would involve strong interactions between stakeholders across scales – the focal scale (the catchment) and the scale above (the state of Victoria and its capital city, Melbourne). This is part of the general resilience of the system: its levels of trust, leadership, adaptability and the responsiveness of governance. Maintaining the resilience of the whole, interconnected system at the catchment scale may well depend on whether some necessary local-scale transformational changes can be achieved (Folke *et al.*, 2010).

The unfolding story of the GB illustrates that developing policy and the management of a social–ecological system amounts to understanding how that system works across scales and domains, knowing about the threshold effects that can lead to significant and perhaps irreversible shifts in the system, and knowing how the controlling – often slowly changing – variables on which these threshold levels occur will respond to external shocks. The cross-scale dynamics in both its biophysical and social domains play a significant role in determining the trajectory of the whole, interlinked system.

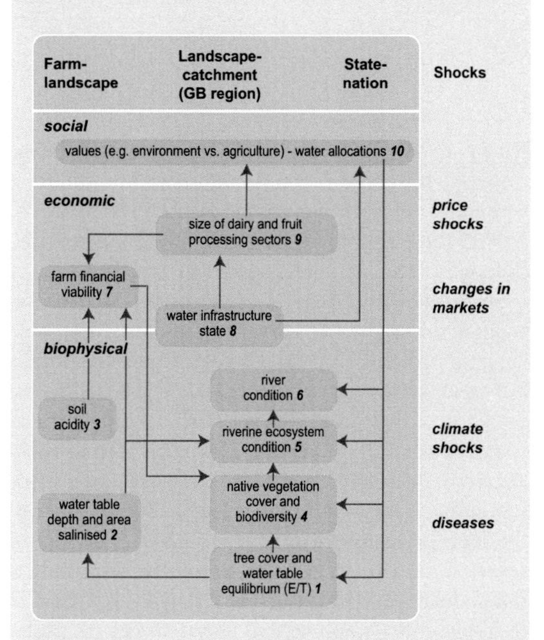

Figure 3.b1.1 Ten slow variables with identified thresholds in the panarchy that constitutes the Goulburn–Broken Region. The arrows between boxes indicate possible cascading threshold effects (Walker *et al.*, 2009).

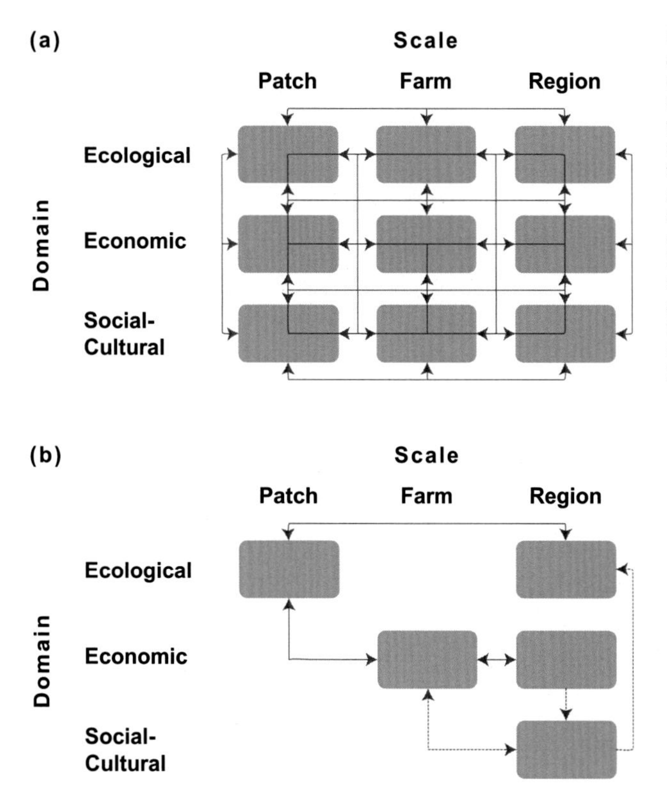

Figure 3.4 A conceptual model of how one regime shift at one spatial scale and one domain can cascade and cause shifts in other domains (Kinzig *et al.*, 2006). The model builds on four case studies on different continents, where the general lesson is that ecological regime shifts mostly occur at either small or slightly larger scales, while the knock-on effects on the economic system often come at a different scale, such as when a field-scale collapse of agricultural production affects household-level economies. If this happens to enough individual households, whole villages (a socio-cultural domain) might need to migrate from a region. While (a) shows all the possible shifts and interactions, (b) synthesises information from a set of case studies showing that shifts in a domain often occur at different scales (Kinzig *et al.*, 2006).

Kinzig *et al.* (2006) reviewed a set of similar case studies to the one in Box 3.1 to illustrate how different thresholds interact and offer a conceptual model of how passing one threshold frequently induces a cascade effect that ultimately leads to crossing one or more additional thresholds (Figure 3.4). They also suggest that these cascading regime shifts not only cross domains, i.e. social, ecological and economic, but also occur at different scales. The different domains seem to run the risk of shifts happening at specific scales. For example, ecological regime shifts are more likely on a patch or landscape scale, while economic or social shifts are more likely at the scale of the farm or region (Kinzig *et al.*, 2006). This makes sense, since a farm is an economic or social entity rather than an ecological one. We are also used to thinking about cultural and social forces on scales larger than a patch. A manager who focuses too strongly on one domain or scale is likely to miss the possibility of interactions with regime shifts and the likelihood that a new, resilient and possibly less desirable system will emerge. These findings suggest that concentrating on regime shifts in a single domain at a single scale is likely to be misleading with respect to the larger dynamics of the entire system, its implications and future trajectories.

3.4.3 Regime shifts may also involve truly linked social–ecological systems

When regime shifts happen in interlinked social–ecological systems it might be impossible to determine exactly what domain caused the regime shift. Instead, it is the interaction among a set of different social, ecological and economic variables that collectively govern the dynamics of a system and make it change to a new state. One way to think about this is to first identify the alternative regimes of interest, and then look at how the feedbacks among a smaller set of variables that control system behaviour are different in these regimes. Enfors and Gordon (2007) used this approach to examine how a village landscape in

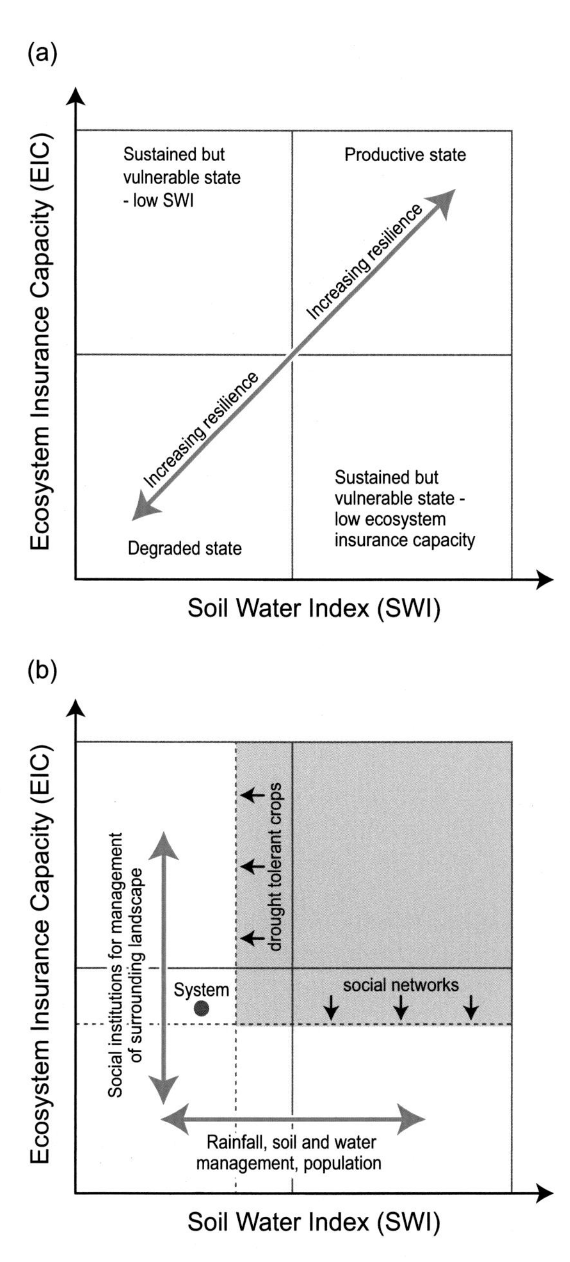

Figure 3.5 Conceptual framework for analysing resilience in smallholder farming systems, based on research in Tanzania (Enfors and Gordon, 2007). The two suggested key variables of the system, the Soil Water Index (SWI) and Ecosystem Insurance Capacity (EIC), are placed on the axes of the diagram. The area between the axes represents the space within which the system could move between two alternative stability domains. One is identified as productive and the other as degraded. In the former state (a), enough food and other provisioning ecosystem services are generated to support the people living in the system over time. In the latter, agricultural products and other ecosystem services are still being produced, but management erodes the productive potential. Rainfall, land management, population and institutions governing resource management influence the position of the system, as can be seen in (b). The position of the thresholds could be moved to some extent, affecting the relative position of the system and increasing or decreasing the space for the productive and degraded domains.

Tanzania went through a regime shift from a 'productive' to a 'degraded' state (Figure 3.5).

A productive state is here defined as a system where adequate biophysical resources are generated over time to support the people living in the system, and the system develops along a trajectory in which management of the land results in positive feedbacks that maintain or even improve the productive potential in the system (Figure 3.6). In a degraded state, management practices trigger a negative spiral of feedbacks that degrade the productive potential of the resource base over time (Figure 3.6). The system develops along a trajectory in which the vulnerability of local communities increases. The variables chosen were not in a single, specific domain, but combined socio-ecological variables where biophysical as well as socio-economic factors mutually drive change and reinforce each other. Enfors and Gordon (2007) suggested two interlinked variables: the Soil Water Index (SWI), which is related to per capita water availability in the soil for crop production; and Ecosystem Insurance Capacity (EIC), which is the capacity of ecosystems to generate goods for people in times of drought. For example, when the harvest fails, other ecosystem goods can become more important, such as collecting wild fruits, leaves and herbs to supplement diets, making charcoal to sell on the local market and making handicrafts. Figure 3.6 shows the feedbacks that sustain or degrade these variables.

Another suggested social–ecological regime shift of relevance to water has been hypothesised as taking place at the river basin scale. The river basin is one of the clearest system boundaries in water-related issues and it can therefore be useful to think about options for detecting possible thresholds and regime shifts at that scale (Cumming, 2011). One idea is that river basin closure (Falkenmark and Molden, 2008), i.e. where all of the blue water in the basin becomes committed to societal uses, will result in a regime shift. The rationale would be that once all the water has already been committed, a substantial change in social organisation is required for continued development (Cumming, 2011).

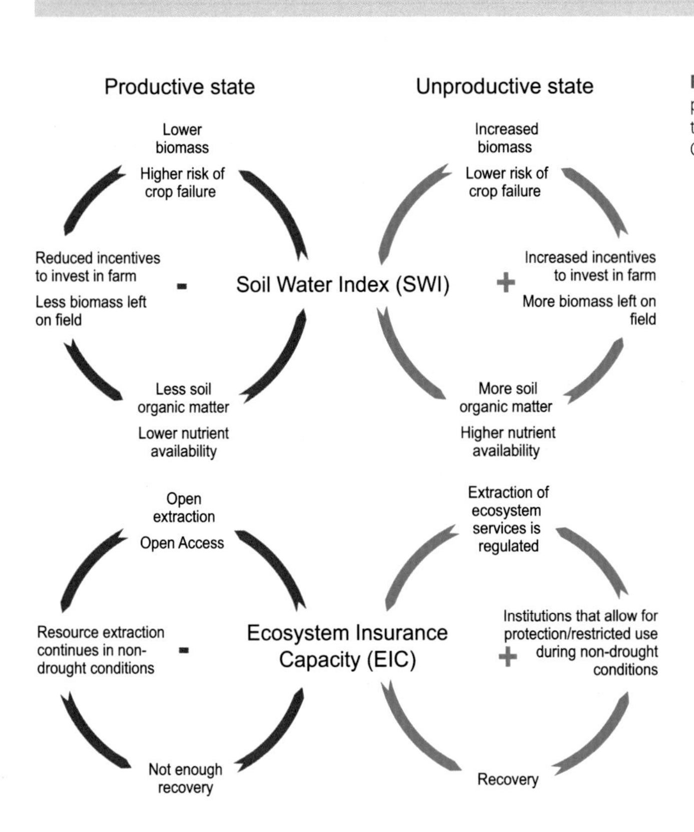

Figure 3.6 Feedback loops that sustain either productive states (+) or degrading states (−) in relation to the Soil Water Index and Ecosystem Insurance Capacity (Enfors and Gordon, 2008).

3.5 Enhancing resilience in the face of regime shifts

There are different ways to avoid regime shifts. Attempts could be made to avoid changing the drivers that are triggering the regime shift. However, this is very difficult in some cases. For example, even if there were drastic cuts in emissions of greenhouse gases, we are already committed to a certain degree of climate change and therefore will continue to be vulnerable to climate-change-driven regime shifts. Similarly, as is stated in Chapter 1, we are committed to feeding at least 9 billion people, which means that abandoning agriculture on a large scale is not an option.

3.5.1 Focus on slow variables

One way to improve management is to focus more on the variables that are slower relative to other system dynamics in which change may go unnoticed over long periods of time. For example, despite some early indications in the nineteenth century that clearing woody vegetation could lead to an increased risk of salinisation in Australia, Australians in general were unaware of this phenomenon until salinisation started to become widespread (Anderies *et al.*, 2006). Identifying and monitoring key slow-changing hydrological processes, such as water table rise in Australia, can be used to predict the likelihood of a regime shift.

3.5.2 Predicting thresholds is important but difficult

The ability to manage systems to avoid regime shifts in a world of high levels of variability would be improved if we could predict how close an ecosystem is to critical thresholds. Devising methods to provide early warning of regime shifts is a significant scientific challenge, and research suggests that changes in pattern formation and increasing variance could be used to detect when and where systems are more likely to experience regime shifts (Brock and Carpenter, 2006; Kéfi *et al.*, 2007). Detecting regime shifts in empirical data also requires the development and use of statistical methods specifically designed to identify evidence of abrupt shifts in long-term data (Scheffer and Carpenter, 2003). Identifying and understanding thresholds, and monitoring ecosystems to know when these are close, however, is often complicated. There are indications that by the time we become aware that we are close to a threshold, it is already too late (Biggs *et al.*, 2008).

3.5.3 Building resilience through diversity and spatial heterogeneity

The other way to reduce vulnerability to regime shifts is to improve the management and build the resilience of ecosystems (Gordon *et al.*, 2008). Land-use change, particularly in agriculture, has driven biodiversity loss and simplified landscapes. Management of agricultural systems has tended to focus on maximising yields, thereby emphasising the production of a single ecosystem service at the expense of other such services. Changing the focus of agricultural management on to how reliably and profitably food can be produced while also producing other ecosystem services has been a key recommendation of several large-scale assessments of ecosystems and agriculture (Molden, 2007; Foley *et al.*, 2005; Millennium Ecosystem Assessment, 2005). This change in focus could help increase resilience to regime shifts by maintaining or enhancing functional and response diversity (Elmqvist *et al.*, 2003). Functional diversity is the species diversity that maintains a specific ecosystem function, while response diversity is the diversity of responses that different species have to variations. Both are particularly important to reorganisation after disturbance (Elmqvist *et al.*, 2003; Folke *et al.*, 2004).

One of the major ways in which regime shifts can become irreversible is if the ecological processes that maintain a regime vanish from the landscape. For example, fragmentation of the Australian landscape altered the hydrology, which resulted in increased dryland salinity. Fragmentation also reduced plant reproduction and dispersal, however, reducing the ability of systems to recover from salinisation once hydrology was restored (Cramer and Hobbs, 2005). This illustrates the importance of sustaining habitats for species that can connect over the landscape – so-called mobile links – by providing ecological functions such as seed dispersal (Lundberg and Moberg, 2003).

Managing spatial heterogeneity in landscapes can also play a vital role in identifying critical areas that have a disproportionately high influence on the risk of regime shifts at a larger scale. Variations in ecological processes across a landscape mean that some locations will be more vulnerable to regime shifts than others, or contribute more to regime shifts than others. This can allow managers to focus their rehabilitation or monitoring efforts where they are likely to have the greatest effect. For example, relatively small areas in catchments that combine high soil phosphorus concentrations and high runoff potential are disproportionately responsible for the majority of phosphorus runoff into freshwater lakes (Nowak *et al.*, 2006). Water quality can therefore be substantially improved by managing these 'critical source areas' rather than managing an entire catchment (Sharpley *et al.*, 2001).

At the global scale, some regions are source areas for moisture recycling, i.e. they provide proportionally quite large amounts of evaporation which sustain rainfall in other regions (Chapter 4; see also van der Ent *et al.*, 2010). The clearing of forests in source areas can thus have quite large impacts on rainfall, since evaporation decreases and weakens moisture recycling, while the clearing of forests in other regions might have no influence at all on rainfall.

3.5.4 Understanding landscape dynamics

It is also important to understand landscape dynamics. Management and policy often assume ecosystems or water flows to be relatively stable, and therefore neglect to plan or manage for disturbance and reorganisation (Gunderson and Holling, 2001). Management that adapts to variations in external drivers can increase resilience. For example, grazing management that accounts for variability in rainfall rather than considering only average conditions can increase rangeland productivity (Janssen *et al.*, 2004). Alternatively, extreme events can be used to engineer a regime shift. For example, rainy periods associated with El Niño can be used in combination with grazing reduction to restore degraded ecosystems, while grazing reduction alone would be insufficient to achieve this goal (Holmgren *et al.*, 2006).

3.5.5 Managing the spatial disconnects between causes and effects of change

The regime shifts identified in Table 3.1 were created by agricultural modification of the hydrological cycle and present a challenge to ecosystem governance. Through blue water flows, local or continental vapour flows, or changed climate variability and seasonality, these modifications often transmit the consequences of change to a location that is spatially and temporally separate from where the initial change occurred. Coping with these disconnects can be improved by governance systems that connect local agricultural management practices to the scales at which key ecological processes operate (Folke *et al.*, 2005). Since regime shifts are often a surprising outcome of slow change, governance that is able to learn how to effectively anticipate, avoid and respond to abrupt ecological change will be better prepared for future surprises (Folke *et al.*, 2005). Achieving this goal requires better biophysical understanding of the key feedback processes that connect regime shifts at different scales (Dekker *et al.*, 2007; Scheffer *et al.*, 2005). Ecologists and hydrologists need to be involved in ensuring that this happens, and especially in expanding the scales at which they operate in order to provide better linkages between local ecological processes and global change interactions. While this is a complex and difficult endeavour, one way to start would be to identify the regions on which to focus efforts, based on where regime shifts are thought to be most likely.

3.6 Traps: highly resilient but undesirable situations

Most of the time we consider regime shifts to be undesirable, since they are often surprising and relatively rapid, and can result in large, unexpected losses of benefits for many in society. So far in this chapter we have mainly considered these types of undesirable shifts. However, some sociological–ecological regimes can be undesirable but still highly resilient. They can, for example, be situations that result from a previous regime shift or from general degradation. Examples include low productivity in the livestock system in Uganda (Box 3.2) and a degraded catchment in the Andes (Box 3.3). These degraded or undesirable but resilient systems are sometimes referred to as traps.

Resilience theory suggests the existence of two main types of trap: poverty traps and rigidity traps (Gunderson and Holling, 2001). They share the same characteristic of being difficult to escape from, but differ vastly in the amount of resources and connections that are available to attempt change. This section explores the features that inhibit change in some water-related cases in which societies either seem to have become stuck in degrading development trajectories which there is a desire to steer away from, or simply lack the desire to develop.

Box 3.2 Re-greening the Uganda 'Cattle Corridor'

Alain Vidal, CGIAR Challenge Program on Water and Food Denis Mpairwe, Makerere University

Donald Peden, International Livestock Research Institute

The CGIAR (Consultative Group on International Agricultural Research) Challenge Program on Water and Food works with its partners in the Uganda Cattle Corridor, an area that covers about one-third of the country and extends from the southwest to its northern and northeast borders. There has been severe land degradation in this region, linked to inappropriate cultivation, overgrazing and widespread charcoal production, even though there is potentially sufficient rainfall for sustainable land management. Rainfall on exposed, bare soil causes the expansion of clay particles, sealing the surface and preventing infiltration. Most precipitation then runs off the land, causing erosion and sedimentation of water bodies. Termites have repeatedly consumed young grass seedlings despite numerous efforts to reseed the highly degraded pastures. This situation exemplified a typical one in which the ecosystem has passed a seemingly irreversible threshold and was trapped in a non-productive, degraded state from which recovery appeared impossible.

Ugandan Animal Science researchers at Makerere University, however, took an idea from Ethiopia and convinced cattle keepers in the community to corral their animals together at night in order to concentrate manure. Doing this for 2 weeks before reseeding enabled the re-establishment of grass after repeated years of failed efforts. Apparently, termites prefer to eat manure rather than seedlings. Once grass cover was established, rainfall infiltration greatly improved, pasture production increased from nil to about 3000 kg/ha dry weight and soil erosion virtually ceased. The use of manure served as a trigger to flip the ecosystem back into a more productive and sustainable state of production. As a hypothesis, the researchers suggest that established pasture released organic matter into the soil through the shedding of root biomass associated with the onset of annual dry periods. These roots seem to sustain the termites and mitigate their need to feed on live vegetation.

Box 3.2 (*cont.*)

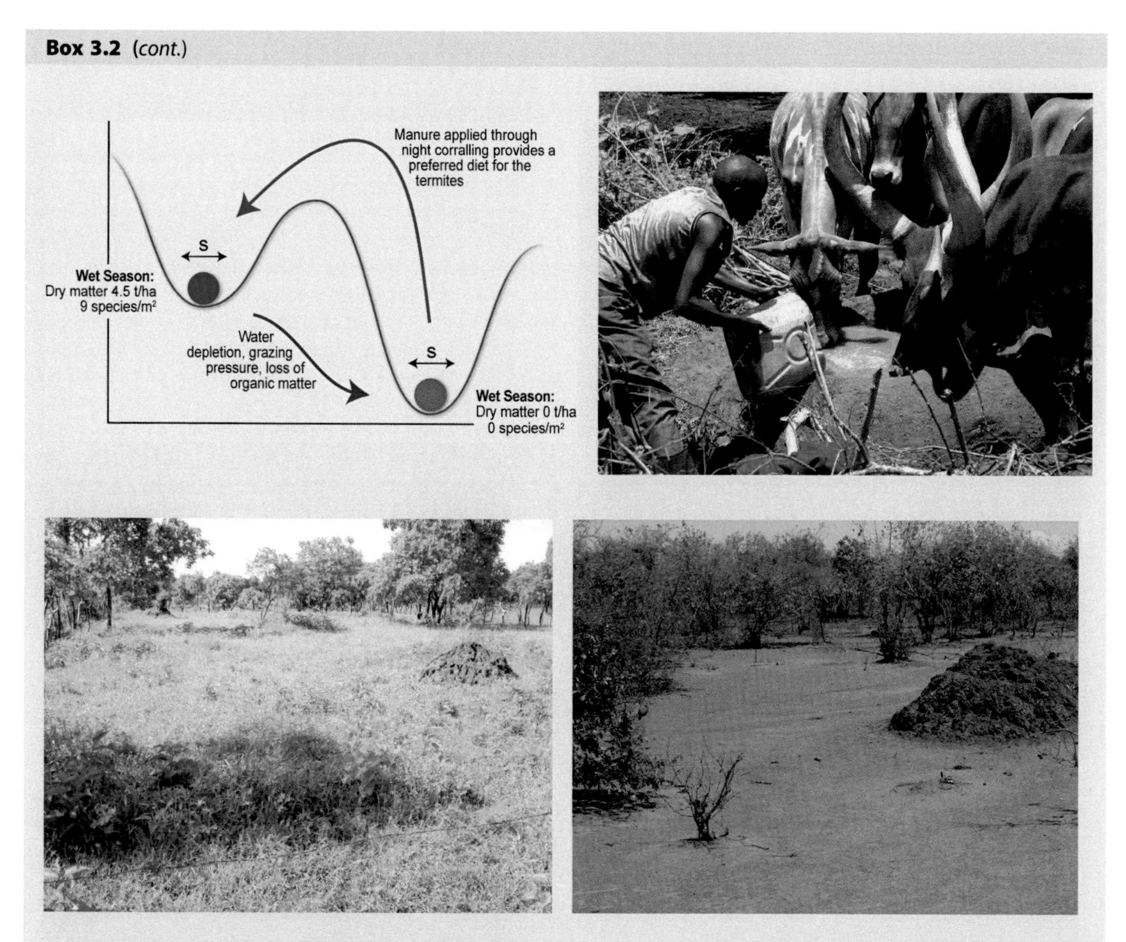

Figure 3.b2.1 (a) Diagram showing how the Uganda Cattle Corridor system was moved from a degraded 'attraction basin' to a more desirable one; (b) Ugandan cattle drinking; (c) typical wet season, in the productive state; (d) typical wet season in the degraded state.

The restoration of upslope vegetative pasture also resulted in reduced surface water runoff and evaporation, preventing sedimentation of valley tanks – a water harvesting practice – and enabling the maintenance of higher quality and a greater volume of reservoir water. Within the reservoir, the research team found that water plants such as *Nymphaea* and *Lemna* allowed aeration, which increased the efficiency of nitrification thereby increasing nitrogen levels in the soil. Furthermore, the use of *Lemna* is particularly suitable because it reduces evaporative water loss by up to 20% compared to open water sources and absorbs sediments. In response to the development of these technologies, local communities are in the process of mandating the protection of riparian vegetation and water quality. Local livestock keepers are now investing their own resources in the development and maintenance of common property pasture and water resources. In terms of resilience analysis, this re-greening of the Uganda Cattle Corridor is a good example of how transformative management of an agro-ecosystem can reverse a non-linear and seemingly irrevocable shift.

This transformation opens up opportunities to improve ecosystem services and animal production. Maintaining this positive outcome will require improved vegetation management, especially through restrictions on grazing pressure and charcoal production which were probably the shocks that caused the shift in ecosystem structure and function. One key to resilience lies in the maintenance of the food supply for termites through better vegetation management – implying the need for enhanced ground cover vegetation which can convert water depletion by runoff and evaporation to water depletion by transpiration. This is the key water intervention achieved by integrated termite management and vegetation management.

Box 3.3 Restoring ecosystem services in the Andes

Alain Vidal, CGIAR Challenge Program on Water and Food
Marcela Quintero, International Centre for Tropical Agriculture (CIAT)

The CGIAR Challenge Program on Water and Food has explored widely, together with its partners in the Andes, how benefit-sharing mechanisms can be established to optimise ecosystem services. Fuquene Lake, about 150 km north of Bogota, collects the water from the Rio Ubate. Communities manage a range of high altitude Andean production systems, including multiple cropping and livestock, from 2000–3500 m above sea level. This affects water quality by producing high levels of sediment as well as nitrates and phosphates, which are deposited in the lake accelerating its eutrophication and reducing the surface area of the water. Downstream municipalities, which depend either partially or totally on water from the Suarez River, which begins at the outlet of the lake, and navigators, are concerned about the future of the lake. Agriculture and cattle raising have expanded the agricultural frontier and degraded the ecosystem, in particular the *paramo* – a high Andean alpine-like ecological zone composed of high altitude wetlands. A change from traditional agricultural methods to conservation agriculture, especially for potato production, was chosen as a mechanism to decrease sediment and nutrient flows. Research findings showed that conservation agriculture has helped to restore *paramo* soils, especially those characteristics which determined the original capacity of buffering and filtering water in the upstream part of the basin. (The water volumetric content of the soil deep horizon significantly increased from 48% to 54%.) Moreover, these practices were found to be an extraordinarily effective way to increase the soil carbon stock (accumulated organic matter also significantly increased) and to reduce the net greenhouse-gas emissions produced by the conventional crop-livestock system (Quintero, 2009).

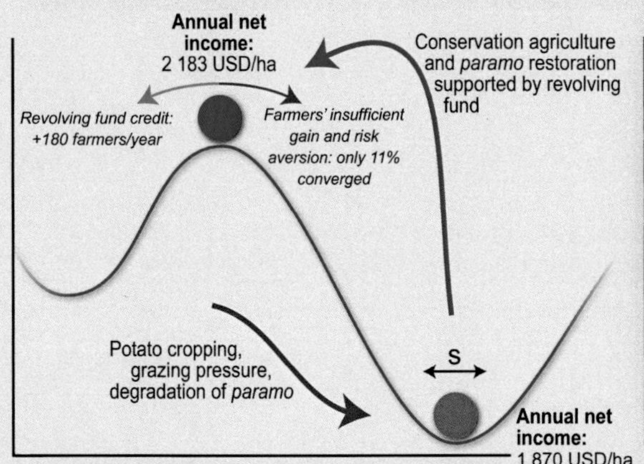

Figure 3.b3.1 (a) The graph represents how the *paramo* system can move from a degraded attraction basin to a more desirable one, (b) stakeholders discussing, (c) view over degraded *paramo* landscape.

Box 3.3 (*cont.*)

Conservation tillage results in an increase in social benefits but the economic gains remain modest. A 17% increase in net revenues in Fuquene farms would not be enough to overcome the risk aversion of farmers and other adoption barriers, or to encourage them to make additional investments to cover the up-front extra costs of conservation agriculture, such as the cultivation of oats as a cover crop. In an agricultural development, it is usually thought that even when interest rates are low, most traditional farmers need at least a 15–20% increase to make a change worthwhile. This fact may explain why the practice has not yet been widely adopted in the Fuquene catchment – such practices have been implemented on about 1800 (11%) of the 16 933 ha under potato production in the catchment. A revolving fund provides credits to farmers willing to implement conservation tillage in their potato-based production systems, however, and since 2005 it has incorporated about 180 small farmers per year.

This combination of institutional (the revolving fund) and technical (conservation tillage) innovations provides a financial incentive to those farmers willing to switch to conservation agriculture, helping to reverse the non-linear degradation of the fragile land and water ecosystems and to shift them back to a 'safe operating space' where they can continue to provide their services to local communities. However, as shown in Figure 3.b3.1, negative feedbacks such as risk aversion and other adoption barriers maintain the social and ecological systems of the Fuquene basin at the limit of such a safe operating space.

3.6.1 Poverty traps: running on the treadmill with no returns

A poverty trap – however measured – is about staying poor, not just about being poor at a moment in time (Barrett *et al.*, 2011). It involves self-reinforcing mechanisms that cause poverty to persist. Poverty traps reflect a lack of options to develop or deal with change. They occur in situations of low resource availability, low diversity and low connectedness to other systems (Gunderson and Holling, 2001). This is often a typical situation in, for example, many dryland environments with scarce natural capital, such as nutrients and water, infrastructural capital, such as roads and telephones, and human capital, such as educational opportunities; and where many people are remote both in a geographical sense and in terms of decision-making at larger scales (Reynolds *et al.*, 2007). In such situations, the trap can be a result of poor rural people's heavy dependence on and over-exploitation of finite natural resources such as water, nutrients and soil for their livelihoods (Barrett *et al.*, 2011). It can also happen in situations of failure of social institutions, for example, where society is traumatised by conflict, there is a lack of trust among citizens or individuals lack extended networks (Gunderson and Holling, 2001). Barrett *et al.* (2011) suggest that a lack of informed adaptive management can also cause poverty traps.

There are many examples of water availability, variability and management playing a role in both sustaining poverty traps and providing mechanisms for their transformation. Many regions with an arid/semi-arid climate coincide with regions with high levels of poverty. Water variability, such as recurrent droughts and dry spells, can play an important role. It is quite common for rural farmers to build up capital during relatively good, wet years, only to lose many assets in dry years when several factors coincide simultaneously, such as yield reductions, rising food prices and dramatic falls in livestock prices (Enfors and Gordon, 2008). Farmers might then have to sell other assets and reduce their normal small-scale business activities due to lack of investment capital, and there can be a slowdown in other investment strategies such as an inability to pay school fees.

Since poverty traps are about being locked into persistent degraded or low-production conditions, with poor capital and networks, external support is generally needed to get out of a poverty trap. However, simply providing money, technical expertise, infrastructure and public education is seldom sufficient. Successful investments depend on the capacity of people within the social–ecological system to use investment to create new and continuing opportunities (Folke *et al.*, 2010). However, low levels of connectedness can prevent the mobilisation of ideas and resources to solve problems in these situations, and new ideas can have difficulty emerging or gaining impetus (Gunderson and Holling, 2001). Work to develop social capital is therefore important to ensure that transformations out of poverty traps can take place. Box 3.2 illustrates how it was possible to transform a degraded livestock environment into a much more productive one, and that this would not have been possible without investing in a community organisation and its members. Box 3.3 shows successful catchment restoration, which only happened due to investment in both technical and institutional innovations. The institutional investment was particularly helpful in overcoming people's risk aversion, which initially was a major factor hampering development.

3.6.2 Rigidity traps: where past successes are the failures of tomorrow

In contrast to poverty traps, rigidity traps arise in situations where resource availability is high and connectedness provides great impetus (Gunderson and Holling, 2001). While this might sound like a good situation, it can exist at the expense of creativity and adaptability. Hegmon *et al.* (2008) suggest that societies might become rigid if they limit buffering options, narrow socially accepted differences, become too attached to tradition, technology or place or are too path dependent. This can happen in systems that are highly successful. An initially successful company can for example resist disturbances such as internal criticism among staff about working conditions, or fail to monitor external changes in its surroundings that might help predict changes in long-term markets. An underlying reason can be initial success, which makes the company complacent or over-confident. People and institutions in these situations can try to resist change and persist with their current management and governance systems despite clear indications that change is essential (Holling and Meffe, 1996). In some cases, the rigidity comes from the unintended consequences of many actions, such as a build up of bureaucratic systems, while in others it is intentional rigidity developed by leaders who want to maintain their power (Hegmon *et al.*, 2008). This behaviour constrains the ability of people to respond to new problems and opportunities. One example is the identity that people have as farmers in the Murray Darling River in Australia, which is so deeply embedded in the culture (Walker *et al.*, 2009) that it prevents them from diversifying away from farming even though it is necessary to avoid irreversible salinisation in a region currently struggling with severe salinisation problems (Box 3.1).

Many cases that have ended up in rigidity traps involve the development of large-scale irrigation with large-scale downstream effects and vast fixed costs. The Colorado River basin is one example. Water resources have been overcommitted and over-allocated due to unrealistically high assumptions about annual river flows, leading to large reductions in river flows with enormous implications downstream. The river today is chronically overstressed, the population has grown by 700% in the past 70 years, and wetlands have shrunk to one-tenth of their original size. The Colorado River is also one of the most highly regulated rivers, with rigid infrastructure and institutions. Many dams and reservoirs have been built to supply water for hydropower, irrigation schemes and huge water transfers to California. There have been many efforts to change the situation but people and institutions are resisting change and persisting with current management and governance systems, which has eroded adaptive capacity.

This rigidity trap is a consequence of strongly bureaucratic systems that maintain a historical situation. The 'Law of the River' in this case is one example of inflexible property rights that can also inhibit change. These rights include the 'priority concept', in which the first water user has eternal rights to that water and newer rights, such as environmental flows, will only receive what is left over after all prior demands have been satisfied. A lot of trust is placed in the promotion of supply-side management as a possible solution, through technological fixes such as desalination plants and cloud seeding. There is little chance of creating incentives that are strong enough to cause change. For example, emerging water markets will have difficulty valuing environmental flow requirements and dam removals are not being seriously considered.

Hegmon *et al.* (2008) have studied the rigidity and transformation of three societies in the American southwest in 1000–1450 CE: the Mimbres, Mesa Verde and Hohokam. They argue that people in the Hohokam region, where large populations were centred around irrigation systems, created – consciously or not – a particularly rigid way of life that offered few alternatives, so that when conditions changed in negative ways linked to a worsening climate, deteriorating health and decaying social and physical infrastructure, people stayed put for generations until the social and physical infrastructure finally disintegrated. They suggest that people may literally have felt trapped, perceiving no way to make changes and no place to go, and so they stayed put while things fell apart around them.

3.6.3 A need for transformation: when you are in a hole, stop digging

When asked how to break out from a trap, Brian Walker, one of the most highly cited scholars in resilience-related work, advised: 'When you are in a hole, the first thing you should do is stop digging.' However, much more is needed to have a substantial and lasting impact on these resilient systems. It is sometimes argued that when societies or groups find themselves trapped in an undesirable regime, it may prove necessary to configure an entirely new 'stability landscape', which is dominated by substantially different feedbacks and key processes. In the metaphor of the sailing boat in the introduction to this chapter, once it has capsized it is not necessarily the best solution to right it and keep sailing. Perhaps too much has been lost. Perhaps it is better to sell it and do something else. The capacity to create such a new regime is known as transformability: the capacity to create new beginnings from which to evolve a new way of living when existing ecological, economic or social structures become untenable (Walker and Meyers, 2004). There will always be a tension between maintaining the resilience of a desirable current regime and simultaneously building the capacity for transformation should the need arise.

Table 3.2 summarises some of the aspects involved in poverty and rigidity traps and shows the differences between them. Just as the character of the traps differs, so do the potential solutions for how to get out of them. What creates transformative capacity may differ between poverty and rigidity traps. If you are in a poverty trap and have few resources, external investment is usually needed for transformation to be possible. This investment should focus on strengthening social capacity that enables the mobilisation of alternative resources beyond current livelihoods, and improving the processes for using them. This also means that local, small-scale investment is seldom enough. There is a need to coordinate across scales (Barrett and Swallow, 2006). Since rigidity traps are often caused by mental gridlocks, we suggest that some type of 'wake-up call' is needed. Using times of crises as windows of opportunity might be particularly relevant here. The development of alternative mental models or identities and a stronger vision are also important. The processes that can enable transformation are further discussed in Chapter 8.

Summary

The role of water in regime shifts

This chapter addresses the involvement of water in abrupt, unexpected regime shifts in social–ecological systems. Regime shifts are large, abrupt and persistent changes in structure and function, where the cost in

Table 3.2 Key differences between poverty and rigidity traps, and their relation to water

	Poverty trap	Rigidity trap
Governance and investments	Low social capital, low level of networking skills Low levels of leadership Low internal resources and investments	Highly bureaucratic and/or hierarchical structures High financial rewards Investment in infrastructure
Monitoring	General lack of monitoring, low skills	Monitoring of wrong or too narrow set of variables
Dealing with disturbances	Continual, often recurring disturbances (e.g. droughts, conflicts or pests) Difficulty in keeping resources in bad years	Takes advantage and mobilises in good years, unprepared for bad years
General water issues involved	Land degradation, loss of infiltration and water holding capacity Low initial endowments/natural capital Often green water related	Overuse of irrigation water Technological fixes Often blue water related
Identity	Strong identity (as a farmer in our cases)	Strong identity as a productive person/society

terms of money, effort or scale of change needed to shift a system back – even if it is possible – is larger than the change it took to trigger the regime shift. Some of these changes are relatively easy to reverse if the pressures on the system are reduced. Others are difficult to reverse or even irreversible.

Resilience in social–ecological systems is characterised by the existence of reinforcing processes and stabilising feedbacks. A regime shift implies that one system of feedbacks is replaced by a different system of feedbacks. System dynamics are shaped by internal variables and external forces acting on the system. Internal variables may undergo slow change as part of the system's dynamics. These slower processes are seldom monitored, even though this would be useful to obtain early warning of potential regime shifts.

Water-related disturbances and feedbacks

Water's many different roles and functions in the life-support system mean that it is profoundly involved in processes of and responses to regime shifts, as both a state variable and a control variable. We have demonstrated that regime shifts can occur across the full hydrological cycle and at several different scales, from the local to the global. Many regime shifts are cross-scale in nature, where for example a change in global markets can affect local conditions in unexpected ways, which can even cascade back to affect global biophysical phenomena. One major example is the case of the Borneo rainforest shifting from being a carbon sink to a carbon source.

Disturbances may originate from blue water systems, through the depletion of river flow, groundwater resources or water quality; from green water systems, through soil water recharge, melting permafrost or productivity depletion; or from vegetation systems, through deforestation, land-use change and alterations in consumptive water use or moisture feedback. Feedbacks between water and processes in which water is involved are of crucial importance to resilience.

Even though the details are poorly understood, it has become increasingly clear that three water-related threshold processes have interacted with processes that generate feedbacks that can either dampen or reinforce change. Three sets of tipping points have been identified, generated by: land-use management issues, such as deforestation-linked moisture feedbacks, land mismanagement feedbacks and salinisation; water overuse, such as groundwater collapse, river basin closure and river basin depletion; and global change pressures, such as regional processes, sea level rise and saltwater intrusion, drastic rainfall regime change and glacial melt.

Building resilience to regime shifts

To avoid water-related regime shifts, it is not enough to manage water by itself. Water management needs to be combined with stewardship of the whole landscape, including ecological and social processes. It is the interactions among social, ecological and hydrological variables that both create vulnerability to unwanted regime shifts and help build resilience against them.

An understanding of diversity and dynamics in a broad sense is important when designing interventions to build resilience to regime shifts. This includes aspects such as finding critical source areas at different scales, and analysing how disturbances can be used as sources of renewal. The ability to manage systems to avoid regime shifts in a world of high variability would be improved if we could predict how close an ecosystem is to critical thresholds. However, by the time we have identified that we are close to a threshold, it could already be too late because the feedbacks may already have started to switch. Finding operational levels to which management can respond, rather than seeking true thresholds, which are extremely costly and difficult to find, could be a way of defining thresholds of probable concern.

Traps and how to get out of them

Systems can sometimes be trapped on undesirable but resilient pathways of development. We distinguished between poverty traps, characterised by low resource availability, low diversity and low levels of connectedness, and rigidity traps, characterised by high levels of resource availability and connectedness but also rigid resistance to change.

A poverty trap involves self-reinforcing mechanisms that cause poverty to persist. These reflect a lack of options to develop or deal with change. In rigidity traps, people and institutions can try to resist change and persist with their current management and governance systems, despite clear recognition that change is essential.

To break out of poverty traps, it is often important to build social and institutional capital which can be mobilized for use in other types of more direct investment, e.g. in infrastructure and public education. With rigidity traps, it is often important to develop an alternative vision of the future in order to break mental gridlocks that might otherwise hinder change, despite the capital and networks that exist.

References

Adger, W. N., Eakin, H. and Winkels, A. (2008). Nested and teleconnected vulnerabilities to environmental change. *Frontiers in Ecology and the Environment*, **7**, 150–157.

Anderies, J. M., Ryan, P. and Walker, B. H. (2006). Loss of resilience, crisis, and institutional change: lessons from an intensive agricultural system in southeastern Australia. *Ecosystems*, **9**, 865–878.

Antle, J. M., Stoorvogel, J. J. and Valdivia, R. O. (2006). Multiple equilibria, soil conservation investments, and the resilience of agricultural systems. *Environment and Development Economics*, **11**, 477–492.

Barrett, C. B. and Swallow, B. M. (2006). Fractal poverty traps. *World Development*, **34**, 1–15.

Barrett, C. B., Travis, A. J. and Dasgupta, P. (2011). On biodiversity conservation and poverty traps. *Proceedings of the National Academy of Sciences*, **108**, 13907–13912.

Bestelmeyer, B. T., Trujillo, D. A., Tugel, A. J. and Havstad, K. M. (2006). A multi-scale classification of vegetation dynamics in arid lands: What is the right scale for models, monitoring, and restoration? *Journal of Arid Environments*, **65**, 296–318.

Biggs, R., Simons, H., Bakkenes, M. *et al.* (2008). Scenarios of biodiversity loss in Southern Africa in the 21st century. *Global Environmental Change*, **18**, 296–309.

Bossio, D. (2007). Conserving land-protecting water. In *Water for Food, Water for Life: A Comprehensive Assessment of Water Management*, ed. Molden, D. London: Earthscan, pp. 551–583.

Boyer, E. W., Howarth, R. W., Galloway, J. N. *et al.* (2006). Riverine nitrogen export from the continents to the coasts. *Global Biogeochemical Cycles*, **20**.

Brock, W. A. and Carpenter, S. R. (2006). Variance as a leading indicator of regime shift in ecosystem services. *Ecology and Society*, **11**, 9.

Carpenter, S. R. (2003). *Regime Shifts in Lake Ecosystems: Pattern and Variation. Excellence in Ecology.* Oldendorf/Luhe, Germany: Ecology Institute.

Carpenter, S. R. (2005). Eutrophication of aquatic ecosystems: bistability and soil phosphorus. *Proceedings of the National Academy of Sciences of the United States of America*, **102**, 10002–10005.

Centre for Research on the Epidemiology of Disaster (CRED) (2012). EM-DAT: The OFDA/CRED International Disaster database version 12.07. Universitē catholique de Louvain, Brussels.

Clarke, C. J., George, R. J., Bell, R. W. and Hatton, T. J. (2002). Dryland salinity in south-western Australia: its origins, remedies, and future research directions. *Australian Journal of Soil Research*, **40**, 93–113.

Cowling, S. A., Betts, R. A., Cox, P. M. *et al.* (2004). Contrasting simulated past and future responses of the Amazonian forest to atmospheric change. *Philosophical Transactions of the Royal Society of London Series B-Biological Sciences*, **359**, 539–547.

Cramer, V. A. and Hobbs, R. J. (2005). Assessing the ecological risk from secondary salinity: a framework addressing questions of scale and threshold responses. *Austral Ecology*, **30**, 537–545.

Cumming, G. S. (2011). The resilience of big river basins. *Water International*, **36**, 63–95.

Curran, L. M., Trigg, S. N., McDonald, A. K. *et al.* (2004). Lowland forest loss in protected areas of Indonesian Borneo. *Science*, **303**, 1000–1003.

D'Almeida, C., Vörösmarty, C. J., Hurtt, G. C. *et al.* (2007). The effects of deforestation on the hydrological cycle in Amazonia: a review on scale and resolution. *International Journal of Climatology*, **27**, 633–647.

Dawson, T. E. (1998). Fog in the California redwood forest: ecosystem inputs and use by plants. *Oecologia*, **117**, 476–485.

Dekker, S. C., Rietkerk, M. A. X. and Bierkens, M. F. P. (2007). Coupling microscale vegetation–soil water and macroscale vegetation–precipitation feedbacks in semiarid ecosystems. *Global Change Biology*, **13**, 671–678.

del-Val, E., Armesto, J. J., Barbosa, O. *et al.* (2006). Rain forest islands in the Chilean semiarid region: fog-dependency, ecosystem persistence and tree regeneration. *Ecosystems*, **9**, 598–608.

Diaz, R. J. (2001). Overview of hypoxia around the world. *Journal of Environmental Quality*, **30**, 275–281.

Douglas, E. M., Niyogi, D., Frolking, S. *et al.* (2006). Changes in moisture and energy fluxes due to agricultural land use and irrigation in the Indian monsoon belt. *Geophysical Research Letters*, **33**.

Elmqvist, T., Folke, C., Nyström, M. *et al.* (2003). Response diversity, ecosystem change, and resilience. *Frontiers in Ecology and the Environment*, **1**, 488–494.

Enfors, E. I. and Gordon, L. J. (2007). Analysing resilience in dryland agro-ecosystems: a case study of the Makanya catchment in Tanzania over the past 50 years. *Land Degradation & Development*, **18**, 680–696.

Enfors, E. I. and Gordon, L. J. (2008). Dealing with drought: the challenge of using water system technologies to break dryland poverty traps. *Global Environmental Change*, **18**, 607–616.

Falkenmark, M. and Molden, D. (2008). Wake up to realities of river basin closure. *International Journal of Water Resources Development*, **24**, 201–215.

Feddema, J., Oleson, K., Bonan, G. *et al.* (2005). A comparison of a GCM response to historical anthropogenic land cover change and model sensitivity to uncertainty in present-day land cover representations. *Climate Dynamics*, **25**, 581–609.

Foley, J. A., Coe, M. T., Scheffer, M. and Wang, G. (2003). Regime shifts in the Sahara and Sahel: interactions between ecological and climatic systems in northern Africa. *Ecosystems*, **6**, 524–532.

Foley, J. A., DeFries, R., Asner, G. P. *et al.* (2005). Global consequences of land use. *Science*, **309**, 570–574.

Folke, C., Carpenter, S., Walker, B. *et al.* (2004). Regime shifts, resilience, and biodiversity in ecosystem management. *Annual Review of Ecology Evolution and Systematics*, **35**, 557–581.

Folke, C., Carpenter, S. R., Walker, B. *et al.* (2010). Resilience thinking: integrating resilience, adaptability and transformability. *Ecology and Society*, **15**, 20.

Folke, C., Hahn, T., Olsson, P. and Norberg, J. (2005). Adaptive governance of social-ecological systems. *Annual Review of Environmental Resources*, **30**, 441–473.

Folke, C., Jansson, Å., Rockström, J. *et al.* (2011). Reconnecting to the biosphere. *AMBIO: A Journal of the Human Environment*, **40**, 719–738.

Fraser, E. and Rimas, A. (2011). The psychology of food riots. Foreign Affairs. Available at: http://www.foreignaffairs.com/articles/67338/evan-fraser-and-andrew-rimas/the-psychology-of-food-riots.

Fu, C. B. (2003). Potential impacts of human-induced land cover change on East Asia monsoon. *Global and Planetary Change*, **37**, 219–229.

Galaz, V., Moberg, F., Olsson, E.-K., Paglia, E. and Parker, C. (2010). Institutional and political leadership dimensions of cascading ecological crises. *Public Administration*, **89**, 361–380.

Galloway, J. N., Dentener, F. J., Capone, D. G. *et al.* (2004). Nitrogen cycles: past, present, and future. *Biogeochemistry*, **70**, 153–226.

Genkai-Kato, M. and Carpenter, S. R. (2005). Eutrophication due to phosphorus recycling in relation to lake morphometry, temperature, and macrophytes. *Ecology*, **86**, 210–219.

Gordon, L., Dunlop, M. and Foran, B. (2003). Land cover change and water vapour flows: learning from Australia. *Philosophical Transactions of the Royal Society of London Series B: Biological Sciences*, **358**, 1973–1984.

Gordon, L. J., Peterson, G. D. and Bennett, E. M. (2008). Agricultural modifications of hydrological flows create ecological surprises. *Trends in Ecology and Evolution*, **23**, 211–219.

Gunderson, L. H. and Holling, C. S. (eds) (2001). *Panarchy: Understanding Transformations in Human and Natural Systems*. Washington DC: Island Press.

Hegmon, M., Peeples, M. A., Kinzig, A. P. *et al.* (2008). Social transformation and its human costs in the prehispanic US southwest. *American Anthropologist*, **110**, 313–324.

Holling, C. S. (1973). Resilience and stability of ecological systems. *Annual Review of Ecology and Systematics*, **4**, 1–23.

Holling, C. S. and Meffe, G. K. (1996). Command and control and the pathology of natural resource management. *Conservation Biology*, **10**, 328–337.

Holmgren, M., Stapp, P., Dickman, C. R. *et al.* (2006). Extreme climatic events shape arid and semiarid ecosystems. *Frontiers in Ecology and the Environment*, **4**, 87–95.

Hughes, L. (2003). Climate change and Australia: trends, projections and impacts. *Austral Ecology*, **28**, 423–443.

Hutchings, J. A. (1996). Spatial and temporal variation in the density of northern cod and a review of hypotheses for the stock's collapse. *Canadian Journal of Fisheries and Aquatic Sciences*, **53**, 943–962.

Hutyra, L. R., Munger, J. W., Nobre, C. A. *et al.* (2005). Climatic variability and vegetation vulnerability in Amazonia. *Geophysical Research Letters*, **32**.

Janssen, M. A., Anderies, J. M. and Walker, B. H. (2004). Robust strategies for managing rangelands with multiple stable attractors. *Journal of Environmental Economics and Management*, **47**, 140–162.

Kabat, P., Ludwig, F., van der Valk, M. and van Schaik, H. (eds) (2012). *Climate Change Adaptation in the Water Sector*. London: Routledge.

Karlsson, J. M., Lyon, S. and Destouni, G. (2011). *Arctic Hydrology Shifts with Permafrost Thawing and Thermokarst Lake Changes*. AGU 2011 Fall Meeting, 2011. San Francisco: AGU.

Kéfi, S., Rietkerk, M., Alados, C. L. *et al.* (2007). Spatial vegetation patterns and imminent desertification in Mediterranean arid ecosystems. *Nature*, **449**, 213–217.

Kinzig, A. P., Ryan, P. A., Etienne, M. *et al.* (2006). Resilience and regime shifts: assessing cascading effects. *Ecology and Society*, **11**, 1. Available at: http://www.ecologyandsociety.org/vol11/iss1/art20/.

Los, S. O., Weedon, G. P., North, P. R. J. *et al.* (2006). An observation-based estimate of the strength of rainfall-vegetation interactions in the Sahel. *Geophysical Research Letters*, **33**.

Ludwig, J. A., Wilcox, B. P., Breshears, D. D., Tongway, D. J. and Imeson, A. C. (2005). Vegetation patches and runoff-erosion as interacting ecohydrological processes in semiarid landscapes. *Ecology*, **86**, 288–297.

Lundberg, J. and Moberg, F. (2003). Mobile link organisms and ecosystem functioning: implications for ecosystem resilience and management. *Ecosystems*, **6**, 87–98.

Millennium Ecosystem Assessment (2005). *Ecosystems and Human Well-being: Synthesis*. Washington DC: Island Press.

Miller, G., Mangan, J., Pollard, D. *et al.* (2005). Sensitivity of the Australian monsoon to insolation and vegetation: implications for human impact on continental moisture balance. *Geology*, **33**, 65–68.

Molden, D. (ed.) (2007). *Water for Food, Water for Life: A Comprehensive Assessment of Water Management*. London: Earthscan.

Nowak, P., Bowen, S. and Cabot, P. E. (2006). Disproportionality as a framework for linking social and biophysical systems. *Society and Natural Resources*, **19**, 153–173.

Oyama, M. D. and Nobre, C. A. (2003). A new climate-vegetation equilibrium state for tropical South America. *Geophysical Research Letters*, **30**.

Page, S. E., Siegert, F., Rieley, J. O. *et al.* (2002). The amount of carbon released from peat and forest fires in Indonesia during 1997. *Nature*, **420**, 61–65.

Peters, D. P. C. and Havstad, K. M. (2006). Nonlinear dynamics in arid and semi-arid systems: interactions among drivers and processes across scales. *Journal of Arid Environments*, **65**, 196–206.

Peters, D. P. C., Pielke, R. A., Bestelmeyer, B. T. *et al.* (2004). Cross-scale interactions, nonlinearities, and forecasting catastrophic events. *Proceedings of the National Academy of Sciences of the United States of America*, **101**, 15130–15135.

Quintero, M. (2009). Effects of conservation tillage in soil carbon sequestration and net revenues of potato-based rotations in the Colombian Andes. Masters thesis, University of Florida.

Reynolds, J. F., Smith, D. M. S., Lambin, E. F. *et al.* (2007). Global desertification: building a science for dryland development. *Science*, **316**, 847–851.

Rietkerk, M., Dekker, S. C., de Ruiter, P. C. and van de Koppel, J. (2004). Self-organized patchiness and catastrophic shifts in ecosystems. *Science*, **305**, 1926–1929.

Sankaran, M., Hanan, N. P., Scholes, R. J. *et al.* (2005). Determinants of woody cover in African savannas. *Nature*, **438**, 846–849.

Scheffer, M., Carpenter, S., Foley, J. A., Folke, C. and Walker, B. (2001). Catastrophic shifts in ecosystems. *Nature*, **413**, 591–596.

Scheffer, M. and Carpenter, S. R. (2003). Catastrophic regime shifts in ecosystems: linking theory to observation. *Trends in Ecology and Evolution*, **18**, 648–656.

Scheffer, M., Holmgren, M., Brovkin, V. and Claussen, M. (2005). Synergy between small- and large-scale feedbacks of vegetation on the water cycle. *Global Change Biology*, **11**, 1003–1012.

Scheffer, M. and van Nes, E. H. (2007). Shallow lakes theory revisited: various alternative regimes driven by climate, nutrients, depth and lake size. *Hydrobiologia*, **584**, 455–466.

Sharpley, A. N., McDowell, R. W., Weld, J. L. and Kleinman, P. J. A. (2001). Assessing site vulnerability to phosphorus loss in an agricultural watershed. *Journal of Environmental Quality*, **30**, 2026–2036.

Snyder, P. K. (2010). The influence of tropical deforestation on the Northern Hemisphere climate by atmospheric teleconnections. *Earth Interactions*, **14**, 1–34.

Sternberg, L. D. L. (2001). Savanna-forest hysteresis in the tropics. *Global Ecology and Biogeography*, **10**, 369–378.

van der Ent, R. J., Savenije, H. H. G., Schaefli, B. and Steele-Dunne, S. C. (2010). Origin and fate of atmospheric moisture over continents. *Water Resources Research*, **46**, W09525.

Vörösmarty, C. J., McIntyre, P. B., Gessner, M. O., *et al.* (2010). Global threats to human water security and river biodiversity: adjusted human water stress threat. Available at: http://databasin.org/datasets/6cb375aa273449b2b4353b9e265d5557 (accessed 18 January 2013).

Walker, B. and Meyers, J. A. (2004). Thresholds in ecological and social–ecological systems: a developing database. *Ecology and Society*, **9**, 3.

Walker, B. H., Abel, N., Anderies, J. M. and Ryan, P. (2009). Resilience, adaptability, and transformability in the Goulburn–Broken catchment, Australia. *Ecology and Society*, **14**(1), 12. Available at: http://www.ecologyandsociety.org/vol14/iss1/art12/.

Walker, B. H., Anderies, J. M., Kinzig, A. P. and Ryan, P. (2006). Exploring resilience in social-ecological systems through comparative studies and theory development: introduction to the special issue. *Ecology and Society*, **11**, 12.

Zheng, X. Y. and Eltahir, E. A. B. (1997). The response to deforestation and desertification in a model of West African monsoons. *Geophysical Research Letters*, **24**, 155–158.

Zickfeld, K., Knopf, B., Petoukhov, V. and Schellnhuber, H. J. (2005). Is the Indian summer monsoon stable against global change? *Geophysical Research Letters*, **32**.

Logged and terraced areas stand next to primary and secondary growth rainforest in the Bandan River Area of Sarawak, Malaysia, with a great difference in soil moisture, ecosystem services and water retention.

Crucial functioning of and human dependence on the global water system

The analysis in this chapter goes deeper into the core functions of the global water system (GWS), the role of water as the bloodstream of the biosphere and the key interactions between scales, including remote linkages or so-called teleconnections. It discusses how various changes in the Earth System could slowly make the GWS less resilient to shocks such as droughts, the key role of bidirectional interactions between water, land and vegetation, and local, regional and planetary boundaries. The chapter highlights three large-scale interactions that indicate long-term responses by the socio-ecological system to land-use changes in terms of so-called tipping points. It finally presents a comparative analysis of water and food security for a number of large river basins in Asia, Africa and South America.

4.1 The role of water as the bloodstream of the biosphere

Based on the principles outlined in Chapter 2, this chapter focuses on the roles of water as, at the same time, a state and a control variable in the GWS, interlinking different components and regions within the Earth System. We look at the different drivers of change, and how they affect water and water-related impacts and feedbacks as well as resilience at all scales.

The chapter first describes the role of water in biomass production and in all ecosystems, making it the 'bloodstream of the biosphere'. Water links the different sub-systems of the Earth System through the different components of the hydrological cycle: evapotranspiration, advection, condensation, precipitation, infiltration, groundwater recharge and runoff (see Figure 4.1).

Solar radiation is the primary driver of the hydrological cycle, causing transpiration by plants, evaporation from land and open water surfaces, and the transportation of atmospheric moisture. When condensing, atmospheric moisture forms clouds and precipitation. Precipitation infiltrates the soil, recharges groundwater and flows along river networks to the ocean.

According to a recent multi-model study (Haddeland *et al.*, 2011), which simulated the hydrological cycle at the end of the twentieth century, global terrestrial mean precipitation amounts to about 125 000 km^3 per year, albeit with large variations between the models. This is the total water resource for humans and the environment, partitioned into green water (more than 70 000 km^3) which is evaporated or transpired back to the atmosphere, and blue water – the remainder – which becomes runoff or recharges groundwater, eventually returning to the sea.

Water plays a key role in linking the different components of social–ecological systems at all scales by:

- controlling energy fluxes between land and the atmosphere, damping temperature fluctuations and regulating the climate, making it appropriate for living systems;
- mediating biogeochemical reactions;
- dissolving and transporting substances (in particular nutrients) between different environmental compartments;
- sustaining and regulating biomass growth for food and bioenergy production, as well as other ecosystem services;
- serving a range of socio-economic, cultural, religious and other purposes, e.g. for energy production, transportation, health and recreation.

There are a number of bidirectional interactions between water and vegetation in ecosystems. On the one hand, water is central to photosynthetic

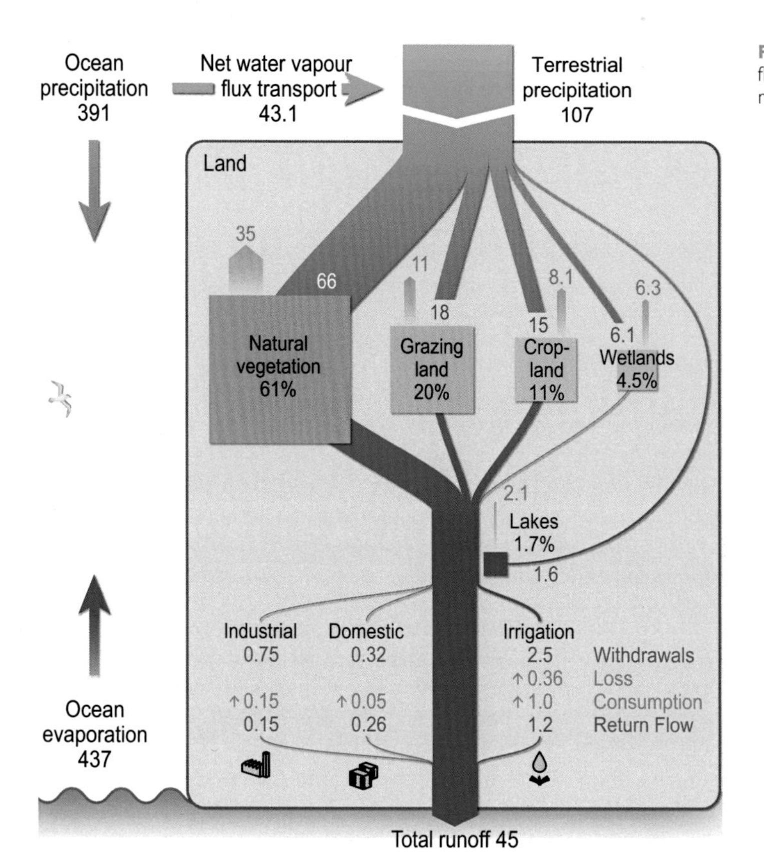

Figure 4.1 The hydrological cycle, fluxes and flows, with numbers calculated from the LPJmL model (see Chapter 5).

biomass production and for transporting nutrients to and within plants. On the other hand, vegetation, among other factors, controls the partitioning of rainfall on the land surface into blue and green water, returning large amounts of water to the atmosphere via transpiration from where it will serve again as rainfall for downwind regions (so-called moisture recycling, see Savenije, 1995), and slowing down the runoff to the sea.

Because water maintains and stabilises biological processes, it has been termed the 'bloodstream of the biosphere' (see Box 4.1; Figure 4.2; Ripl, 2003). In this role, water has no substitute. The Water, Energy and Food Security Nexus Conference in Bonn, Germany, in 2011, picked up on this image and made 'water running through the veins of our economy' a headline of the conference's thematic profile.

4.1.1 Water sustains all ecosystem services on which humans depend

This section explains how water acts as a control variable, for example, with respect to food production and other ecosystem services. More details on the interactions between water and food are provided in Chapter 5, and on water and ecosystem services in Chapter 7.

Previous assessments of water availability and human water appropriation have focused mostly on blue water (e.g. Alcamo *et al.*, 2007; Islam *et al.*, 2007; Shiklomanov, 2000; Vörösmarty *et al.*, 2000), but green water use for crop production has also been quantified recently (e.g. Gerten *et al.*, 2011; Liu and Yang, 2010; Mekonnen and Hoekstra, 2010; Siebert and Döll, 2010). Consumptive blue water uses are estimated at less than 1500 km^3 per

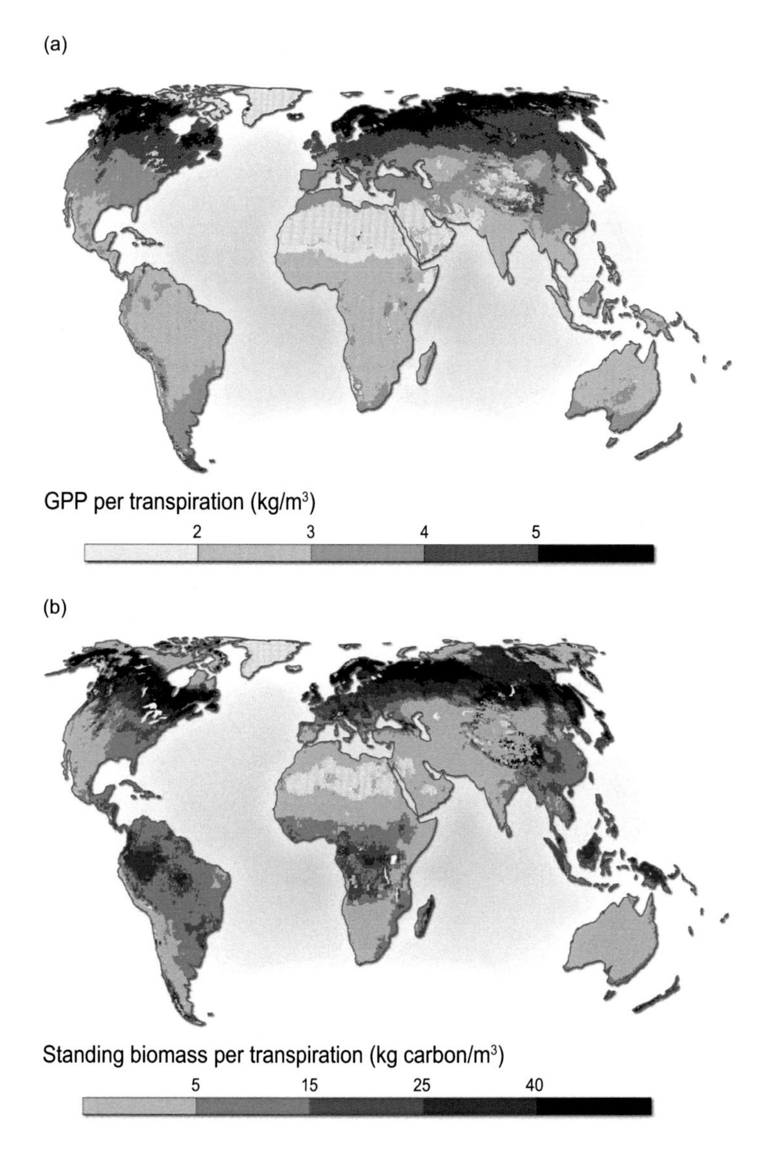

Figure 4.2 Water as the bloodstream of the biosphere. The interaction between water and carbon fluxes and stores in biomass production, or the 'water intensity' of carbon assimilation and storage. (a) Gross primary production (annual carbon 'flow') per transpiration. (b) Vegetation carbon stock in vegetation per transpiration per year.

year, while green water use for crop production alone amounts to 5000–6000 km^3 per year – and about twice this amount when green water used in livestock production (evaporation from grazing land) is also taken into account (Hoff *et al.*, 2010).

Box 4.1 Water has shaped the face of our planet

Wilhelm Ripl Technical University of Berlin (Professor Emeritus)

At the Earth's surface the liquid medium water is involved in dissipating the solar energy pulse, distributing it in space and time. This dynamic process dampens temperature peaks and creates the thermal and hydrological conditions that support all natural processes. Evaporation and precipitation form a more or less open water cycle almost without loss of matter, which cools hot spots and warms cooler areas.

Box 4.1 *(cont.)*

Solar-driven cycles

A set of solar-driven, dissipative water and matter cycles – with water as the most important dynamic agent – have shaped the face of the planet and constitute the key to life. Together, the cyclical processes of evaporation and precipitation, of dissolution and crystallisation and finally disintegration of the water molecule with recombination through respiration, are three essential cyclical water processes together dampening the solar pulse. The dynamic agent water sorts solid and dissolved matter by controlling buoyancy under hydrodynamic conditions and the solubility properties of practically all compounds in water. There are hardly any structures that cannot be considered hydromorphic in a direct or indirect way (Ripl, 1992, 1995, Figure 4.b1.1).

However, since humankind has ruled the fate of this Earth, the natural way of self-organisation has been steadily disrupted in an incremental fashion. If sustainable development is ultimately to be desired by societies and believed to be the main goal of humankind, some improvement in the human control of water cycles and material transport processes seems inevitable. The current situation is characterised by randomisation of previously ordered matter and processes – mainly through the use of non-renewable energy and the distortion of the global cooling system through omnipresent interference with water cycle and vegetation cover. Accelerated, irreversible matter transport processes to the sea, from originally fertile localities covered in vegetation, are still increasing.

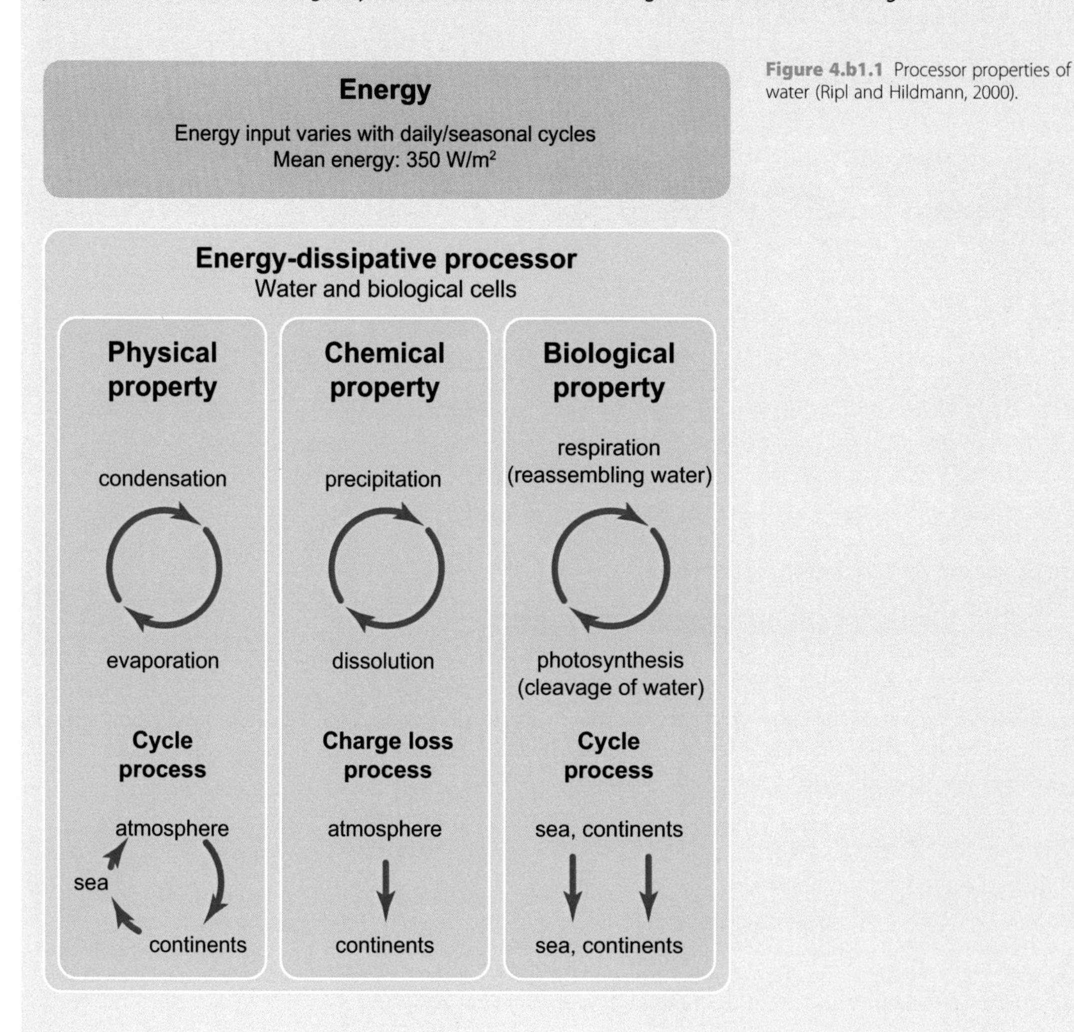

Figure 4.b1.1 Processor properties of water (Ripl and Hildmann, 2000).

Box 4.1 (*cont.*)

Development of energy-dissipating structures

Processes in nature have been altered from net productivity and reproduction of the most energy-efficient (pioneer) organisms to a community-based diversity of organisms, closing matter cycles locally *in situ* and producing sustainable structures by minimising losses (Odum, 1983). Open systems turned into more closed and stable systems due to more resource-economic adaptations to local and phase-related behaviour. These better optimised structures could grow and spread, while less efficient processes with more open structures and higher losses were forced to cease.

Breaking up short-circuited matter cycles

The opening up by humankind of the course that processes will take, such as, for example, the conversion of local evapotranspiration processes into the passage of water through the soil, along with the increased use of groundwater – both responsible for the loss of nutrients and minerals – has had adverse effects on sustainable life processes. An increasing irreversible transportation of life-supporting matter from top-soils – minerals and nutrients taken away in drainage ditches and by one-directional sewage-treatment plants in an irreversible manner to the sea – constitutes a highly efficient factor in the widespread breakdown of vegetation cover and consequent desertification. Recycling from the sea will only occur after hundreds of millions of years, when the seabed is eventually converted again into continents.

The even larger amounts of green water required to sustain other ecosystems and their services have not received much attention. An exception is the work by Rockström *et al.* (1999). New process-based analyses using the LPJmL model (see Chapter 5) arrive at total annual green water fluxes of about 20 000 km^3 for forest biomes, and 22 000 km^3 per year for other natural biomes (see Figure 4.3). The stewardship of these ecosystems and their water requirements is important, as they yield various types of ecosystem services which are critical for the functioning of the Earth System and for human welfare (Costanza *et al.*, 1997). In this context, ecosystems are seen as 'natural water infrastructure' (Diamond *et al.*, 2006).

The so-called provisioning services of ecosystems (Millennium Ecosystem Assessment, 2005) include the production of food (see Chapter 5), biofuel (see Box 2.1) and other biomass such as timber and fibres. Another category of services is 'regulating services' – carbon sequestration and storage, nutrient cycling and moisture recycling.

Much like food production, carbon sequestration and storage in natural and anthropogenic ecosystems – a climate-regulating function – requires different amounts of water per unit of service, depending on location, climate and biome type. Rockström *et al.* (2012) calculate that sequestering 1.5 Gt of carbon per year in soils (restoring soil carbon by best management practices for maize type vegetation) would require an additional 1000 km^3 of green water flow, and sequestering 1.5 Gt carbon per year in vegetation (converting grassland to forest) an additional 1300 km^3, both of which would lead to a subsequent reduction in downstream blue water availability.

The essential role of water in regulating nutrient availability (nutrient cycling) in ecosystems has been severely compromised by land-use changes. Liu *et al.* (2010) demonstrate the large nitrogen losses from the agricultural ecosystems that have replaced natural ecosystems, and the subsequent downstream nitrogen accumulation. If these nutrients reach the sea, they are essentially lost to terrestrial ecosystems and their services.

Terrestrial ecosystems return moisture from precipitation back to the atmosphere, regulating the climate and enhancing water availability in downwind areas (Keys *et al.*, 2012). Ecosystems depend on and at the same time regulate water availability, on which all human activities depend.

There have been initial attempts to associate economic values with all the different types of ecosystem services (Kumar, 2010), but no comprehensive assessment of water dependency or the biophysical or economic water productivity of different types of ecosystem services.

An integrated assessment of water in the Earth System requires a change in perspective from water

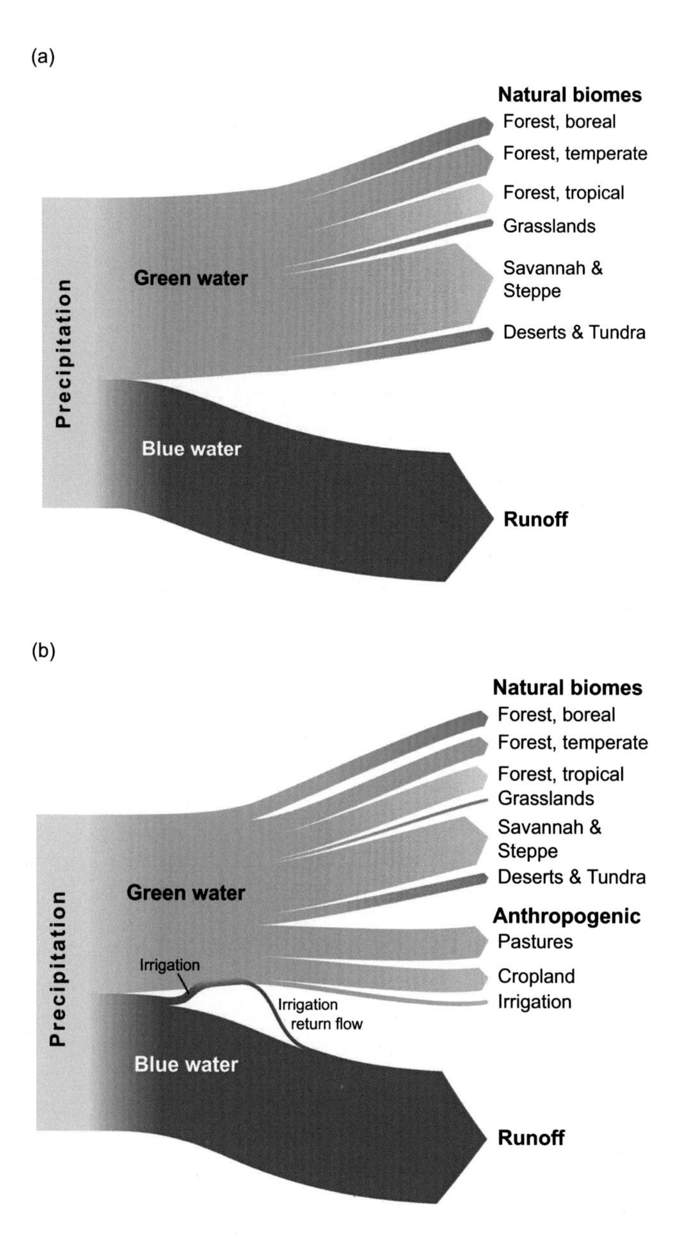

Figure 4.3 LPJmL-based arrow graphs of global green and blue water fluxes through different biomes for (a) potential natural vegetation and (b) current land cover.

as a sector to water as an 'intrinsic part of interconnected social–ecological systems' (Olsson and Galaz, 2009), with water driving change (water as a control variable) and at the same time being affected by changes (water as a state variable). Such an assessment needs to address the close bidirectional interactions between water, land, energy and food, as well as environmental security (Hoff, 2011).

4.2 The Global Water System as part of the Earth System

A system is a functional unit composed of a set of sub-systems and components, which together generate a level of organisation that is fundamentally different from the level of each individual part. Water links the different components of the Earth System, but it can also be interpreted as a system in itself. The term 'global water system' was coined and the Global Change Programs initiated a Global Water System Project (GWSP, see http://www.gwsp.org) to tackle research questions on the global aspects of water within the Earth System, its management and its governance (Alcamo *et al.*, 2008).

The Global Water System (GWS) is the global suite of water-related human, physical, biological and biogeochemical components and their interactions (see Figure 4.4). It plays a key role in the abiotic and biotic dynamics of the Earth System and in human society. Human activities are significantly and rapidly changing this system. Human-induced changes to the water system are now global in extent (from Global Water System Project, 2005). The principles addressed in Chapter 2 on the Earth System also apply to the GWS. Changes in the GWS feed back to and cascade through other components of the Earth System.

Like the Earth System as a whole, the GWS is characterised by: (a) interactions and feedbacks; and (b) teleconnections, which link water-related causes and effects across large distances, in particular through atmospheric moisture transport, virtual water transfers through trade and water uses as a result of FDI in land.

4.2.1 Remote water connections between regions

As is described in Chapter 2 for the Earth System, in the GWS, 'responses to different human pressures become feedbacks, which in turn can lead to further pressures'. So water is affected by global change, but at the same time also a driver of global change. For example, water in all its anthropogenic and ecosystem uses is one of the key resources affected by climate change. In turn, changes in water use and water availability cumulatively affect ecosystems and climate on a local to regional and possibly even a global scale. Some prominent examples of large-scale water-related interactions in the Earth System with potential non-linear responses and feedbacks are:

1. **The drying and savannisation of the Amazon rainforest** in response to a combination of local and external drivers, such as intensification of soy (feed) and biofuel production, including for export, pushing the agricultural frontier through direct and indirect land-use change further into the rainforest, and cutting new corridors of development through the forest (Fearnside, 2007), causing rapid deforestation – on average 16 000 km^2 annually in the past decade (INPE,

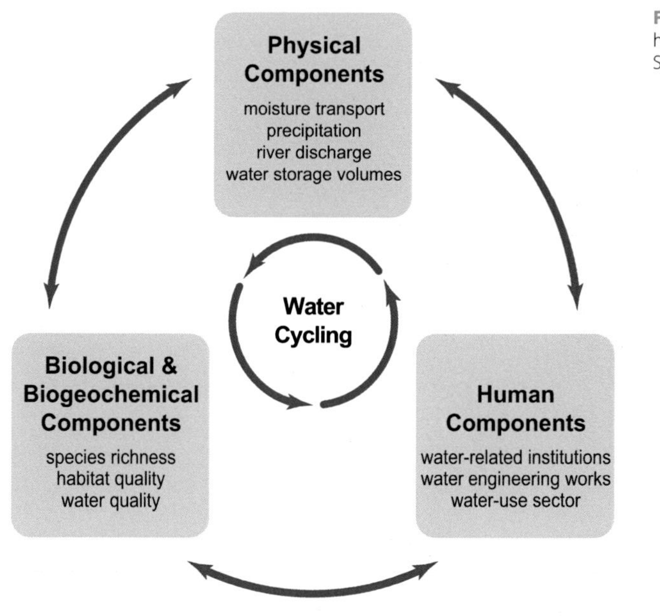

Figure 4.4 The global water system with the hydrological cycle in the centre (from the Global Water System Project, 2005).

2012) and interacting with climate change. If a threshold of about 40% deforestation or of 3.5–4° warming is exceeded, a sudden shift may occur to a new savannah-type ecosystem (Marengo *et al.*, 2009; Nobre and Borma, 2009) or a more seasonal, monsoonal, climate (Malhi *et al.*, 2009). Such changes in vegetation and evapotranspiration could cause feedbacks through reduced moisture recycling and a subsequent reduction in biomass production, which could eventually change Amazonia from a C-sink to a C-source with potential repercussions for the global climate system. Amazonia contains about 10% of the global carbon stored in land ecosystems and accounts for 10% of global primary productivity (Marengo *et al.*, 2009).

2. **The greening of the Sahel:** a greening of large parts of the Sahel has been observed in response to a combination of local and external drivers. This greening, or increase in vegetation and NPP, may partially be a response to climate variability (Figure 4.5). Additional potential drivers include changes in land and vegetation tenure, national policies and the involvement of international donors (Reij *et al.*, 2009). Despite the generally higher levels of evapotranspiration of more dense vegetation, the greening of the Sahel has been accompanied by an increase in river discharge at all scales, probably as a result of reductions in soil water storage (Amogu *et al.*, 2010). Using a similar line of argument to positive land–atmosphere feedbacks, Ornstein *et al.* (2009) propose a forced greening of the Sahara through large-scale irrigation and afforestation in order to enhance moisture recycling, which in turn could generate a stable, more humid regional climate, increased bioproductivity and carbon sequestration.
A similar proposal for intensifying land-water interactions was made for the Negev, suggesting the use of fossil brackish groundwater for afforestation and subsequent carbon sequestration (Issar, 2010). Similarly, Ripl (2003) suggests 'winning back manageable land from the desert'.

3. **A weakening of the Indian monsoon:** India, which is often called an 'irrigation economy'

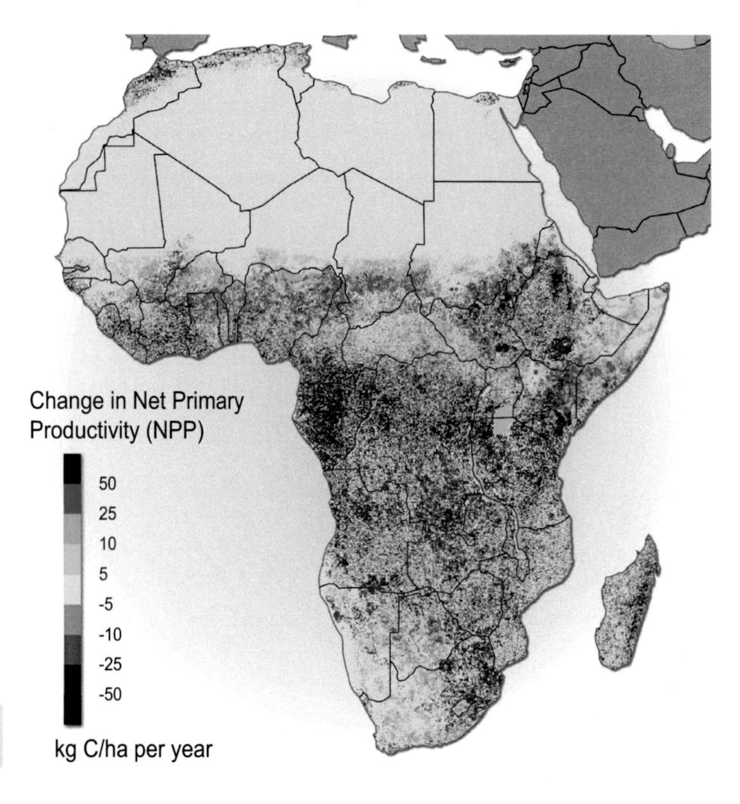

Figure 4.5 Change in vegetation in Africa, calculated as the change in net primary productivity (NPP), between 1981 and 2003. These data show increases in vegetation in the broader Sahel region, in contrast to other parts of the continent (FAO, 2008).

(e.g. Shah *et al.*, 2006), has hundreds of millions of people who depend for their survival on irrigation water from overexploited aquifers. Excessive groundwater pumping and application on irrigated fields adds more than 300 km^3 of evaporated water annually to the atmosphere, which, according to Douglas *et al.* (2009), could lead to a weakening of the Indian monsoon. Complementing such a scenario, Chase *et al.* (2003) hypothesise additional weakening of the Indian monsoon in response to deforestation. However, the direct effect of global warming on atmospheric circulation could act in the opposite direction to stabilise monsoon systems. If the combination of these different drivers were to weaken the Indian monsoon, this would have severe impacts on agriculture across South Asia, potentially increasing the demand for irrigation even further.

The mechanisms involved in the Asian monsoon and its potential for destabilisation are described in more detail in Box 4.2.

Taking these types of interactions and feedbacks between the atmosphere, water, land and social–ecological systems into account is a major challenge for water management and governance, because they may occur unexpectedly, cause non-linear changes and reach across different scales and sectors. Moreover, they can have effects in distant regions through teleconnections.

4.2.2 Remote water connections between regions

Teleconnections are chains of cause-and-effect that operate through several intermediate steps and across long distances. In the GWS, teleconnections link local water resources to external drivers of change across

Box 4.2 Irrigation impacts on water and energy budgets

Ellen M. Douglas, University of Massachusetts, Boston

Globally, agricultural water use in the form of crop irrigation comprises 70% of all human water withdrawals. Irrigation water use can alter the hydrological cycle in several ways: by reducing base flow to rivers; by increasing physical evaporation from soils and standing water, and through transpiration from vegetation; by adding to the greenhouse effect, since water vapour is also a greenhouse gas; by changing cloud coverage and depth; through changes in vegetation distribution and surface albedo and roughness; and by subsequent feedbacks to precipitation and runoff, and contributions to soil moisture and groundwater storage.

Increases in vapour flows

Gordon *et al.* (2005) noted that increases in water vapour flows are correlated with intensive food production on the Indian subcontinent, and suggest that expanding irrigation in this area could increase the risk of changes in the Asian monsoon system and possibly affect food production capacities in other regions, such as sub-Saharan Africa, as well. Boucher *et al.* (2004) concluded that the addition of water vapour in dry regions results in a non-linear increase in precipitable water at the regional scale, which is compensated by a decrease in other regions through changes in convection. Chase *et al.* (2003) and Pielke *et al.* (2003) documented changes in the global and Indian monsoon system. Extensive agricultural conversion in southern Florida has resulted in reduced precipitation and redistributed latent heat flux and atmospheric water vapour (Marshall *et al.*, 2004a; Marshall *et al.*, 2004b). Such changes can result in increased vulnerability for human populations.

Altered partitioning between latent and sensible heat

The presence of irrigation affects the partitioning of surface energy (net solar energy) between latent heat (the energy required for evaporation and transpiration) and sensible heat (the energy that changes air temperature). Under wetter conditions, latent heat fluxes typically dominate the energy budget, while under drier conditions sensible heat fluxes dominate. In humid climates, temperatures are moderated by latent heat fluxes as vegetation converts soil moisture into water vapour during photosynthesis. With much less water and vegetation, arid and semi-arid regions are usually hotter because most of the surface energy is transformed to sensible heat, which increases air temperature. However, large-scale land-use change and agricultural irrigation can alter this pattern.

Box 4.2 *(cont.)*

Effects in the United States

In the United States, land-cover change over the past 290 years has led to a weak warming along the Atlantic coast and a strong cooling of more than 1°C over the Midwest and Great Plains region, along with some reduction in precipitation due to changes in large-scale moisture advection (Roy *et al.*, 2003). Ozdogan *et al.* (2010) used model simulations over the continental United States to demonstrate that agricultural irrigation increases latent heat flux by 9 W/m^2 (or 12% of latent heat flux (Qle) from non-irrigated crops) and decreases sensible heat flux by 8 W/m^2 (11%), when averaged over all irrigated cropland. They also found that including irrigation in a land surface modelling scheme improved both energy and water budget predictions, which could ultimately lead to improved weather prediction and climate models. The partitioning of surface energy also affects the thickness of the planetary boundary layer (PBL), which is the layer of atmosphere closest to the Earth's surface. More sensible heat means a deeper (thicker) PBL, which allows for more convective activity. The thickness of the PBL can range from hundreds of metres to 3 km, and is typically deeper in the daytime than at night and in the summer than in winter.

Effects in India

The Indian subcontinent is a particularly interesting region for studying human–land–atmosphere interactions because it is home to one-sixth of the world's population, most of whom rely heavily on the summer monsoon rains for their survival. India leads the world in total irrigated land. Irrigation withdrawals represent 80–90% of all water use in India. Approximately 60% of irrigated food production depends on irrigation from groundwater (Shah *et al.*, 2000). Irrigated agriculture in India has been a key component of economic development and poverty alleviation (Bansil, 2004). From 1951 to 1997, gross irrigated areas expanded four-fold from 23 million to 90 million hectares, resulting in rapidly declining groundwater levels in as many as 15 of India's 28 states (Bansil, 2004). While the benefits of this expansion in agricultural productivity have been immediate, the environmental and social costs over the long term are becoming increasingly apparent as well. For instance, a great deal of the water used for irrigation in India is drawn from deep groundwater bore wells. As a result, groundwater stores in India, estimated at 432 billion m^3 (CWC, 1998), have been declining by 20 cm/year in as many as 15 Indian states (Bansil, 2004). Groundwater stores in most of these states are predicted to dry up by 2025 or sooner (Jha, 2001).

Changes in precipitation patterns

In India, there is also evidence that intensive irrigation has led to changes in precipitation patterns. Lohar and Pal (1995) report that mean monthly rainfall in West Bengal in 1983–1992 was less than half that observed in 1973–1982. The doubling of the area covered by summer paddy crops – mostly along the coast – over this timeframe is a possible cause. Two-dimensional numerical simulations indicate that wetter soils along the coast reduce the temperature gradient between the land and the sea, weakening sea-breeze circulation and reducing convective rainfall, which is a significant source of localised heavy precipitation in the coastal area. De Rosnay *et al.* (2003) reported a 9.5% increase in latent heat fluxes due to irrigated agriculture in India. Douglas *et al.* (2006) compared vapour fluxes (estimated evaporation and transpiration) from pre-agricultural and contemporary land covers and found that mean annual vapour fluxes have increased by 17% (340 km^3) with a 7% increase (117 km^3) in the wet season and a 55% increase (223 km^3) in the dry season. Two-thirds of this increase was attributed to irrigation, with groundwater-based irrigation contributing 14% and 35% of the vapour fluxes in the wet and dry seasons, respectively.

Possible destabilisation of the summer monsoon

In addition to the obvious environmental and socio-economic impacts that could result from changes in the Indian monsoon, Douglas *et al.* (2009) used the Regional Atmospheric Modelling System (RAMS) to show that agricultural intensification and irrigation can modify surface moisture and energy distribution, which alters the boundary layer and regional convergence, meso-scale convection, and precipitation patterns over the Indian monsoon region (see Figure 4.b2.1). They reported a statistically significant decrease in area-averaged sensible heat flux of nearly 12 W/m^2 between the pre-agriculture and irrigated agriculture scenarios. Changes in latent heat fluxes (−26% to +24%) and sensible heat fluxes (−77% to +8%) were found when model outputs were averaged over Indian states. Decreases in sensible heat in the states of Punjab (87.5 W/m or 77%) and Haryana (65.3 W/m^2 or 85%) were found to be statistically significant at

Box 4.2 (*cont.*)

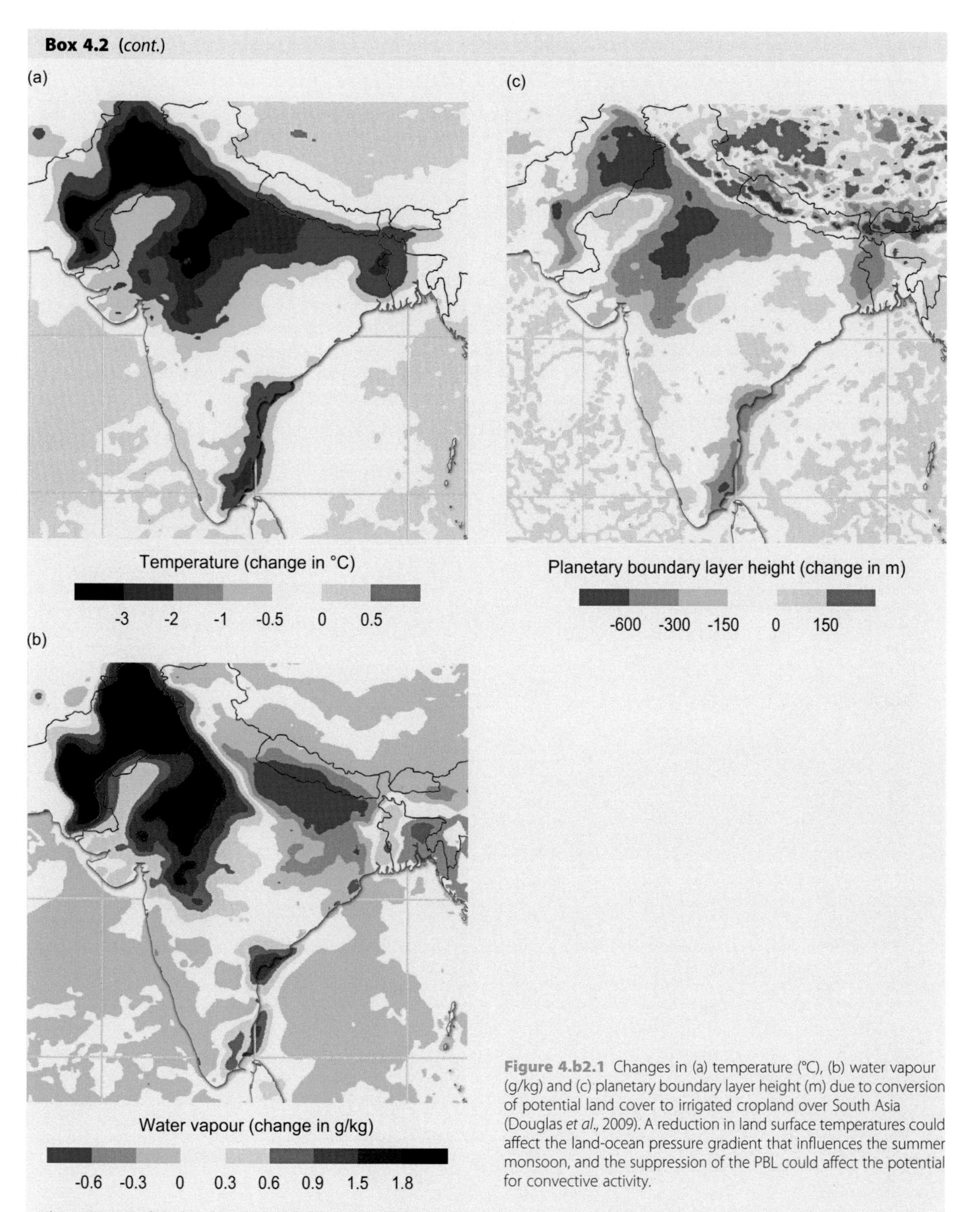

Figure 4.b2.1 Changes in (a) temperature (°C), (b) water vapour (g/kg) and (c) planetary boundary layer height (m) due to conversion of potential land cover to irrigated cropland over South Asia (Douglas *et al.*, 2009). A reduction in land surface temperatures could affect the land-ocean pressure gradient that influences the summer monsoon, and the suppression of the PBL could affect the potential for convective activity.

the 95% confidence level. Zickfeld *et al.* (2005) show that any perturbation in the radiative budget over the subcontinent could weaken the driving pressure gradient and potentially destabilise the summer monsoon circulation. Continued land surface changes due to agricultural intensification could therefore have a profound impact on the monsoon system.

Box 4.3 Land surface feedbacks and tipping points across scales in the Mediterranean

Holger Hoff, Stockholm Environment Institute and Potsdam Institute for Climate Impact Research

The Mediterranean basin is unique in that it has a very high rate of internal moisture recycling: more than 75% of the precipitation originates from evapotranspiration within the basin. In summer, large amounts of moisture accumulate in the atmosphere over the sea.

Two potential tipping points

A first tipping point of the land–atmosphere system may have already been crossed along the Spanish Mediterranean coast and other parts of the northern Mediterranean, where a decrease in summer rainstorms and an increase in droughts has been observed for some time (Figure 4.b3.1). This seems to be related to historical and more recent land-use changes, i.e. deforestation and the drying of coastal wetlands, and their replacement by agricultural land and more recently also by urban land and land burned by forest fires. The new land cover does not provide enough moisture (it provides more sensible heat instead) to the air masses moving inland with the sea breeze, which means that precipitation over the mountains ceases. The system 'closes' and the atmospheric moisture returns to the sea instead. Reduced runoff causes an aridification of the landscape up to about 80 km inland. Vegetation responses to this reduction in water availability act as a positive feedback by reducing evapotranspiration even further. The resulting drier vegetation is also more fire prone.

The accumulation of moisture in the atmosphere through this cascade of effects could also lead to atmospheric instabilities and an increase in large storms. Furthermore, it could contribute to heating the region and the sea by

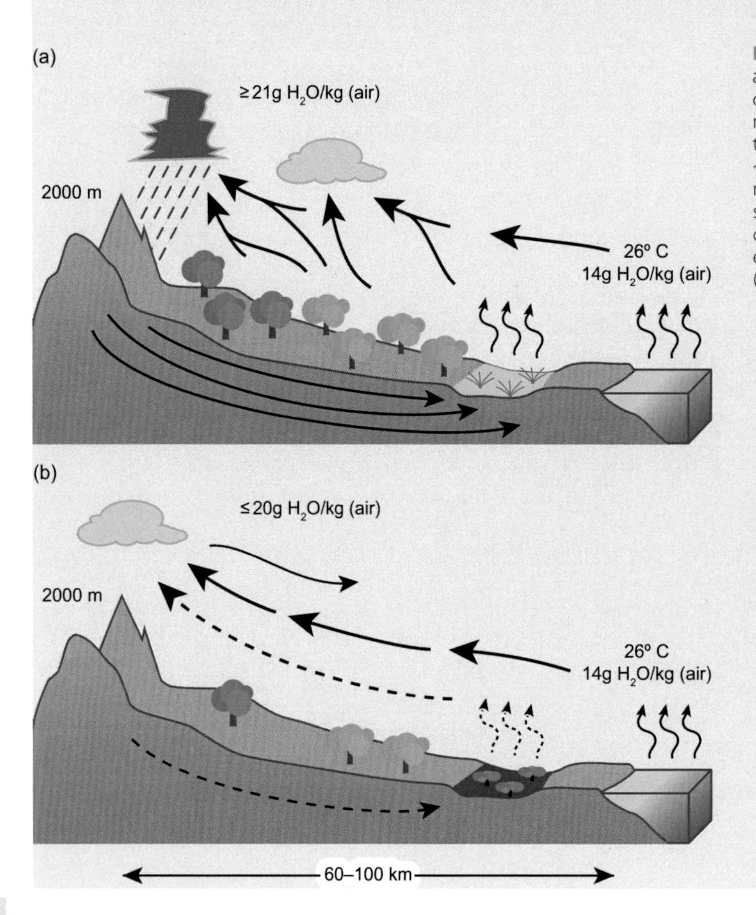

Figure 4.b3.1 The summer rainstorm cycle along the Western Mediterranean. (a) Past conditions supplied enough additional moisture to the marine air mass (sea breeze) to trigger precipitation almost every day at ~1000 m altitude, allowing for water recirculation within the coastal system, i.e. surface water and aquifers. (b) Current conditions with increased heating and less evapotranspiration along a drier surface (adapted from Millán, 2010).

Box 4.3 (*cont.*)

acting as a greenhouse gas. An increase in the frequency and intensity of torrential rainfall events in the autumn and winter has been observed in the coastal zones of the Mediterranean, driven in particular by higher sea surface temperatures.

A second tipping point might be reached when these torrential rainfall events and the erosion associated with them cause severe land and ecosystem degradation or desertification. The previously reduced vegetation cover could amplify this erosion further. A critical first tipping point – and even the second tipping point – may have been crossed long ago along the African Mediterranean coast and Almeria.

Further upscaling of these effects might occur through changes in the evaporation-to-precipitation ratio in the western Mediterranean, which could increase the salinity of the Mediterranean and eventually change the outflow of saltier water to the Atlantic (the salinity valve).

There are many open questions when observing this system:

- How closely is the observed change along the Spanish Mediterranean coast related to changes in the hydrological cycle in other parts of the basin?
- How certain is the propagation and an upscaling of the local effects in Spain?
- What actions in terms of improved land and water use can stop or reverse these negative trends?
- Can afforestation and irrigation – and the use of surface irrigation instead of drip irrigation – on agricultural land increase the moisture of the land–atmosphere system and lead to a re-greening of the landscape?
- Is this already happening with the reforestation linked to land abandonment in parts of the northern Mediterranean?
- Would that regeneration also be an option for the southern Mediterranean, or has that region gone too far towards a drier, degraded state, making a return to moister conditions impossible?
- Finally, is the situation in the southern Mediterranean very different because of the lack of coastal mountains in large parts of the area?

regions. These external drivers generally act in combination with local pressures.

The main water-related teleconnections in the GWS are:

- liquid water flows along large – often transboundary – river basins, or large-scale water transfers across basin boundaries;
- atmospheric vapour transport;
- virtual water flows through trade or related to FDI. Virtual water is the amount of water required to produce a commodity, in particular crops or livestock, which is saved by a country when substituting local production with imports.

Teleconnection 1: water flows along large rivers and water transfers between basins

Downstream sections of river basins depend to a varying extent on runoff generated upstream. Some heavily utilised or dry basins nowadays also depend on transfers of water from other basins.

One example is the Nile basin. Egypt is completely dependent on rainfall in upstream countries and the political will of these countries to release most of the resulting discharge for downstream use. In the past, this has mostly been ensured by the lack of capacity in the upstream countries – in particular Ethiopia – to capture and use a significant fraction of the Nile discharge, in combination with Egypt's pressure on international donors not to fund blue water infrastructure in Ethiopia but this situation is now changing.

Inter-basin water transfers have been implemented or are planned on an ever larger scale, such as in the American West, resulting for example in the depletion of the Colorado River before it reaches the sea, between southern and northern China and between eastern and western India. These mega-projects often have severe social and environmental impacts with knock-on effects for local and regional water systems. In addition, their economic rationale is not always clear.

Teleconnection 2: atmospheric moisture transport

Atmospheric moisture transport links even more distant regions than river flows or inter-basin transfers. Evapotranspiration in one region eventually becomes precipitation in another. This process has been called 'moisture recycling' (Savenije, 1995).

Large-scale land-use changes in a source region for atmospheric moisture can change water availability in 'downwind' regions. The travel distances of moisture evaporated from land surfaces are typically several thousand kilometres before the moisture returns to the land surface as precipitation (van der Ent and Savenije, 2011).

Past European land-use changes (see Figure 2.9) may have caused a reduction in precipitation in Asia, in particular in China, given the strong dependence of China's precipitation on atmospheric moisture influx from the west (van der Ent *et al.*, 2010). Amazonian deforestation and the resulting lower level of evapotranspiration (see above) may eventually reduce precipitation in the Parana–La Plata basin. Marengo *et al.* (2009) show large-scale southward atmospheric moisture transport in this region. Indian monsoon changes due to the expansion of irrigated agriculture (see Douglas *et al.*, 2009; previous section) might affect precipitation and accordingly the length of the growing season in sub-Saharan Africa. Camberlin (1997) and Janicot *et al.* (2009) confirm a strong positive correlation between Indian summer monsoon activity and rainfall over East Africa.

Land-use changes in the upper Nile, such as the draining of the large Sudd wetlands, could change precipitation patterns in Central Africa and eventually also in West Africa, through several moisture recycling loops (van der Ent *et al.*, 2010). Box 4.4 elaborates further on the subject of moisture recycling.

Teleconnection 3: international and domestic virtual water flows with trade and FDI

Trade enables countries to overcome local scarcities. For example, the Middle East and North Africa region relies heavily on external water resources (Fader *et al.*, 2011; Yang *et al.*, 2007) to produce the food required to feed its rapidly growing population. Large flows of virtual water are associated with the international trade in agricultural commodities, which consume large amounts of water during production (e.g. about 1 m^3 of water for 1 kg of cereals or for 0.1 kg of meat). A more detailed analysis of virtual water content can be found in Gerten *et al.* (2011).

Hoekstra and Chapagain (2009) calculate a global annual flow of virtual water linked to exports of agricultural products of 1263 km^3 (including 306 km^3 due to re-exports), of which 986 km^3 is associated with crops and 276 km^3 with livestock products. The LPJmL-based virtual water content of 11 crop functional types, in combination with ComTrade data, add up to 393 km^3/year of virtual water flows linked to exports (Fader *et al.*, 2011). As trade – including trade in agricultural commodities – grows rapidly (WTO, 2008), so do virtual water flows. Ramirez-Vallejo and Rogers (2004) project a further increase in virtual water flows linked to trade under future trade liberalisation, including a doubling of net imports by sub-Saharan Africa in 2020.

There are also major flows of food and associated virtual water within countries – often across large distances, such as from northern to southern China and from western to eastern India. Most of these flows occur from surplus to deficit areas, but in the case of China and India the food surplus areas are also the drier parts of the country, with water availability augmented or planned to be augmented by large-scale real water transfers (Ma *et al.*, 2006; Verma *et al.*, 2009).

As described in Chapter 2, trade externalises resource exploitation and overexploitation, and possibly degradation, to other countries or regions. Examples of the use of local water resources to meet external demands include:

- about 20% of the drying up of the Aral Sea has been attributed to water abstractions for cotton exports to the EU (Hoekstra, 2009);
- the depletion of the Ogallala aquifer is in part for export food production (see Box 4.8);
- exports of coffee, cocoa, biofuels, etc., from African countries, some of which suffer from water scarcity and malnutrition (Hoff *et al.*, 2013).

To derive trade-related external water impacts, the consumptive water use of export crops has to be compared with the level of water scarcity and opportunity costs in the producing region, in order to move from a 'shoe size' to a genuine 'footprint' analysis (Pfister and Hellweg, 2009).

Rapidly expanding FDI in agricultural land (Anseeuw *et al.*, 2012; Friis and Reenberg, 2010) in particular in African countries, has an effect on local resources similar to virtual water trade (Rulli *et al.*, 2013). In fact, FDI is to some extent a response to recent price instability on international commodity markets. Land- and/or water-scarce countries aim to

Box 4.4 The origin and fate of atmospheric moisture over continents

Ruud J. van der Ent, Delft University of Technology

Hubert H.G. Savenije, Delft University of Technology

The precipitation that falls on land originates in part from terrestrial evaporation. This feedback of moisture from land to the atmosphere and from the atmosphere back to the land is called moisture recycling (see van der Ent and Savenije, 2011; van der Ent *et al.*, 2010). Figure 4.b4.1a shows the proportion of precipitation that comes from terrestrial evaporation, Figure 4.b4.1b shows the proportion of the evaporation that returns as precipitation on land, and Figure 4.b4.1c indicates the average distance an evaporated water particle travels through the atmosphere before it precipitates.

On average, 40% of the precipitation on land originates from terrestrial evaporation and 57% of all terrestrial evaporation returns as precipitation on land. Obviously, there is a lot of spatial heterogeneity. Many interesting

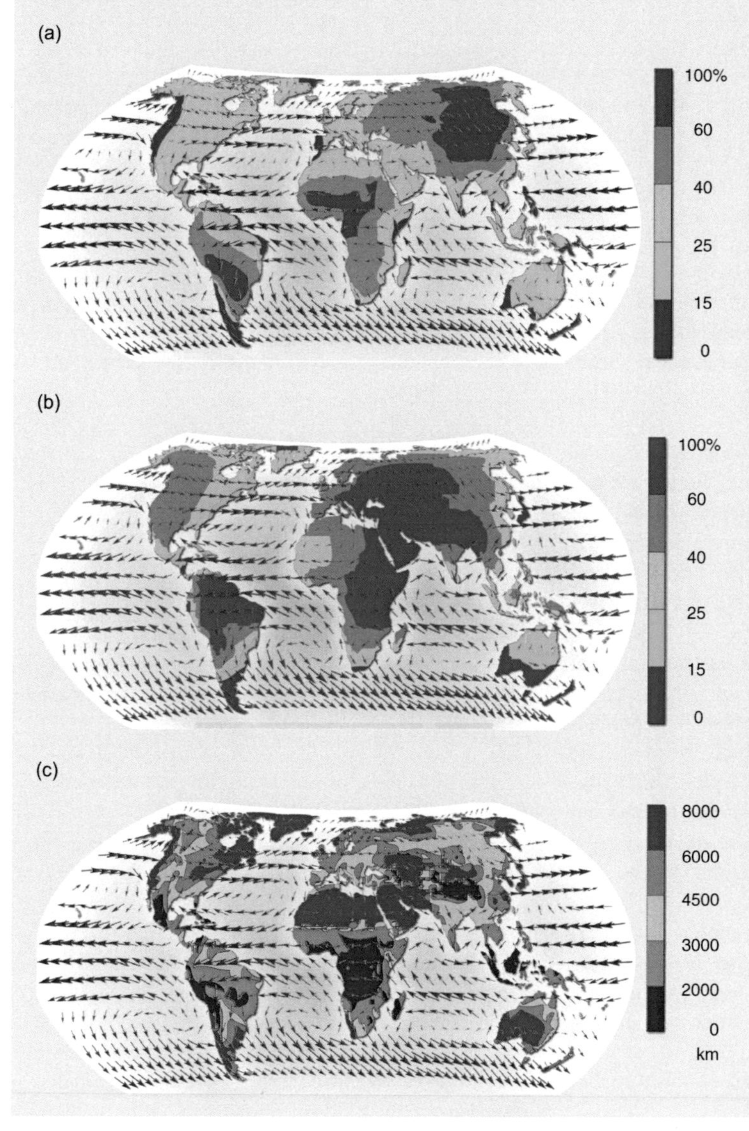

Figure 4.b4.1 Different metrics identifying the sources and sinks of atmospheric moisture and indicating moisture recycling strength. ρ_c is the fraction of the precipitation that originates from terrestrial evaporation – shown in (a), ε_c is the fraction of the evaporation that returns as precipitation over land (b), λ_ρ, is a local metric indicating the average distance a water particle travels through the atmosphere from evaporation to precipitation (if upwind conditions are constant; c). The arrows indicate the horizontal moisture flux field. All data represent averages over the period 1999–2008. Input data were taken from ERA-Interim reanalysis (Berrisford *et al.*, 2009). The metrics ρ_c, ε_c and λ_ρ follow a moisture accounting procedure which is further described in van der Ent and Savenije (2011); van der Ent *et al.* (2010).

Box 4.4 (*cont.*)

features can be observed from Figure 4.b4.1. For example, 80% of China's water resources originate from terrestrial evaporation (Figure 4.b4.1a, $\rho_c \approx 0.8$ in the source regions of China's major rivers), which makes sense since 60% of the evaporation over large parts of the Eurasian continent returns as precipitation on land (Figure 4.b4.1b, $\varepsilon_c > 0.6$). Furthermore, evaporation from the Amazon region (Figure 4.b4.1b) sustains the precipitation in the Río de la Plata basin (Figure 4.b4.1a). Similarly, in Africa it can be seen that the Great Lakes region and subsequently the Congo basin are a major source of rainfall in the Sahel. The local moisture recycling strength in Central Africa is generally high. As a rule, precipitation does not travel more than 2000 km through the atmosphere (Figure 4.b4.1c, $\lambda_p < 2000$ km). Very high local moisture feedback is also observed just east of the Andes and around the Tibetan Plateau (ρ_c = high, ε_c = high, λ_p = low), where strong moisture advection towards the mountains in combination with orographic lift sustains high rates of local moisture recycling.

Evaporation has a direct link to land use and water management. These results therefore provide a first order estimate of the potential impact of land-use change on climate. For example, the results suggest that deforestation in the Amazon and Congo is likely to lead to less rainfall downwind, whereas water conservation could have a positive multiplier effect. In order to provide a more direct link to land use, future moisture recycling research should make a distinction between: fast moisture feedback, or evaporation from interception; retarded moisture feedback, or transpiration and soil evaporation; and continuous evaporation, or open water evaporation.

overcome their domestic resource limitations by acquiring so-called unused land to grow food, feed, fibre or biofuel. This acquisition of land and water can turn into 'land grabbing' or 'water grabbing' when the local population is deprived of its formal or informal customary water rights and access to water (Skinner and Cotula, 2011).

All these teleconnections – long distance river flows, inter-basin transfers, atmospheric vapour transportation and virtual water flows linked to trade or FDI – expose local water resources to external drivers, such as consumption patterns and land and water scarcity in other regions. Water management and governance have not yet addressed these teleconnections and external drivers. It has been argued (Wichelns, 2010) that non-sustainable water use is driven not by external demand, but by local policies and decisions on water use, which need to be addressed at the local to national scale. This argument, however, assumes that local institutions, in particular in developing countries, are strong enough to keep up with international market forces and actors.

The way forward will be to reconcile local and external or international agendas, e.g. by regulating FDI so that it improves – or at least does not degrade – local livelihoods and environmental conditions (Hoff *et al.*, 2012).

It is possible to find 'chains of teleconnections' in the GWS, such as:

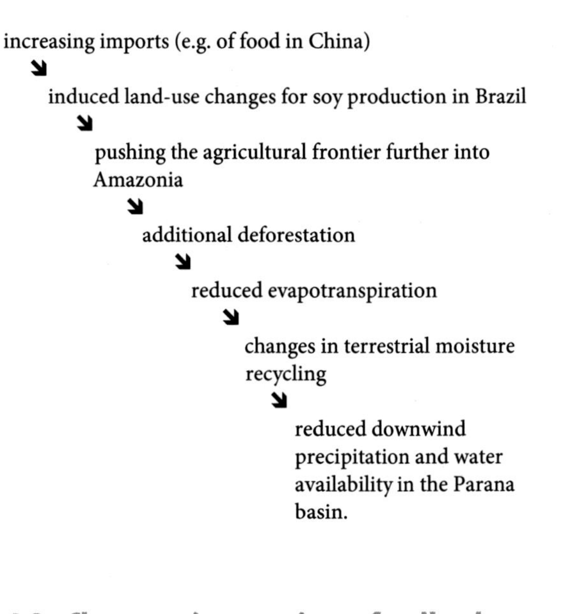

4.3 Changes, interactions, feedbacks, teleconnections and resilience in the Global Water System

The changes in the Earth System described in Chapter 2 also affect all components of the GWS. Some of these changes are local in nature (e.g. land-use change and increasing water demand) but in their sum also affect the GWS, while others are truly global (e.g. increasing CO_2 concentrations and climate change).

Through the feedbacks and teleconnections described above, these changes cascade through the different water-dependent social–ecological systems up to the global scale and affect their resilience. Slow drivers such as land degradation may push the social–ecological systems closer to critical thresholds, so that they become less resilient to shocks such as droughts – with drastic results including food crises and migration. IWRM and governance need to adapt to the accelerating dynamics and turbulence in the Earth System and the GWS.

4.3.1 Land-use change, water and resilience

Land-use change happens locally but in some cases is also driven externally, as a result of globalisation, trade and FDI (von Braun, 2009) and increasing demands for agricultural products in other regions. Examples of land-use change for export production include soy plantations in Latin America, oil palm plantations in South East Asia or flower farms in sub-Saharan Africa.

Land-use changes, in particular towards intensive agriculture, often result in simpler landscape configurations, reduced agrobiodiversity and loss of redundancy in ecosystem functions. These changes may result in land degradation, the erosion of long-term productivity and the need to be compensated by continuous additional inputs of e.g. fertiliser, biocides or water (Kumar, 2010). In combination with reduced livelihood choices, such land-use change can increase the vulnerability of social–ecological systems to climatic, market or other shocks or calamities and erode resilience (Falkenmark and Rockström, 2008). If accompanied by land degradation, soil water and carbon storage are also reduced. An example of such a – largely externally driven – landscape degradation is the Borneo rainforest, which has been reduced and fragmented by the production for export of palm oil, among other things. In the original multifunctional rainforest, El Niño events and associated droughts were essential as triggers for regenerating the rainforest and its biodiversity. In the simplified palm oil landscapes, El Niño events can lead to large wildfires. The 1997 wildfires in Borneo, for example, may have caused CO_2 emissions equivalent to about one-third of the annual global emissions from fossil fuels. Borneo has shifted from a carbon sink to a carbon source (Folke *et al.*, 2011).

Land-use change, for example, from natural ecosystems to agricultural land and other 'anthropogenic biomes' (Ellis and Ramankutty, 2008) changes green and blue water stocks, flows and productivity. It also redirects water flows between the different types of ecosystem services, with associated trade-offs in water productivity levels and the benefits derived, e.g. from upstream recharge areas and downstream discharge areas, or even between different countries.

Figure 4.6 shows the increase (green and blue) or decrease (orange and red) in evapotranspiration due to anthropogenic land-cover change from potential natural vegetation to the current land use (after Ramankutty *et al.*, 2008).

Evapotranspiration normally decreases when forests are replaced by crops, causing less moisture recycling and possibly a down-regulation of the hydrological cycle at the regional scale, which could, for example, lead to the savannisation of Amazonia. On the other hand, evapotranspiration increases when irrigation is applied, through the 'pumping' of additional surface or groundwater into the atmosphere (Gordon *et al.*, 2005; Rost *et al.*, 2008).

Land use affects water flows and stocks, and at the same time is controlled by water availability, e.g. when the low productivity of dryland agriculture forces farmers to expand into more marginal land or shift from rainfed to irrigated agriculture. Irrigation may initially increase resilience, for example, to droughts, but this effect is likely to reverse once aquifers are depleted due to overexploitation (see the example of India's irrigation economy above, where millions of livelihoods depend on rapidly depleting aquifers), or when climate change reduces total water availability. The associated social–ecological systems have become dependent on the higher yields or multiple cropping that irrigation has brought – and on cheap water and energy for pumping. Migration as a response to such a dilemma shifts the pressure on resources and overexploitation to other regions, and often to cities.

The effects of land-use changes are not limited to the region in which they occur, but may also affect water availability and eventually resilience in other regions through atmospheric moisture recycling, see teleconnection 2 above and Keys *et al.* (2012).

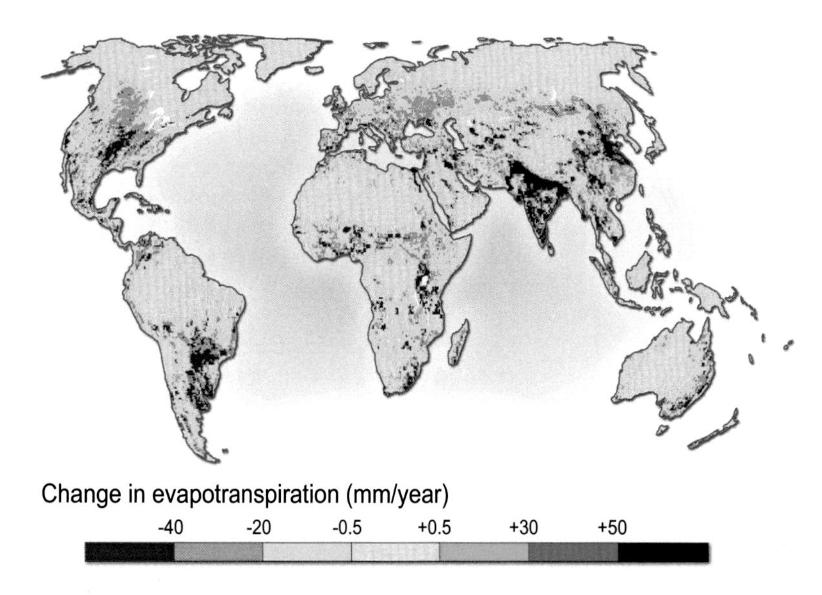

Figure 4.6 LPJmL-based global map of anthropogenic evapotranspiration changes from potential natural vegetation to current land cover (land-cover data from Ramankutty *et al.*, 2008).

4.3.2 Climate change, water and resilience

Climate change causes changes in temperature, water availability and variability, as well as changes in crop water demand and productivity, glacier runoff and sea levels. At the same time, variability and extremes, such as floods or droughts, are expected to increase (IPCC, 2011).

Figure 4.7 shows the projected change in water availability over most of this century, as simulated by LPJmL, for an ensemble average of 17 global climate models for constant atmospheric CO_2 concentrations. While laboratory and field experiments show an increase in plant water use efficiency and, accordingly, a reduction in transpiration with increasing CO_2 concentration (the fertilisation effect), large uncertainties remain about the strength of this 'water-saving' effect at the landscape scale. Despite this uncertainty, we can safely assume that future water scarcity will be mitigated somewhat by increasing atmospheric CO_2 concentration.

Climate-change-driven sea level rise puts a large proportion of the coastal population at risk in terms of flooding and degraded water resources due to saltwater intrusion into surface and groundwater bodies. Figure 4.8 categorises major estuaries according to the number of people affected by inundation due to effective sea level rise, which includes eustatic sea level rise, land subsidence in response to the extraction of groundwater or hydrocarbons, and fluvial sediment deposition.

Higher temperatures in the long run reduce snow and glacier water storage in regions such as the Himalayas or the Andes, resulting in a reduced buffer capacity of associated social–ecological systems in dry seasons and dry years.

Extreme events, such as floods and severe storms, have increased in recent decades according to multiple data sources not only in number, but also the number of people affected and the costs (CRED, 2012; Figure 4.9).

However, it is not clear whether the observed increase in extremes can be attributed to climate change or other drivers such as landscape transformation, or is within natural levels of variability. Nor is it clear whether future climate change will lead to further increases in extreme events in the future. The IPCC (2011) states with *medium confidence* that some regions of the world have experienced more intense and longer droughts, in particular in southern Europe and West Africa, and also with *medium confidence* that droughts will intensify in the twenty-first century in some seasons and areas.

In any case, slow-changing climate variables that cause e.g. temperature increase, aridification or loss of snow water storage – in combination with other pressures such as land degradation – can reduce resilience against shocks such as droughts or dry spells, which due to their different respective timescales pose different types of threats

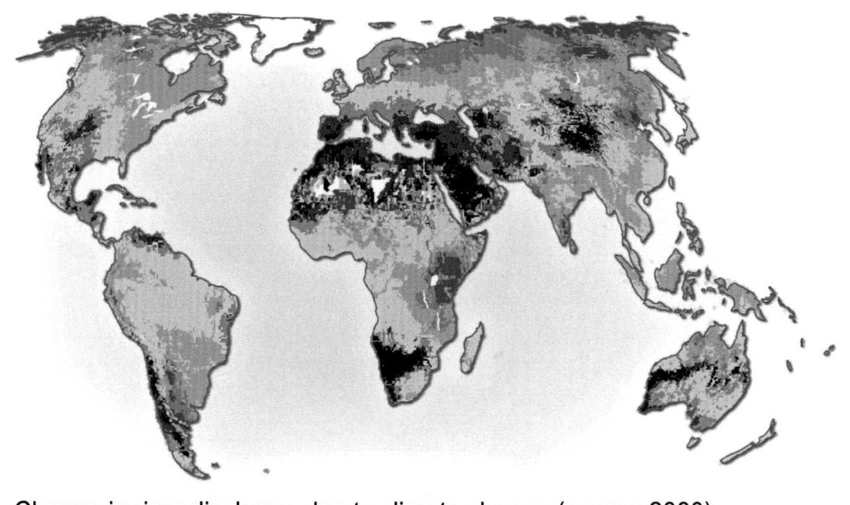

Figure 4.7 Change in river discharge due to climate change: blue water availability per cell by the 2080s compared to around 2000 according to LPJmL with current levels of atmospheric CO_2 concentrations.

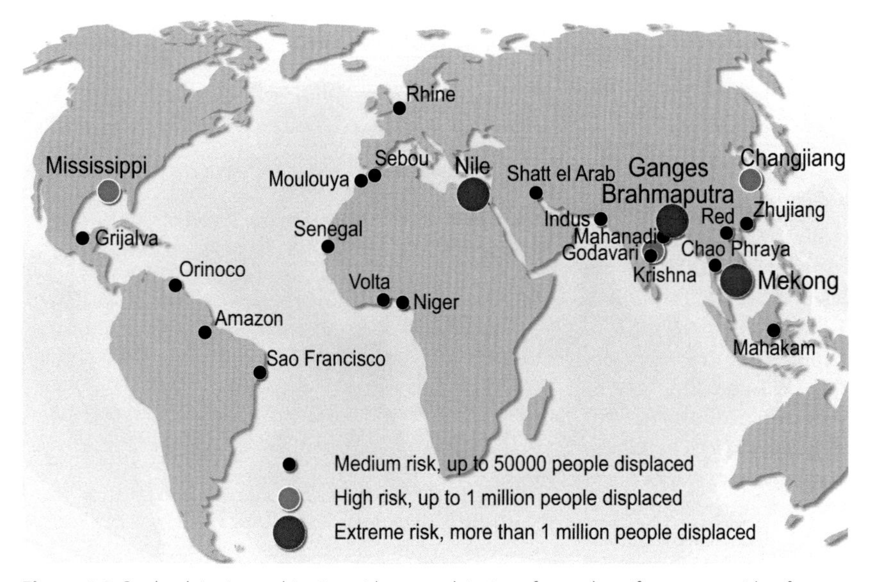

Figure 4.8 Sea level rise in combination with overexploitation of coastal aquifers presents risks of water resources degradation in estuaries of large river basins (Parry *et al.*, 2007).

to social–ecological systems. Feedbacks from climate change through increasing plant water demand may add pressure to already stressed systems (see the drying Amazon or weakening Indian monsoon above).

Box 4.5 demonstrates the potentially severe consequences for agriculture in Australia of the **decreasing reliability** of rainfall, and the need to build resilience in water systems and to be prepared for an uncertain future and potential shocks.

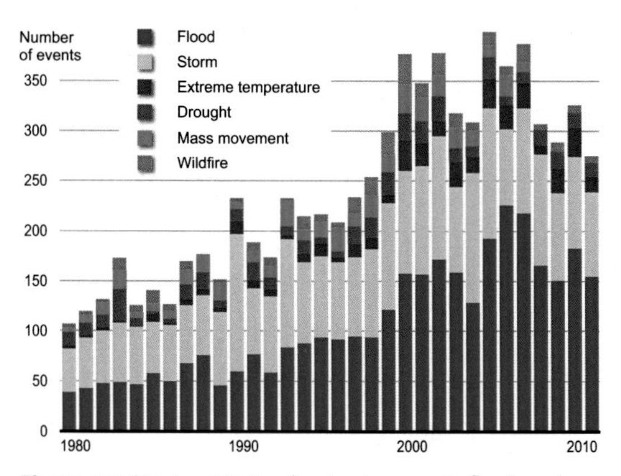

Figure 4.9 Disaster statistics, showing increases in floods and other climate-related extreme events (CRED, 2012).

4.3.3 Increasing water demands and resilience

Water demand, withdrawal and consumptive use are growing rapidly in many parts of the world, compounded by water quality degradation, in response to population increases and economic development, changes in lifestyles and diets, and technological development. The resulting overexploitation is demonstrated e.g. in lower groundwater tables, reduced river discharge, compromised environmental flows and basin closures (Falkenmark and Molden, 2008). Box 4.6 highlights areas of groundwater overexploitation, in particular in India, China, western Asia, North Africa, the United States and Mexico.

Groundwater serves as storage which buffers dry seasons or droughts and at the same time avoids evaporative losses and reduces vulnerability to pollution, compared to surface water storage. The loss of these functions through overexploitation – which can also result in irreversible saltwater intrusion and the compaction of aquifers – is likely to reduce the resilience of groundwater-dependent social–ecological systems.

Like groundwater, surface water and its respective aquatic ecosystems can be overexploited and degraded in terms of structure, diversity, functions and services (Arthington *et al.*, 2009), and with it the resilience of the social–ecological systems that depend on it. Box 4.7 summarises current knowledge on environmental flows and the methods for assessing and implementing them.

Box 4.5 Coping with a chaotic climate in Australia

Will Steffen, The Australian National University and Climate Commissioner

Large hydrological variability

Australia is the 'land of drought and flooding rains', a continent with a historically high variability in its hydrological cycle that has created water challenges for its inhabitants for 60 000 years. Indigenous Australians used their high level of mobility to follow the resources that moved across the landscape in response to a variable climate. European and Asian Australians, with a vastly shorter period of experience on the continent and a more sedentary lifestyle, are still learning to cope with this variability.

The challenge now is to understand not only how modes of natural variability – ENSO, Indian Ocean Dipole (IOD), Southern Annual Mode (SAM) and Pacific Decadal Oscillation (PDO) – work and how they interact with each other, but also how the underlying trends of global climate change are interacting with these existing modes of natural variability to produce novel patterns of rainfall and water availability. Although some progress is being made with understanding how some of the physical processes that influence Australia's rainfall are changing (e.g. Frederiksen *et al.*, 2011; Nicholls, 2010; Timbal *et al.*, 2010), much remains uncertain and the level of confidence in future predictions is low for much of the continent.

Rainfall reliability

In the face of such uncertainty, the agricultural sector, for example, is developing new ways of gaining insights into changing rainfall patterns to help it cope with the future. The concept of rainfall *reliability*, rather than variability, has been developed to help the cropping sector cope with change in water availability (Steffen *et al.*, 2011). Reliability refers to the distribution of rainfall through a particular period relevant to a user group, for example, a growing season (e.g. whether a location receives a certain amount – say 25 mm – per month for the period). Figure 4.b5.1 shows the change in the reliability of autumn rainfall over the past 100 years. The pattern is complex, but one location – the southwest corner of Western Australia – shows a clear decrease in reliability.

Rainfall amounts have changed

This decrease in rainfall reliability has been accompanied by a decrease in the absolute magnitude of rainfall, which together mean that overall water availability is decreasing and the availability of water throughout the year is becoming less predictable.

Box 4.5 (*cont.*)

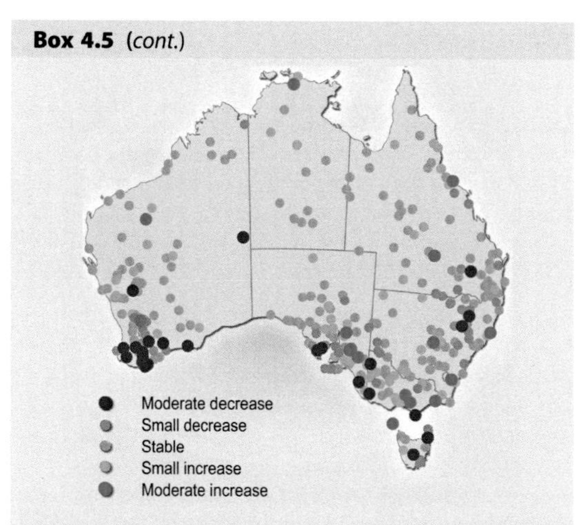

Figure 4.b5.1 A time series analysis applied to the reliability of autumn rainfall over the past 100 years (Steffen *et al.*, 2011).

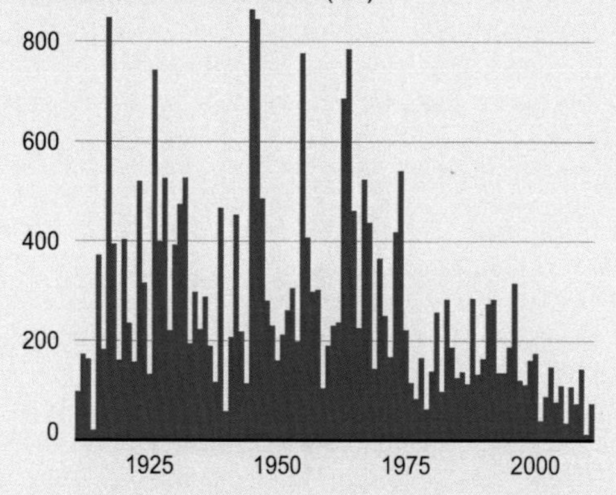

Figure 4.b5.2 The inflow to Perth's water supply dams, 1911–2011 (Water Corporation, 2012).

Interestingly, because these trends are now known with some certainty, the ability of people in the region to cope with these changes is surprisingly high. Figure 4.b5.2 shows the change in inflow to the water supply dams of the city of Perth over the past century, with a sharp, stepped decrease beginning in the mid-1970s. The city has coped well so far, with a combination of increases in efficiency in the water system, demand management and the construction of two desalination plants.

The agricultural sector in the region, which is dominated by wheat production, has also coped well. Despite the drop in rainfall from the mid-1970s, wheat yields actually increased over the same period because of a range of changes in technology and management that were adopted, some for reasons unrelated to the rainfall change but some directly in response to it. An example of the latter is the use of previously waterlogged areas for crop production, while an example of the former is the widespread adoption of no-till agriculture, which conserves soil moisture (Steffen *et al.*, 2006). However, following the very dry conditions in 2010, production of wheat and other winter crops in 2010–2011 was 43% lower than for the previous season (ABARES, 2011).

The Millennium Drought, 1997–2009

Where reliability, variability or longer-term trends are not changing in such clear patterns, however, it is much more difficult to cope with changes in water availability. Unfortunately, this is the case for much of Australia, including the more heavily populated south-eastern region. A crippling drought from 1997 to 2009, which threatened urban water supplies in Melbourne, Sydney, Brisbane and Canberra and led to severe water use restrictions in those cities, was suddenly and dramatically broken in mid-2010 by a period of heavy rainfall and flooding along much of the east coast. December 2010 was the wettest December ever recorded in eastern Australia (BoM, 2011). These rapid and dramatic swings in water availability have undermined efforts to secure water resources for the future in some areas, as actions planned to respond to a drying climate have now been called into question.

Towards resilience to chaotic shocks

Only in a few places in Australia – the southwest corner of Western Australia and the southeast corner of the continent – can coping responses be made to directional changes in rainfall or water availability with a high degree of confidence. In most places, the country faces a highly uncertain future with changes in water availability of both uncertain magnitude and uncertain direction (wetter or drier). This implies that building the general resilience of water systems to chaotic shocks and uncertain trends is more appropriate than trying to adapt to directional changes that have high levels of uncertainty. Australia is still the land of drought and flooding rains, but they are perhaps becoming less predictable and more extreme.

Box 4.8 describes that hydro-climatological conditions in combination with socio-economic trends (through changes in water demand) can affect the resilience of water-scarce regions. The North China Plains have much higher population pressure than

Box 4.6 Groundwater use and groundwater stress today and in the future

Petra Döll, University of Frankfurt

It is not known with certainty how much water is withdrawn from groundwater and how much from surface waters. Uncertainty is highest in the globally dominant water-use sector – irrigation. Shah *et al.* (2007), for example, estimate that 26.5 million ha of land is under groundwater irrigation in India, while the corresponding value in Siebert *et al.* (2010) is 36.8 million ha. For Turkey, the respective estimates are 0.7 and 1.7 million ha. Based on a detailed study of Siebert *et al.* (2010), it is estimated that about 40% of irrigation water is derived from groundwater. Irrigation-related withdrawals from groundwater amounted to about 1300 km^3/year in 1998–2002 and accounted for about 90% of total groundwater withdrawals (Döll *et al.*, 2012).

Net groundwater abstraction

To assess the environmental impacts of human water use, net water abstraction from groundwater and surface water bodies must be considered, rather than water withdrawals. Net abstractions are defined as the difference between water withdrawals from the storage compartment and return flows to the compartment, with net abstraction from groundwater also taking into account the return flow of surface water that is used for irrigation. Positive values of net abstraction from groundwater (NAg) or net abstraction from surface water (NAs) indicate groundwater or surface water storage losses, respectively, whereas negative values indicate storage gains. Examples of areas with high positive NAg and high negative NAs are the High Plains of the central USA, the westernmost part of India (among others the states of Gujarat and Rajasthan) and the region around Beijing in north-eastern China, where return flows of irrigation water pumped from groundwater increase surface water flows and storage (Figure 4.b6.1 a and b, Döll *et al.*, 2012). Both NAg and NAs are high in most of the Ganges basin in India and the Po basin in Italy, while in most of Spain, both NAg and NAs are positive but NAg dominates. Global NAg was about 250 km^3/year, and global NAs about 1200 km^3/year in 1998–2002 (Döll *et al.*, 2012).

Groundwater depletion

The ratio of NAg (from Döll *et al.*, 2012) to groundwater recharge (from Döll and Fiedler, 2008) is an indicator of groundwater stress and groundwater depletion (Figure 4.b6.1c). Taking into account the spatial resolution of the analysis (grid cells of 0.5° by 0.5°, i.e. of 2000–3000 km^2), the negative impacts of groundwater withdrawals may occur even for small groundwater stress values of 0.01 to 0.1, in particular if groundwater withdrawals are restricted to parts of the grid cell. At this level of groundwater stress, base flow to surface waters may already be reduced significantly, affecting freshwater ecosystems and surface water supply security. Groundwater tables may have declined such that groundwater-dependent vegetation and groundwater supply is negatively affected. Where groundwater stress is larger than 1, groundwater depletion is likely to occur, with continuously declining water tables. This is the case in semi-arid areas with intensive irrigation in the United States and Mexico, around the Mediterranean, in western Asia, in India and in northern China (Figure 4.b6.1c).

Trends

Groundwater irrigation, that is, areas equipped for irrigation by groundwater, has increased in recent decades both in absolute terms and in terms of the percentage of the total irrigated area (Siebert *et al.*, 2010). This trend is likely to continue in the future, and to be exacerbated by climate change. One reason is that climate change is likely to increase total irrigation water requirements due to temperature increases and changed precipitation patterns on 60–70% of the currently irrigated land (Döll, 2002). The other reason is that the variability of surface water supplies is likely to increase due to increased precipitation variability, making groundwater an even more attractive source of water. However, in areas with high groundwater stress, irrigation water use may decrease in the future, as declining water tables make groundwater use economically inefficient or societies value a good ecological state of the environment higher than the benefits of irrigation.

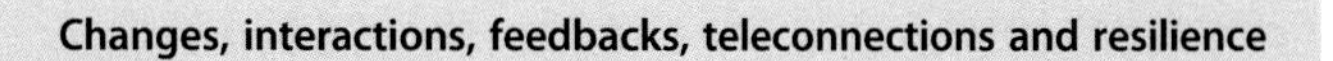

Box 4.6 (*cont.*)

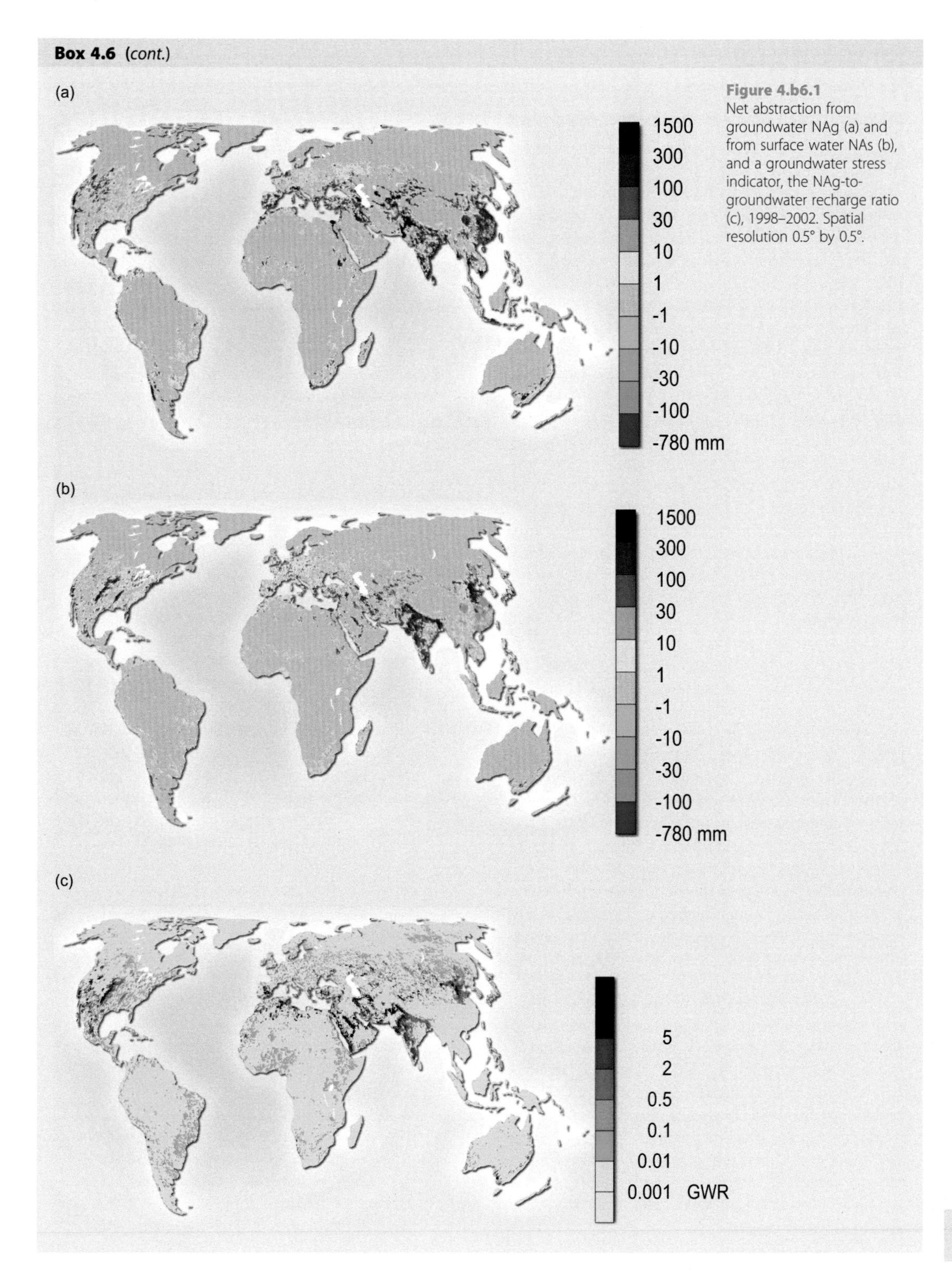

Figure 4.b6.1
Net abstraction from groundwater NAg (a) and from surface water NAs (b), and a groundwater stress indicator, the NAg-to-groundwater recharge ratio (c), 1998–2002. Spatial resolution 0.5° by 0.5°.

Box 4.7 Sustaining flows in the world's rivers

Jay O'Keeffe, Rhodes University

Michael E. McClain, UNESCO-IHE Institute for Water Education

River systems are at the heart of human society. Where would London be without the Thames? Cairo without the Nile? Delhi without the Yumana? Rome without the Tiber? Everywhere we look, great civilisations have been built up around rivers, which provide not only the water supply and waste disposal on which people depend, but also avenues for cultural, spiritual, navigational and ecological focus. Nonetheless, these rivers are increasingly being degraded by excessive use: polluted, over-abstracted, canalised and disconnected from their floodplains. Incredibly, previously perennial rivers are being managed so that they end up with no flow remaining in the channel – even some of the largest rivers in the world, including the Yellow, Indus, Colorado and Murray, no longer reach the sea for periods of time (Le Quesne *et al.*, 2010). A river channel without water is completely useless – it provides no goods or services for people, and no habitat or resources for its biodiversity. By any criteria, and despite the demands of upstream communities, such management is senseless.

This growing realisation has resulted, over the past 35 years, in the development of the concept of environmental flows: the quantity, timing and quality of water flows and levels required to sustain freshwater and estuarine ecosystems and the human livelihoods and well-being that depend on these ecosystems (Brisbane Declaration, 2007). In contrast to many aspects of IWRM, which are complex and not clearly defined (e.g. Biswas and Tortajada, 2004; Kerr, 2007), the assessment and implementation of environmental flows is a clear process with well-developed and described methodologies at different levels of detail and a well-defined end-point, which is accessible and amenable to policymakers, managers and engineers, as well as to scientists and conservationists.

The underlying principle of environmental flows is that people have to use the resources that rivers provide, that rivers are robust and resilient in the face of reasonable exploitation, and that it is possible to define modified flow regimes which will continue to provide for the needs of people and the rest of their biodiversity in a sustainable manner. There are many well-thought out methods for assessing these flow requirements (Tharme, 2003), depending on the resources, including information and expertise, available in different regions.

Environmental flows have become a standard part of developing water resource policy worldwide. Le Quesne *et al.* (2010) list policy and legislative developments, which include a requirement for environmental flows in Japan, China, Pakistan, India, Australia, New Zealand, South Africa, Kenya, Tanzania, Mexico, Chile, Ecuador, Columbia, Costa Rica, Puerto Rico, Brazil, the countries of the Mekong basin, the EU member states and many US states. In fact, they are 'aware of no major nation in which environmental flows are not now being discussed and/or incorporated into high-level water policy decision-making'. Of course, this process is still at an early stage of implementation, but Hirji and Davis (2009) describe eight implementation projects in countries other than the United States and Australia, both of which have a number of rivers with operating environmental flows.

Methods for assessing environmental flows

The methods currently in use for assessing environmental flows can be grouped into three main categories:

- Hydrology-based or look-up table methods: The original, Montana, method (Tennant, 1976) prescribes flows as a percentage of average wet and dry season flows corresponding to seven classes of ecological condition, from optimum (60–100%) to severe degradation (0–10%). The more detailed Index of Hydrological Alteration, a software program developed by the US Nature Conservancy, examines over 60 ecologically relevant statistics derived from daily hydrological data (Richter *et al.*, 1996). These are effective reconnaissance level assessment methods, which can be applied quickly and cheaply if the required hydrological data are available. However, they lack specific ecological, water quality, geomorphological or social justifications for the recommended flows, and are therefore difficult to defend if there is disagreement.
- Habitat assessment methods such as the Instream Flow Incremental Methodology (IFIM) (Stalnaker *et al.*, 1995), of which the principle model is PHABSIM, use data from multiple rated cross-sections to simulate hydraulic habitat conditions through a section of river, and match these up with the hydraulic habitat preferences of target species, in terms of depths, current velocities and substrate types. PHABSIM produces a database of reductions or increases in preferred habitat with changing discharges –

Box 4.7 (*cont.*)

expressed as weighted usable area (WUA) in m^2 per 300 m of river. This method has been much used in the United States, where it has been successfully legally defended, and is suitable for cases in which there are overriding target species such as salmon or trout for which the flows are being designed.

- Where ecosystem rather than particular species objectives are paramount, holistic methods typified by the Building Block Methodology (BBM, King and Louw, 1998) and the Downstream Response to Imposed Flow Transformation (DRIFT, King *et al.*, 2003) are the most commonly used processes. Holistic methods rely on a group of specialists in fish, invertebrate and riparian or floodplain ecology, water quality, geomorphology and socio-economics examining the consequences of flow changes for the channel shape, substrates, hydraulics, water chemistry, floodplain connectivity and therefore availability of habitat types in time and space. Like IFIM, rated cross-sections or habitat simulation models are used to link habitat availability to hydrology. The outcome may be a specific recommendation, usually with stakeholder involvement, to modify the flow regime to achieve predefined objectives, or a description of the outcomes of a series of flow scenarios from which management choices can be made.

Recently, a regional assessment method known as the Ecological Limits of Hydrological Alterations (ELOHA) has been developed (Poff *et al.*, 2009). Hydrological modelling is used to build a foundation of baseline and current hydrographs for river segments throughout the region. These segments are then classified into flow regime types that are expected to have different ecological characteristics, and are further sub-classified according to important geomorphic features that define hydraulic habitat features. Current flow conditions are analysed to determine how far they have been altered from natural patterns. Finally, ecological response relationships to altered flows are developed for each river type. ELOHA is currently being tested in a variety of rivers worldwide.

Implementing environmental flows

The implementation and persistence of environmental flows depend crucially on political will, public perceptions of their importance and a recognition of the long-term economic benefits of balancing resource use and protection. The success of environmental flows in maintaining or rehabilitating riverine ecosystems and biodiversity can only be measured by long-term monitoring. Preliminary cost–benefit reviews of existing environmental flows are encouraging. In 2006–07, 9 years after the completion of Katse Dam and 4 years after the completion of Mohale Dam and Matsoku Weir – elements of the Lesotho Highlands Water Scheme (LHWS) in Southern Africa, rivers downstream of the structures either achieved or exceeded their target environmental conditions. Environmental flows amounting to between 10.4 and 15.4% of total releases from these structures were recommended by the assessments (IUCN, 2004). The cost of releasing environmental flows and paying millions of US dollars in compensation to downstream villagers amounted to 0.5% of the project costs and did not significantly affect the project's economic rate of return (Le Quesne *et al.*, 2010). The environmental flows have positively affected some 39 000 people downstream of the dams, who depend partly or wholly on the river (Hirji and Davis, 2009).

The LHWS case study is typical of projects in which environmental flows are retrofitted to the river after dams have already been built and water uses allocated. A contrasting case study is the Mara River, which flows from the Mau escarpment in Kenya, from natural forest that has been extensively cleared in the past 15 years, through irrigated farmland, the extensive grazing lands of the Masai into the iconic Masai Mara (Kenya) and Serengeti (Tanzania) conservation areas. The river provides the only perennial water in these conservation areas, from which it flows in a westerly direction, through extensive wetlands, into Lake Victoria at Musoma. The river flows have been modified by erosion and deforestation, irrigation and abstraction, and are threatened by plans for further abstraction and inter-basin transfer. Despite this, the river is not heavily impounded, and the flow patterns remain substantially natural, apart from reduced low flows, particularly during droughts. A team of Kenyan and Tanzanian specialists, contracted by WWF East Africa and the US Global Water for Sustainability Programme (GLOWS), conducted a 2-year BBM assessment of the flows required to maintain the river in a near natural state, facilitated by UNESCO-IHE. The results indicated that, at the environmental flow site on the Kenya/Tanzania border, flows amounting to 35% of the average flow recorded over the 26 years on

Box 4.7 *(cont.)*

record would be sufficient to achieve this objective (LVBC and WWF, 2010). Of course, these are preliminary predictions, with some uncertainty involved, but they indicate that there can still be economic development of the water resources of the Mara River without unacceptable degradation of the natural resources of the basin. In the dry season during drought years there is already a deficit of the required flows, so some storage or demand management would be required during these periods. Additional assessments need to be carried out on the downstream reaches of the river, including the Tanzanian wetlands and the inflows to Lake Victoria. With these caveats, the project has demonstrated that:

- Environmental flow assessments should ideally be carried out before the water resources of a river are substantially allocated. They can then be incorporated into the future plans for development of the river.
- Far from preventing further development, such an assessment can demonstrate that there is potential for further consumptive use without unacceptable degradation of the resources.
- Such an assessment can pin-point when and where there are shortages (during dry seasons in droughts) so that remedial measures, such as on-farm storage, can be taken before the shortages become unmanageable.

the US Central High Plains, but at the same time also a more positive economic development perspective to replace water-intensive agriculture with industry. The term 'ageing' of landscapes, which Ripl (see Box 4.1) uses to describe a biophysical trend, can also be applied in a socio-economic sense to indicate a reduction in the degrees of freedom for development.

In response to the fact that landscapes and their terrestrial ecosystems can no longer provide water in the required amounts and at the required quality in many arid regions, freshwater is increasingly being manufactured from the oceans through seawater desalination (IEA, 2009). Since this process is highly energy-intensive and currently almost completely based on fossil fuels, dependency on desalinated water generates new vulnerabilities to energy price fluctuations as well as additional greenhouse-gas emissions.

Additional demand for water and land for rapidly expanding biofuel plantations (see Box 2.1) adds pressure to the global food market. New biofuels are estimated to have contributed to recent price hikes on global food markets (Rosegrant, 2008), which has implications for food security and the resilience of the poor.

Cities concentrate water demand because of their high population density and generally more resource-intensive lifestyles compared to rural areas. Municipal water demand increases at high rates, as urban population increases and living standards improve. Given decreasing and degrading surface water resources, cities rely on extracting groundwater often from increasingly larger distances. These massive water imports are often accompanied by severe leakage from the city's pipe system, from which rebound effects can occur: groundwater levels may rise again, possibly even above natural levels, causing new flooding problems (Foster *et al.*, 2010).

Like virtual water imports by countries, the effects of virtual and real water imports on the resilience of social–ecological systems is not clear for cities. Overexploitation of the resources of a city's rural hinterland linked to continued urban pressure can provide positive feedbacks, by degrading rural living conditions, causing additional migration to the city, leading to further growth of its population and more resource demands on the hinterland. Potential water-related critical thresholds in urban systems have not been explored in detail.

While the relatively small municipal and industrial blue water demands of less than 200 m^3/capita per year (Lundqvist and Gleick, 1997) are generally met from nearby, food water demands of 1000 m^3/capita per year or more (Gerten *et al.*, 2011) are often met by virtual water imports from distant regions. Berlin provides an example, where the average transport distance of imported crops is more than 4000 km and hence Berlin carries a high external virtual water footprint (Hoff *et al.*, 2013; Figure 4.10).

4.3.4 Virtual water flows and resilience

Virtual water flows linked to trade are not generally driven by water scarcity, but by the comparative advantages of countries in the different production sectors, as well as subsidies and trade policies. Only in the water-scarce Middle East and North Africa countries do virtual water imports correlate closely with water scarcity (Yang *et al.*, 2007).

Box 4.8 A tale of two plains: the consequences of groundwater depletion in the US Central High Plains and the North China Plain

Eloise Kendy, The Nature Conservancy

Bridget Scanlon, University of Texas at Austin

Though far from obvious at first glance, the 450 000 km^2 Central High Plains (CHP) of the United States and the 140 000 km^2 North China Plain (NCP) north of the Yellow River bear many crucial resemblances (Figure 4.b8.1). Over the past five decades, both have mined groundwater to supply irrigation, producing nationally significant agricultural yields and supporting prosperous rural livelihoods. In recent years, water users on both plains have become painfully aware of the alarming rate at which their groundwater reserves are depleting. Clearly, both plains face water-limited futures. Due to their unique hydrologic and socio-economic contexts, however, that is where the resemblance ends.

Groundwater mining

Groundwater mining occurs when groundwater extraction causes the volume of water stored in an aquifer to decrease year after year, as indicated by declining sub-surface water levels. The CHP has had water level declines of as much as 1.5 m/year in individual wells and regional declines of as much as 30 m over a 10 000 km^2 area since irrigation began in the 1950s. This indicates a water deficit – the difference between precipitation and evapotranspiration, which is made up by groundwater mining – of about 75 mm/year. The NCP has water level declines of more than 1 m/year in many places and regional declines of more than 30 m since irrigation with groundwater began in the 1960s. The irrigation deficit is about 200 mm/year in Luancheng County on the NCP.

Precipitation is similar on both plains, in the range of 320–840 mm/year in the CHP and 500–800 mm/year in the NCP. In both regions, precipitation falls primarily in the summer, from June to September (Scanlon *et al.*, 2010).

Differences in groundwater recharge

Natural rates of groundwater recharge differ markedly. Clay soils in the CHP prevent precipitation from percolating into the underlying aquifer except in isolated ephemeral lakes or playas. Therefore, groundwater beneath most of

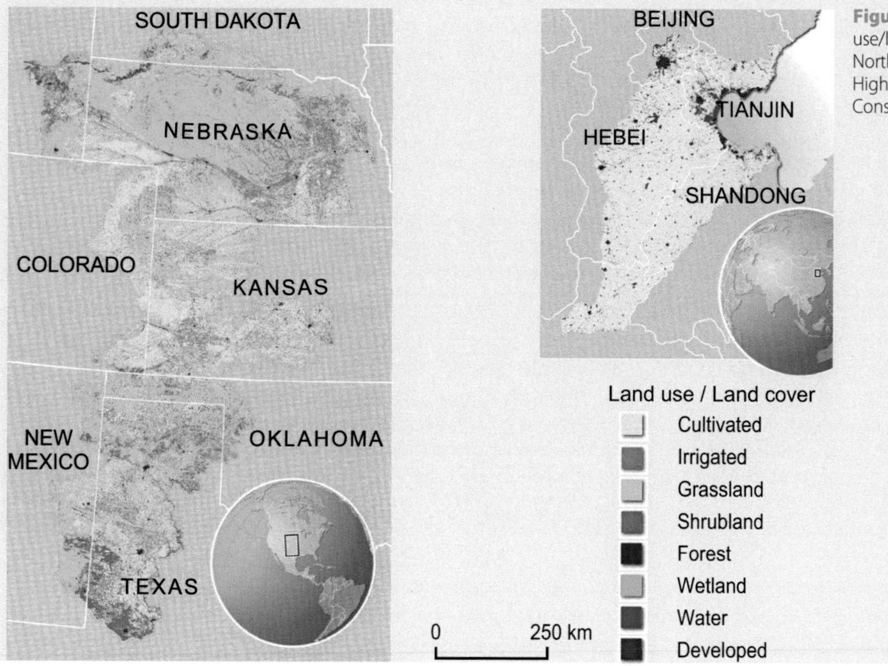

Figure 4.b8.1 Land use/land cover for the North China Plain and the High Plains (MRLC Consortium, 2007).

Box 4.8 (*cont.*)

the CHP is 'fossil' water that filled the aquifer in the past 13 000 years, and is only replenished locally beneath playas (Scanlon *et al.*, 2010; Scanlon *et al.*, 2012). In the NCP, precipitation recharges the aquifer regionally, albeit at lower rates than water is consumed by crops. Prior to the 1950s, runoff from the Taihang Mountains also recharged NCP aquifers but that water is now diverted and consumed before it reaches the NCP (Kendy *et al.*, 2004).

Similarly, natural surface runoff patterns differ. In the CHP, most surface water drains internally to playas that recharge the aquifer. Integrated surface drainage in streams and rivers is limited, and groundwater depletion has converted 'gaining' streams fed by groundwater into 'losing' streams fed only by runoff. In the NCP, the combination of upstream diversions and groundwater depletion has completely dried up formerly navigable rivers, heralding their conversion to cultivated farmland today. Natural, freshwater ecosystems no longer exist on the NCP.

Efforts to reverse aquifer deficits

Both regions have tried to improve irrigation efficiency in order to reverse water table declines. In the CHP, farmers used the 'saved' water to expand their irrigated areas, leading to more water consumption than before. In the NCP, where no additional land was available, water consumption remained essentially unchanged, as did the rate of water level decline because irrigation efficiency improvements simply reduced percolation that previously recharged the aquifer and was not a true loss to the hydrologic system. Efficiency improvements alone could not make irrigation sustainable on either plain (Kendy *et al.*, 2003; Kendy *et al.*, 2004).

For water levels to stabilise, the water budget must balance, that is, groundwater discharge cannot exceed groundwater recharge. Because water resources are limited, both plains can – and will – stabilise their water tables. Whether this happens by necessity after depletion because no usable groundwater remains or by choice before the resource is gone is largely a matter of political will. At some point both plains will arrive at a time when today's rate of water use can no longer continue, and their economies and societies will change accordingly.

In the CHP, 97% of groundwater withdrawals are used for irrigation. Recognising that irrigating the CHP essentially constitutes mining of a non-renewable resource, local groundwater management policy, termed 'managed aquifer depletion', allows depletion of 50% of groundwater storage in 50 years. A conscious decision has been made to continue current irrigation practices until physically they can continue no longer. The current population of less than 1 million is ageing.

In the NCP, urban and industrial uses increasingly compete with irrigation for water. Stabilisation of groundwater levels requires large-scale spatial and temporal reduction of irrigated agriculture. Spatially, the irrigated area would need to decrease. Well-managed cities and industries consume less water than irrigated cropland, and treated urban wastewater can be reused for irrigation, either directly or after recharging depleted aquifers. Temporally, the length of the growing season would need to decrease in places where irrigation water will no longer be available. Currently, irrigation allows farmers to produce two crops per year – winter wheat is planted in late summer and harvested the following spring, and maize, cotton and other crops are grown in the summer. Irrigation-dependent domestic food production stems emigration of surplus rural labour and ensures affordable food prices for everyone, urban and rural alike. This is helping to maintain social stability during the potentially turbulent transition from primarily rural to primarily urban livelihoods. At about 140 million, the population of the NCP is more than 10 times that of the CHP. Megacities such as Beijing, Tianjin and Shijiazhuang are swallowing cropland, as are the township and village enterprises which employ workers in more rural parts of the NCP (Kendy *et al.*, 2007).

Plausible future outlooks

What kind of a future is there for each of these plains? With extremely limited groundwater recharge through playas, irrigation of the CHP will eventually have to stop altogether, to be replaced by lower yielding rainfed crops and rangeland for livestock production, mirroring the shift that is already occurring in parts of the more arid Southern High Plains. In the absence of alternative economic drivers, emigration will decrease the rural population when irrigation is no longer an option. However, the decrease in population pressure holds the promise for restoring natural ecosystems to some extent. Because the groundwater has been mined irreversibly, spring-fed streams will not return but runoff driven streams could potentially support healthy freshwater ecosystems if they were otherwise protected.

The NCP is moving towards a different future. A return to pre-irrigation cropping patterns would synchronise crop production with rainfall, yielding only one crop per year. This would stabilise the water table, but would substantially decrease yields and increase vulnerability to drought. When this happens depends in large part on the

Box 4.8 *(cont.)*

government's emphasis on national food security versus preservation of groundwater resources. People are not leaving the NCP like they are leaving the CHP. Industry may well overtake agriculture as the leading employer and economic driver. The NCP's growing industrial sector will buffer the socio-economic impacts of reduced irrigation, as urban opportunities continue to replace rural livelihoods. Rising incomes and a growing economy may be deployed to improve living standards, including access to treated, well-managed water. However, that is not to say that ecosystem health will improve. Stream flow will not return to rivers, as that would require increasing – rather than just stabilising – groundwater levels and removing all upstream dams.

The NCP and the CHP are probably headed for very different futures, and thus different balances between people and nature. Both are mining groundwater resources unsustainably. Because of differences in hydrology and socio-economic trends, the future CHP is likely to support a reduced rural population with the potential for restoring some freshwater ecosystem functions, while the NCP is likely to support an extremely large population without restoring natural ecosystems.

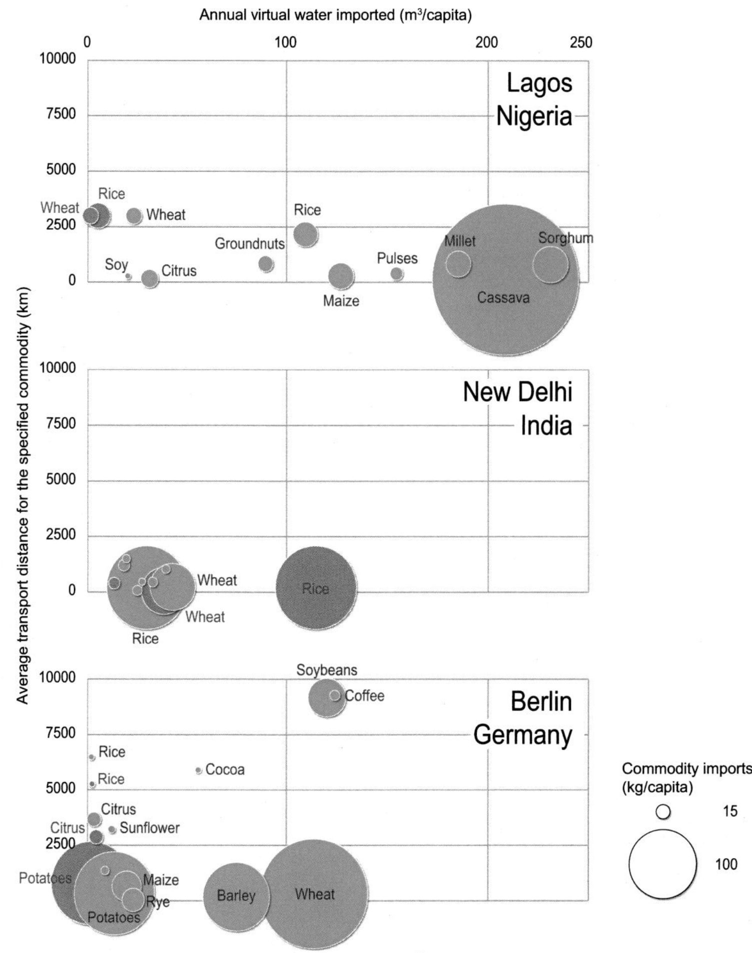

Figure 4.10 Virtual water (green and blue) imports associated with selected crops, and average transport distances, for the cities of Lagos, New Delhi and Berlin (Hoff *et al.*, 2013). The size of the circles indicates the absolute weight of crop imports.

By calculating virtual water imports from LPJmL-based crop virtual water content and combining with ComTrade trade data, and comparing this with LPJmL-derived water scarcities, we found correlation coefficients of −0.51, −0.64 and −0.79, respectively, for per capita net virtual water imports and blue water availability, green-blue water availability and water-limited potential calorie production for the Middle East and North Africa countries.

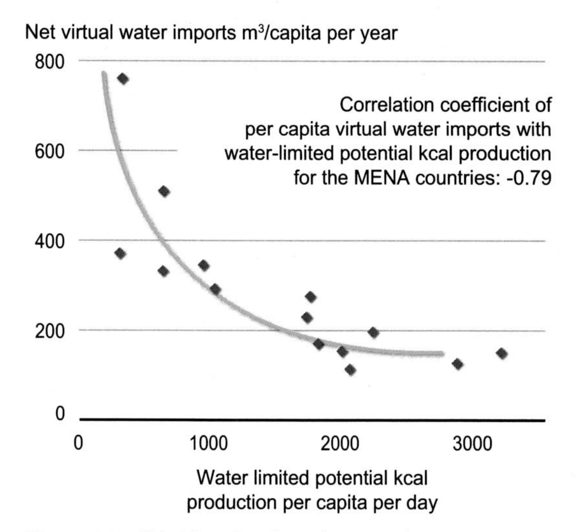

Figure 4.11 LPJmL-based analysis of net virtual water imports vs. water-limited potential kcal production for 14 Middle East and North Africa countries. The data are derived from each country's total blue plus agricultural green water availability in combination with its average crop water productivity (after Gerten *et al.*, 2011).

Figure 4.11 plots net virtual water imports against water-limited potential kcal production per capita for each Middle East and North Africa country. Figure 4.12 shows the relative amounts of domestic blue and green water vs. net virtual water imports for each Middle East and North Africa country (left panel, calculated from exporting countries crop water productivities) and the water savings derived from importing instead of producing locally (right panel, calculated from importing/Middle East and North Africa countries crop water productivities).

It is not clear whether the strong dependence of the Middle East and North Africa countries on external water resources, illustrated in Figure 4.12, and interconnectedness through trade or other mechanisms increase or decrease water-related resilience. On the one hand, the virtual water trade might increase resilience to local droughts or other shocks by geographically expanding the resource base from which a country can draw. It is less likely that distant export producers will be affected by the same shock. On the other hand,

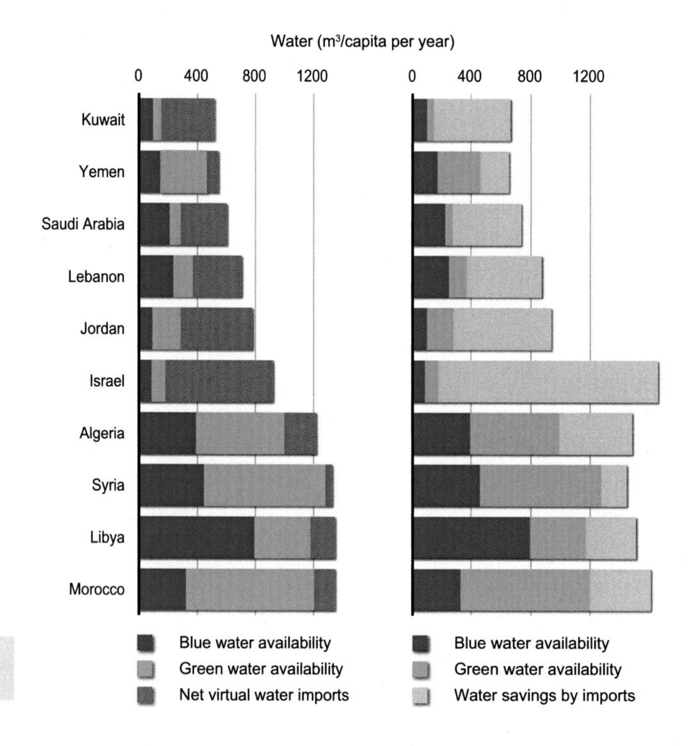

Figure 4.12 LPJmL/ComTrade-based net per capita virtual water imports (brown colour, left panel), water savings by importing (yellow colour, right panel), and green and blue water availabilities per capita (both panels), for a number of Middle East and North Africa countries (after Fader *et al.*, 2011).

political and economic vulnerability might increase if a country relies on a handful of large exporters for external food sources and virtual water supplies. There is also greater exposure to fluctuations in international market prices. Moreover, the substitution of virtual water imports for local production can provide a sense of food security which slows down much needed water sector reforms, such as water pricing and improvements in water productivity.

While virtual water trade patterns are not generally driven by water scarcity, FDI in land is largely driven by water-scarce countries such as China, India and the Arab nations (Anseeuw *et al.*, 2012). The green plus blue water availability for these countries is: 1200 m^3/capita per year (China), 1050 m^3/capita per year (India, Egypt), 300 m^3/capita per year (Saudi Arabia) as calculated by LPJmL (according to Gerten *et al.*, 2011).

Among the preferred target countries for FDI are Ethiopia and Sudan (Anseeuw *et al.*, 2012), both of which suffer from high levels of malnutrition and receive external food aid, while at the same time implementing large-scale export food production. There is a risk of increased vulnerability due to FDI when local water and land users are deprived of their customary rights and access. There is a clear need to regulate these investments (see e.g. FAO, 2010; von Braun and Meinzen-Dick, 2009).

These examples of virtual water trade and FDI demonstrate that the use of local water resources and the resilience of water-dependent social–ecological systems can be subject to strong external drivers, or drivers beyond the water sector, such as land management, trade, investment and climate policies. Cumulatively, the above pressures can drive social–ecological systems towards critical thresholds and reduce water resilience, leaving them more vulnerable to shocks such as droughts or food and energy price shocks. However, to date there are no globally consistent metrics to measure these complex interactions and changes in water systems resilience.

4.4 Water-related indicators for water security and food security

Rockström *et al.* (2009) defined nine interacting key **planetary boundaries** for a safe operating space for humanity (see Chapter 1). One of these planetary boundaries is related to global human consumptive blue water use, which was estimated to be about 4000 km^3/year. In order to operationalise this global boundary for water governance and management on the ground, it needs to be further defined and refined, and downscaled, i.e. disaggregated or made spatially explicit. It also needs to be integrated with green water and eventually to be complemented with location-specific sustainability criteria for maximum water exploitation. Such sustainability criteria could, for example, be: avoiding depletion and degradation of aquifers and surface waters, maintaining environmental flows and not significantly changing atmospheric moisture recycling patterns or interfering with monsoon circulations, or socio-economic and human security-related criteria such as access to safe water, and food and energy security.

Various attempts have been made to quantify and map water scarcity, in particular with respect to the demand on water for food production. Over time, water scarcity metrics have advanced from pure blue water scarcity indicators to more comprehensive ones which also include green water on agricultural land and agricultural water productivity. More recently, they have also begun to address other ecosystem services and in future they will hopefully fully integrate socio-economic aspects of water scarcity. An important reference for blue water scarcity is the water crowding or use-to-availability analysis by Falkenmark and Molden (2008) (Figure 1.7, Chapter 1).

In terms of agricultural production and biomass production in general, several subtropical and tropical regions are severely water limited. Temperate zones tend to be more humid. Subtropical and tropical drylands often combine aridity as a slowly changing variable with high levels of climate variability and frequent, and severe, extreme events. They are prone to water deficits and overexploitation of water resources, with subsequent effects on ecosystems and social–ecological systems and their resilience. Water constraints often coincide with low levels of development, poverty and malnutrition (Brown and Lall, 2006). Grey and Sadoff (2007) confirm that poor countries often have 'difficult hydrology' or are 'hostage to their hydrology'. Chapter 3 of this book suggests that 'drylands are more vulnerable to regime shifts than other regions of the world'. What exactly makes them vulnerable and how can water security and water-related resilience be measured? A number of water-related indicators have been developed in the past, an overview of which is presented in Table 4.1.

Water scarcity also depends on the system boundary or scale of analysis. As is shown by Vörösmarty (personal communication, Figure 4.13), the water-scarce

Table 4.1 Overview of existing water-related indicators

Indicator	Unit	Source
Water scarcity, water stress, water crowding	People per million m^3 or m^3/capita per year (blue water only), use to availability ratio	Falkenmark and Molden (2008)
Basin closure (permanent or temporary)	No more water to allocate, no flow into the sea (blue water only)	Falkenmark and Molden (2008)
Water scarcity, water stress	m^3/capita per year for green and blue water	Rockström *et al.* (2009)
Variability	Coefficients of variation runoff of blue water or aridity	Vörösmarty *et al.* (2005)
Reliability	Water availability meeting demand in space and time	See Box 4.5
Use-to-availability ratio	(Domestic + industrial + agricultural blue water use) / river discharge	Vörösmarty *et al.* (2005)
Withdrawal-to-availability ratio, consumption-to-Q90	Total withdrawals relative to 90th percentile stream flow, i.e. the flow that is exceeded 90% of the time	Alcamo *et al.* (2007)
Human and environmental water scarcity	Water availability relative to human and aquatic ecosystem water demand	Smakhtin *et al.* (2004)
Water-constrained calorie production potential	Based on average annual green and blue water availability and crop water productivity	Gerten *et al.* (2011)
Water security	The availability of an acceptable quantity and quality of water for health, livelihoods, ecosystems and production, coupled with an acceptable level of water-related risks to people, environments and economies	Grey and Sadoff (2007)
Water poverty	Combines information on access to water, water quantity, quality and variability, water uses, capacity for water management and environmental aspects	Sullivan and Meigh (2007)
Human water security threat	Expanded water poverty index with more weighted components	Vörösmarty *et al.* (2010)

population increases with decreasing size of spatial units over which water and population are averaged.

There is more than enough water available globally to feed a growing and wealthier population, but at smaller scales there is an increasing mismatch between population (i.e. water demand) and water availability. Water scarcity approaches zero if total global water resources are compared to total global water demand. A global perspective on water scarcity also encompasses virtual water associated with the trade in agricultural commodities, as well as the role of moisture recycling in transforming consumptive water use or evaporative loss in one region into precipitation and new water resources in downwind regions. This global perspective also emphasises the renewable nature of water, which prevents it from becoming overexploited or scarce at the global scale. Locally, however, overexploitation and temporal or permanent water scarcity are increasing.

We provide here a new and more comprehensive definition of **basin closure**, which goes beyond the conventional blue water limit ('additional commitments cannot be met' or the lack of outflow to the sea, see Falkenmark and Molden, 2008) to include green water on agricultural land and agricultural water productivity. From the available green and blue water, weighted by agricultural water productivity, we calculated the potential water-limited kcal production, and compared that to the generally accepted critical level of 3000 kcal/capita per day. If this level is not reached, we consider a basin 'closed'. This definition of basin closure focuses on water for food. Industrial and municipal demands are uniformly accounted for by a total volume of 200 m^3/capita

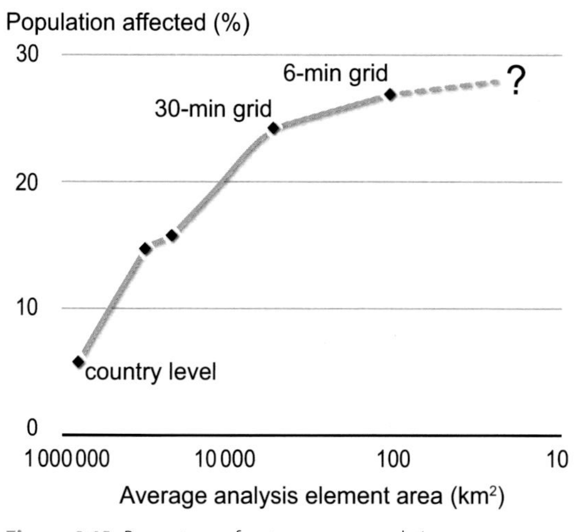

Figure 4.13 Percentage of water-scarce population vs. average element size of the assessment (Vörösmarty, personal communication).

per year (Lundqvist and Gleick, 1997). For environmental flows, we reserve one-third of annual mean discharge.

Cumming (2011) suggests that basin closure poses a critical threshold beyond which a number of new processes and interactions between actors, production systems and land cover may be triggered. Responses to basin closure may include the manufacturing of expensive and energy-intensive, non-conventional water by desalination, real or virtual water imports (see above) or outmigration of population. While desalination and virtual water imports are typical responses to increasing water scarcity in the Middle East and North Africa region and other Mediterranean-type climates, there is no conclusive evidence on the role of water scarcity as a driver of migration. Some evidence for such a last-resort response comes from the Horn of Africa and Syria (Juusola, 2010), where several hundred thousand refugees have tried to escape droughts. About 200 000 refugees fled from the Iraqi marshlands in the 1990s when these were drained (CIMI, 2010) and a similar number of farmers and fishermen migrated to urban areas from the drying Lake Faguibine region in Mali (UNEP, 2011). Niger and Nigeria have experienced massive southward migration of pastoralists in response to droughts (UNEP, 2011). There is further evidence from Kenya of water scarcity driving migration (IRIN, 2009).

There are no consistent and comprehensive indicators of water scarcity, let alone of water-related resilience. Water assessments at the global scale remain simplistic, and the more integrated and detailed analyses are still limited to individual case studies.

4.5 Comparative analysis of water scarcity in a number of large river basins

In order to develop a more comprehensive and globally consistent assessment of water scarcity, we have initiated a comparative analysis of nine large river basins around the world. These rivers, the Ganges, Indus, Limpopo, Mekong, Niger, Nile, São Francisco, Volta and Yellow, were the major focal basins of the first phase of the Challenge Programme on Water and Food (CPWF, http://www.waterandfood.org; see Figure 4.14).

For this comparative analysis of water and food security, we used a combination of top-down and bottom-up approaches: (1) global LPJmL simulations at pixel-based half-degree resolution and (2) basin-by-basin data from the CPWF.

Eventually, this analysis of current and future water scarcity and food security will provide information about:

- the complex interactions and relative magnitudes of internal and external drivers of change;
- the proximity of a basin to a critical basin closure threshold and its resilience;
- upstream-downstream trade-offs between different interventions and investment options, and the potential for regional basin-level cooperation.

4.5.1 Water-basin characteristics

Water availability

Starting from the most common water scarcity indicator, we quantified blue water availability for each basin and complemented this by adding green water availability expressed in m^3/capita per year. Since green water calculations in the CPWF dataset (evapotranspiration over terrestrial areas, ET calculated as the residual between measured precipitation and runoff) are not fully consistent with our calculations, we only use the LPJmL results for this study. Basin populations were extracted from the IIASA gridded

Figure 4.14 The large Challenge Programme on Water and Food (CPWF) river basins covered in the analysis.

population database (Grübler *et al.*, 2007). Blue water availability was calculated by distributing total basin runoff to each basin pixel according to the fraction of total discharge measured for the respective pixel, in order to avoid double counting of discharge along the river course. Municipal, industrial and environmental demands were subtracted from the resulting water availability. Green water was calculated as ET from cropland – the green water resource for food production. Hence, any expansion of cropland would increase that green water resource accordingly. However, future cropland expansion is projected to be very limited (Bruinsma, 2003). The resulting per capita water availability for food production is presented in Figure 4.15a.

Food production potential

Next we used LPJmL according to Gerten *et al.* (2011) to calculate the agricultural water productivity of green plus blue water consumed for the current composition of crops in rainfed and irrigated agriculture. Cai *et al.* (2011) suggest that CPWF water productivity data from the different basins are not comparable, so we only used LPJmL data. Our analysis, presented in Figure 4.15b, confirms that African basins have very low water productivities (see e.g. Cook *et al.*, 2011), in the case of the Nile basin as low as 500 kcal per m^3 of water. By combining green plus blue water availability for food production with agricultural water productivity, we calculated the water-limited potential kcal production/capita per day (see Figure 4.15c).

In order to account for climate variability (for the above calculations we used 30-year averages of annual values) we downscaled the potential kcal production to the tenth percentiles of blue water availability, or the driest three years from each 30-year time series. We found the strongest declines in water-limited potential kcal production in the semi-arid basins, which are also characterised by the highest coefficients of variation in annual water availability, i.e. the Niger, Volta and Limpopo (Figure 4.15c).

Future, more advanced analyses will have to include finer measures of variability in order to better characterise the respective hydro- and agro-meteorological regime (identifying what Grey and Sadoff (2007) call ‘difficult hydrologies’) – using drought indices, flow-duration curves, new measures for the reliability of the rainy season, and so on.

To derive comprehensive indicators of water scarcity, it will also be important to account for reservoir storage per capita, given that this storage can buffer some of the variability. Cook *et al.* (2011) find that the Asian CPWF basins have generally better developed water infrastructure than the African basins, which presumably strengthens their resilience against droughts and other shocks. Similarly, the proportion of irrigated agriculture and the dominant types of irrigation are determinants of a basin’s water productivity and adaptive capacity, and need to be incorporated into comprehensive water scarcity indicators.

In order to assess the importance of external drivers of water scarcity and water-related resilience, we calculated for each CPWF basin:

- the net virtual water imports with agricultural commodities;
- the atmospheric moisture inflow on which each basin’s precipitation depends and the source regions of this moisture.

4.5.2 Basin dependence on virtual water imports

A comparison of imports and exports of food commodities for each basin using national data and the respective proportional contribution of each country to the basin shows that net virtual water imports into the CPWF basins are negligible. More specifically, we calculated virtual water imports and exports from LPJmL crop virtual water contents in combination with bi-national ComTrade data (not including livestock products), weighting each country’s contribution according to its share in the total basin

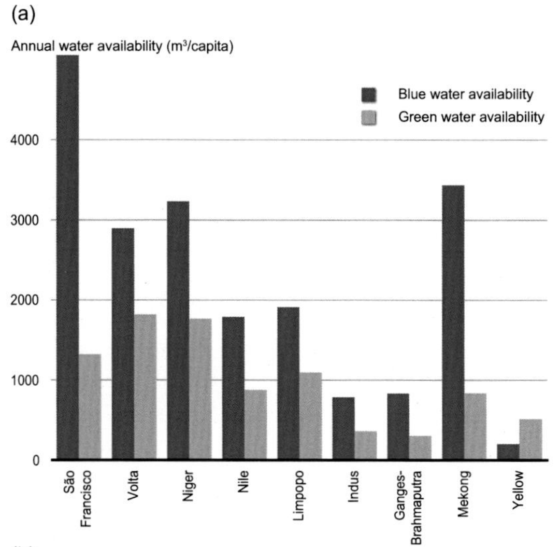

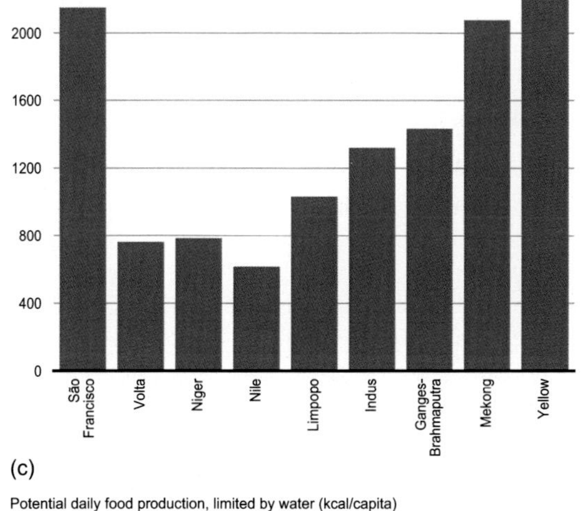

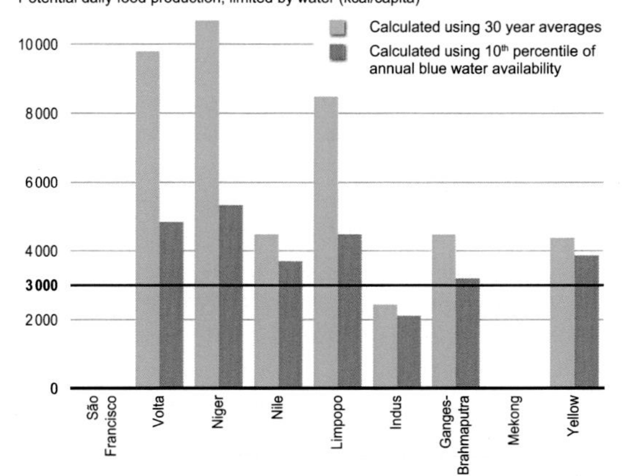

Figure 4.15 Assessment of analysed basins. (a) Blue and green agricultural water availability in m^3/capita per year; (b) agricultural water productivity in kcal per m^3 of green and blue water consumed; (c) water-limited potential kcal production/capita per day, also shows a drop in potential kcal production/capita per day, when using tenth percentile of annual blue water availability instead of 30 year averages – marked with a black horizontal line is the level of 3000 kcal/capita per day above which food security is normally assumed. For São Francisco and Mekong, the values exceed the upper limit of this graph, and are omitted from this chart.

population, assuming that a country's per capita imports and exports are the same for its population inside and outside the basin. Three basins were found to be minor net virtual water importers. Their net virtual water imports per capita per year are very small compared to their internal renewable resources:

Nile basin: 56 m^3
Niger basin: 33 m^3
Volta basin: 7 m^3

All the other CPWF basins are net virtual water exporters, although individual countries within the basin may be net importers. They do not depend on virtual water for their water and food security.

4.5.3 Basin dependence on atmospheric moisture inflow

The atmospheric moisture inflows from other regions on which the basins' precipitation depends – so-called moisture recycling – were calculated using the Delft atmospheric water accounting model or WAM (van der Ent *et al.*, 2010). First, we calculated the fraction of 'internally generated precipitation' originating from atmospheric moisture that was evaporated within the basin – so-called internally recycled moisture. From this we found that CPWF basins' precipitation depends primarily on atmospheric moisture generated outside the basin. Depending on the size and land cover of the basin, only 10–20% of the basins' precipitation is generated from evapotranspiration within the basin (see Figure 4.16a). The extremely long travelling distances of atmospheric moisture between its evaporation from land and precipitation – often several thousand kilometres (van der Ent and Savenije, 2011) – make the 'precipitationshed' from which a basin receives its precipitation quite large (see e.g. Figure 4.16b). This indicates that a significant proportion of

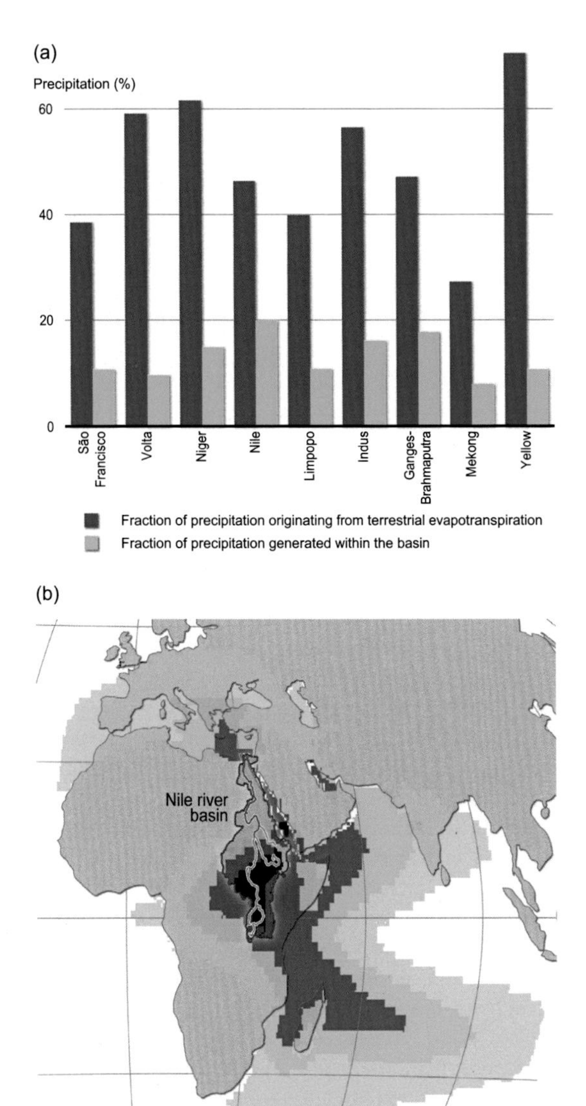

Figure 4.16 (a) Internally generated precipitation, and precipitation originating from terrestrial evapotranspiration as ratio of total basin precipitation. The remainder is precipitation from oceanic evaporation. (b) Precipitationshed for the Nile basin showing the annual evapotranspiration in millimetres that each pixel contributes to Nile basin precipitation. Red line: Nile basin delineation (Nikoli, 2011).

precipitation in the Nile basin originates from evapotranspiration in tropical African land areas, which are prone to land-use change – in particular deforestation.

For each basin, we calculated how much of its precipitation originates from atmospheric moisture evaporated over land, as opposed to evaporated over the sea. The fraction of atmospheric moisture of terrestrial origin varied significantly between the basins, depending on its proximity to the sea and its orientation relative to the prevailing winds. The range was from about 30% to 70% of total precipitation (see Figure 4.16a).

The dependence of the basins' precipitation on evapotranspiration from other land areas may also affect resilience, given that land-use change, in particular deforestation, in the precipitationshed could reduce evapotranspiration (Keys *et al.*, 2012).

We also assessed the global effects of past land-use change from natural to current vegetation and associated changes in evapotranspiration on downwind precipitation, using the LPJmL model in combination with the Delft WAM model (Nikoli, 2011) (see Figure 4.6). We did not, however, address scenarios for future land-use change.

4.5.4 Socio-economic factors in water and food security

To complement the biophysical components of water and food security, we have also begun to calculate selected socio-economic variables per basin. These should eventually be integrated to characterise the resilience of the CPWF basins, which are understood here as socio-ecological systems. GDP per capita can serve as a proxy for the adaptive capacity of the basin population in terms of access to new technologies, but also its capacity to purchase agricultural products and associated virtual water on world markets (Figure 4.17).

The contribution of agriculture to GDP provides an indication of economic diversification, and with that also some measure of the resilience of river basins. Our analysis confirms that of Cai *et al.* (2011), that the contribution of agriculture to GDP – and at the same time the incidence of poverty – is highest in the African CPWF basins (Figure 4.17).

More comprehensive descriptions of water-related resilience will have to address additional dimensions such as the crop diversity index (see Eriyagama *et al.*, 2009), the percentage of the population employed in agriculture as a measure of socio-economic diversity, access to water and sanitation and market access.

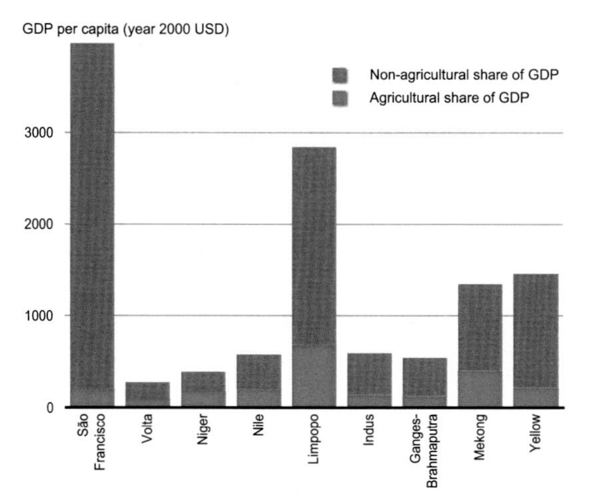

Figure 4.17 GDP per capita and agricultural contribution to GDP, country values weighted according to their share in total basin population.

Market access (see Nelson, 2008) allows rural farmers to sell products and enter development pathways related to commercialisation. It also enables better distribution of food within basins or countries.

4.5.5 Basin dynamics

River basin management and planning require knowledge of trends and development trajectories in order to guide interventions and investment towards improved water and food security and resilience. We analysed and compared the relative importance of a number of internal and external drivers of change:

1. Population change as an internal driver of change affecting water demand and per capita water availability;
2. Climate change as an external driver of change affecting total water availability and crop production, as well as crop water productivity through precipitation and temperature change;
3. Land-use change inside and outside the basin, as an internal and external driver of change affecting evapotranspiration, moisture recycling and eventually basin precipitation.

Population is projected to increase most rapidly in the three African basins: Volta, Niger and Nile. Population will almost triple in each by 2050, which in turn will reduce per capita water availability by a factor of three – assuming constant per capita demand. The lowest pressure from population growth is projected for the Yellow River and Limpopo basins (Figure 4.18a). Potential water-constrained kcal production per capita will shrink accordingly, if climate, productivity and diets remain constant (Figure 4.18b).

When comparing the rates of change of the internal driver, population growth, with those of the external driver, climate change (Figure 4.18a), it is obvious that population growth plus possibly also economic development and changing diets (Figure 2.3) will continue to affect water scarcity more than climate change over the next few decades. However, other climate change effects, in particular changes in variability and extremes, need to be assessed in greater detail as a potentially much more serious threat to water and food security than the change in long-term mean annual water availability.

The sensitivity analysis for land-use change – another driver of change – in the precipitationshed, as described above, also shows relatively small effects on the basins' water availability. Resulting changes in precipitation range from increases of about 30 mm per year in the Asian basins, largely due to increasing evapotranspiration from irrigation in the respective precipitationsheds, to small decreases in precipitation of up to −10 mm per year in the African basins, due to reduced evapotranspiration and loss of natural vegetation (Nikoli, 2011).

4.5.6 Concluding overview

Cook *et al.* (2011) analysed the socio-economic situation, represented by rural poverty and the contribution of agriculture to GDP, in the CPWF basins over time and hypothesised that they are all following a similar development trajectory of decreasing rural poverty in concert with a shrinking contribution by agriculture to GDP (see Figure 4.19). Future research must link these development trajectories to changes in water security and of water-related resilience.

Rural poverty remains high in all the basins in Africa and Asia, but the São Francisco basin in South America has reached a more comfortable position. The fact that it is situated in a more favourable hydro-climatic zone and has plenty of blue water may have contributed to its higher GDP per capita,

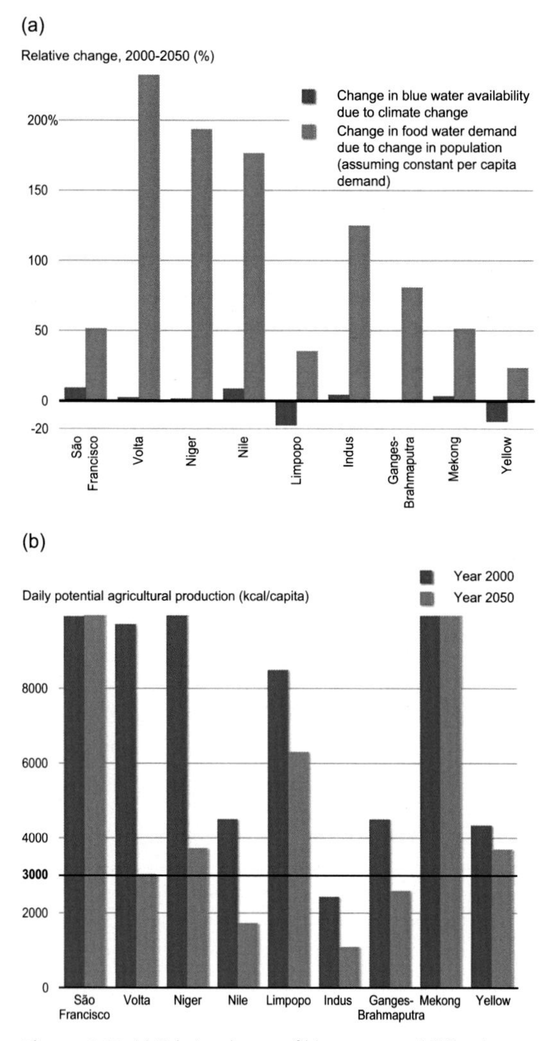

Figure 4.18 (a) Relative change of blue water availability due to climate change (LPJmL ensemble simulation results based on A2 ECHAM scenario) versus the effect of population growth on food water demand; (b) illustrating the situation in the year 2000 and in 2050. The thick line marks 3000 kcal/capita per day, the level above which food security is normally assumed.

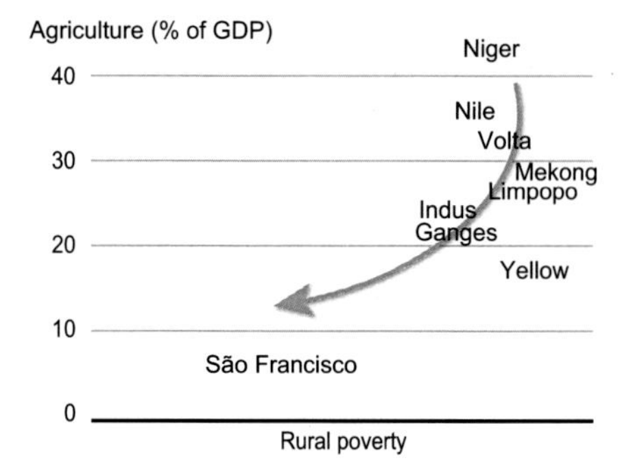

Figure 4.19 Stylised trends in rural poverty and contribution of agriculture to GDP (after Cook *et al.*, 2011).

to which agriculture contributes just over 5%. Next to Mekong, São Francisco has the lowest dependence on terrestrial evapotranspiration and underlying land use, which possibly also improves its resilience. The eight remaining basins all have a high degree of rural poverty and are economically more dependent on agriculture. Most dependent on external moisture influx through the atmosphere is the Yellow River basin, which receives more than 70% of its precipitation from evapotranspiration over terrestrial areas, in this case Eurasia. Comparing their future outlook, all the basins face great challenges as they move along their socio-economic development trajectories.

The Asian basins

The Yellow River, the Indus and the Ganges suffer from severe blue water shortages due to high-level water crowding, and are at the same time also green water (and hence land) scarce, making their food production dependent on irrigation. Population will at least double by 2050 in the Indus and Ganges basins, but population growth will be more limited in the Yellow River basin, which is however – next to Limpopo – most threatened by climate change-driven reductions in blue water availability. Thus, these three basins will all be moving towards water-related critical thresholds, requiring adaptation and/or transformation in their efforts to achieve water and food security. The Mekong basin, on the other hand, has relatively more favourable conditions.

The African basins

The predicament of the three Asian basins is shared by the Nile basin, which is irrigation dependent and is experiencing rapid population growth. Its population is projected to almost triple in the next 40 years. Similarly, the population of Niger and Volta are also projected to triple, in addition to which they also suffer from food water deficiency, making their food security import dependent unless major breakthroughs can be achieved in agricultural water productivity. The Niger basin – on top of these predicaments – is highly vulnerable to droughts. In contrast, the Limpopo basin, where food security is irrigation dependent, is highly vulnerable to climate change, which is expected to increase aridity and vulnerability to droughts, but much less subject to future population pressures compared to the other basins in sub-Saharan Africa.

Summary

The water cycle and vegetation

Water links all of the Earth System through the different components of the hydrological cycle. Solar radiation is the primary driver, causing transpiration by plants, evaporation from moist land and open water surfaces, and the transportation of atmospheric moisture. This vapour forms clouds, and precipitation which infiltrates the soil, recharges groundwater and flows along river networks to the ocean.

There are a number of bidirectional interactions between water and vegetation/land. On the one hand, water is a central component of biomass production and essential for transporting nutrients to and within plants. On the other hand, vegetation influences the partitioning of precipitation between green and blue water at the land surface. Water acts as a control variable with respect to food production and other ecosystem services. Consumptive blue water use is estimated at less than 1500 km^3/year, but green water use for crop production amounts to some four times as much, or twice that amount if the water used in livestock production is also included. However, much larger amounts of green water are required to sustain the other ecosystems, and their services that form the natural water infrastructure of the Earth System. These have not received much attention in the academic literature, although they are critical to the functioning of the Earth System and for human welfare. Examples of such services include carbon sequestration, nutrient cycling and moisture recycling.

Three large-scale water-related interactions in the Earth System have been identified with the potential for non-linear responses and feedbacks: the drying and savannisation of the Amazon rainforest, the greening of the Sahel and the weakening of the Indian monsoon.

Teleconnections

There is a whole set of remote connections, so-called teleconnections, in the GWS: water flows along large rivers and water transfers between basins; atmospheric moisture transportation; and international virtual water flows through trade and FDI. Changes in the Earth System affect all the components of the GWS. Through feedbacks and teleconnections they cascade through the different water-dependent social–ecological systems up to the global scale, affecting their resilience. Slow drivers such as land degradation may push these systems closer to critical thresholds, so that they become less resilient to shocks such as droughts, leading to drastic responses such as food crises and large-scale migration.

Changes, interactions and feedbacks

IWRM and governance needs to adapt to the accelerating and cross-scale dynamics and turbulence in the Earth System and the GWS.

- Land-use change from natural ecosystems to agricultural land and other anthropogenic biomes, as well as agricultural intensification, alters green and blue water stocks, flows and productivity. It also redirects water flows between the different types of ecosystems and ecosystem services, with associated trade-offs in water productivity and the benefits derived, for instance, between upstream recharge and downstream discharge areas, or between different countries through moisture recycling.
- Climate change causes changes in temperature, water availability and crop water demand and productivity, as well as in glacier runoff and sea levels. Variability and disturbances are expected to increase. While it is unclear whether observed increases in water-related extremes such as droughts and floods can be attributed to climate change, their probability increases with climate change. Slow-changing climate variables that cause, for instance, aridification or loss of snow storage, could reduce resilience to shocks such as droughts or dry spells. Feedbacks from climate change through increasing plant water demand are likely to add more pressure to already water-stressed social–ecological systems.
- Water demand, withdrawals and consumptive use are growing rapidly in many parts of the world.

The resulting overexploitation is obvious in lower groundwater tables, such as in the Midwest of the United States, northern China, India and the Middle East, reduced river discharge, compromised environmental flows and basin closures.
- Virtual water flows linked to trade are generally driven not by water scarcity but by the comparative advantages of exporting countries in various production factors, as well as subsidies and trade policies. In the water-scarce countries of the Middle East and North Africa, however, virtual water imports correlate with water scarcity, and an increasing dependence of water-scarce countries on virtual water imports is projected.

Planetary and local freshwater boundaries

Earth System vulnerability has generated analysis of possible planetary boundaries, including a boundary for consumptive freshwater use which has been estimated as 4000 km^3/year. To be operationalised, this boundary needs to be downscaled, that is, made spatially explicit and disaggregated. It also needs to include green water and eventually will have to be complemented by location-specific sustainability criteria for maximum water exploitation. It has been suggested that basin closure poses a critical threshold, which, if crossed, might trigger a number of new processes and interactions between actors, production systems and land cover. Consistent and comprehensive indicators of water scarcity and food security are lacking. Global water assessments are still using simplified indicators, while more in-depth analyses are currently limited to individual case studies and comparisons.

Regional water scarcity exposure

A doubling of population pressure in the coming 40 years in some already water crowded Asian river basins may move them towards water-related critical thresholds, and will require adaptation and/or transformation in an effort to achieve water and food security. In some African river basins, population pressure is currently much lower, but population could even triple during the same period. Foreseeable food and water deficiencies in existing cropland are likely to make their food security import dependent unless major breakthroughs can be achieved in agricultural water productivity.

References

Alcamo, J., Flörke, M. and Märker, M. (2007). Future long-term changes in global water resources driven by socio-economic and climatic changes. *Hydrological Sciences Journal*, **52**, 247–275.

Alcamo, J. M., Vörösmarty, C. J., Naiman, R. J. *et al.* (2008). A grand challenge for freshwater research: understanding the global water system. *Environmental Research Letters*, **3**, 010202.

Amogu, O., Descroix, L., Yéro, K. S. *et al.* (2010). Increasing river flows in the Sahel? *Water*, **2**, 170–199.

Anseeuw, W., Wily, L. A., Cotula, L. and Taylor, M. (2012). *Land Rights and the Rush for Land: Findings of the Global Commercial Pressures on Land Research Project.* Rome: International Land Coalition.

Arthington, A. H., Naiman, R. J., McClain, M. E. and Nilsson, C. (2009). Preserving the biodiversity and ecological services of rivers: new challenges and research opportunities. *Freshwater Biology*, **55**, 1–16.

Australian Bureau of Agricultural and Resource Economics and Sciences (2011). Australian crop report, no. 157. Australian Bureau of Agricultural and Resource Economics and Sciences. Available at: http://adl.brs.gov.au/data/warehouse/pe_abares99001787/ACR11.1_Feb_REPORT.pdf.

Bansil, P. (2004). *Water Management in India.* New Delhi: Concept Publishing Company.

Berrisford, P., Dee, D., Fielding, K. *et al.* (2009). The ERA-interim archive. ERA report series: 1. European Centre for Medium-Range Weather Forecasts, Reading, UK.

Biswas, A. K. and Tortajada, C. (eds) (2004). *Appraising the Concept of Sustainable Development: Water Management and Related Environmental Challenges.* Oxford: Oxford University Press.

Boucher, O., Myhre, G. and Myhre, A. (2004). Direct human influence of irrigation on atmospheric water vapour and climate. *Climate Dynamics*, **22**, 597–603.

Brisbane Declaration. (2007). Environmental flows are essential for freshwater ecosystem health and human well-being. 10th International River Symposium, Brisbane.

Brown, C. and Lall, U. (2006). Water and economic development: the role of variability and a framework for resilience. *Natural Resources Forum*, **30**, 306–317.

Bruinsma, J. (2003). *World Agriculture: Towards 2015/2030: An FAO Perspective.* Rome: Food and Agriculture Organization.

Bureau of Meteorology. (2011). *An extremely wet end to 2010 leads to widespread flooding across eastern*

Australia. Special climate statements: 24. Bureau of Meteorology, Melbourne, Australia.

Cai, X., Molden, D., Mainuddin, M. *et al.* (2011). Producing more food with less water in a changing world: assessment of water productivity in 10 major river basins. *Water International*, **36**, 42–62.

Camberlin, P. (1997). Rainfall anomalies in the source region of the Nile and their connection with the Indian summer monsoon. *Journal of Climate*, **10**, 1380–1392.

Canada-Iraq Marshlands Initiative. (2010). *Managing for Change: The Present and Future State of the Marshes of Southern Iraq*. Canada–Iraq Marshlands Initiative, Washington DC.

Center for Research on the Epidemiology of Disaster. (2012). EM-DAT: The OFDA/CRED international disaster database version 12.07. Université Catholique De Louvain. Available at: http://emdat.be (accessed 6 June 2012).

Central Water Commission. (1998). Water and related statistics. Central Water Commission, New Delhi, India.

Chase, T., Knaff, J., Pielke, R. and Kalnay, E. (2003). Changes in global monsoon circulations since 1950. *Natural Hazards*, **29**, 229–254.

Cook, S., Fisher, M., Tiemann, T. and Vidal, A. (2011). Water, food and poverty: global- and basin-scale analysis. *Water International*, **36**, 1–16.

Costanza, R., d'Arge, R., de Groot, R. *et al.* (1997). The value of the world's ecosystem services and natural capital. *Nature*, **387**, 253–260.

Cumming, G. S. (2011). The resilience of big river basins. *Water International*, **36**, 63–95.

De Rosnay, P., Polcher, J., Laval, K. and Sabre, M. (2003). Integrated parameterization of irrigation in the land surface model ORCHIDEE. Validation over Indian Peninsula. *Geophysical Research Letters*, **30**, 1986.

Diamond, J. M., Ehrlich, P. R., Mooney, H. A. *et al.* (2006). *US Supreme Court: on writs of certiorari to the United States court of appeals for the sixth circuit, nos. 04–1034 and 04–1384, January 12, 2006.*

Döll, P. (2002). Impact of climate change and variability on irrigation requirements: a global perspective. *Climatic Change*, **54**, 269–293.

Döll, P. and Fiedler, K. (2008). Global-scale modeling of groundwater recharge. *Hydrology and Earth System Sciences*, **12**, 863–885.

Döll, P., Hoffmann-Dobrev, H., Portmann, F. T. *et al.* (2012). Impact of water withdrawals from groundwater and surface water on continental water storage variations. *Journal of Geodynamics*, **59–60**, 143–156.

Douglas, E., Beltrán-Przekurat, A., Niyogi, D., Pielke Sr, R. and Vörösmarty, C. (2009). The impact of agricultural intensification and irrigation on land–atmosphere interactions and Indian monsoon precipitation: a mesoscale modeling perspective. *Global and Planetary Change*, **67**, 117–128.

Douglas, E. M., Niyogi, D., Frolking, S. *et al.* (2006). Changes in moisture and energy fluxes due to agricultural land use and irrigation in the Indian monsoon belt. *Geophysical Research Letters*, **33**.

Ellis, E. C. and Ramankutty, N. (2008). Putting people in the map: anthropogenic biomes of the world. *Frontiers in Ecology and the Environment*, **6**, 439–447.

Eriyagama, N., Smakhtin, V. and Gamage, N. (2009). *Mapping Drought Patterns and Impacts: A Global Perspective*. Colombo, Sri Lanka: International Water Management Institute.

Fader, M., Gerten, D., Thammer, M. *et al.* (2011). Internal and external green-blue agricultural water footprints of nations, and related water and land savings through trade. *Hydrology and Earth System Sciences*, **15**, 1641–1660.

Falkenmark, M. and Molden, D. (2008). Wake up to realities of river basin closure. *International Journal of Water Resources Development*, **24**, 201–215.

Falkenmark, M. and Rockström, J. (2008). Building resilience to drought in desertification-prone savannas in sub-Saharan Africa: the water perspective. *Natural Resources Forum*, **32**, 93–102.

Fearnside, P. M. (2007). Brazil's Cuiabá-Santarém (BR-163) highway: the environmental cost of paving a soybean corridor through the Amazon. *Environmental Management*, **39**, 601–614.

Folke, C., Jansson, Å., Rockström, J. *et al.* (2011). Reconnecting to the biosphere. *AMBIO: A Journal of the Human Environment*, **40**, 719–738.

Food and Agriculture Organization (2008). *Global Change in Net Primary Productivity (1981–2003)*. Lada – Land Degradation Assessment in Drylands. Rome: Food and Agriculture Organization. Available at: http://www.fao.org/geonetwork/srv/en/metadata.show?id=37049 (accessed 17 October 2012).

Food and Agriculture Organization (2010). Principles for responsible agricultural investment that respects, rights, livelihoods and resources, discussion note prepared by FAO, IFAD, UNCTAD and the World Bank Group. Rome: Food and Agriculture Organization.

Foster, S., Hirata, R., Misra, S. and Garduno, H. (2010). Urban groundwater use policy – balancing the benefits and risks in developing nations. Strategic Overview Series: 3. World Bank, Washington DC.

Frederiksen, C., Frederiksen, J., Sisson, J. and Osbrough, S. (2011). Changes and projections in Australian winter rainfall and circulation: anthropogenic forcing and internal variability. *International Journal of Climate Change: Impacts and Responses*, **2**, 143–162.

Friis, C. and Reenberg, A. (2010). *Land grab in Africa: Emerging land system drivers in a teleconnected world.* GLP Report: 1. Global Land Project International Project Office, Copenhagen.

Gerten, D., Heinke, J., Hoff, H. *et al.* (2011). Global water availability and requirements for future food production. *Journal of Hydrometeorology*, **12**, 885–899.

Global Water System Project (2005). The Global Water System Project: science framework and implementation activities. ESSP Report: 3. Global Water System Project, Bonn, Germany.

Gordon, L. J., Steffen, W., Jönsson, B. F. *et al.* (2005). Human modification of global water vapor flows from the land surface. *Proceedings of the National Academy of Sciences of the United States of America*, **102**, 7612–7617.

Grey, D. and Sadoff, C. W. (2007). Sink or swim? Water security for growth and development. *Water Policy*, **9**, 545.

Grübler, A., O'Neill, B., Riahi, K. *et al.* (2007). Regional, national, and spatially explicit scenarios of demographic and economic change based on SRES. *Technological Forecasting and Social Change*, **74**, 980–1029.

Haddeland, I., Clark, D. B., Franssen, W. *et al.* (2011). Multimodel estimate of the global terrestrial water balance: setup and first results. *Journal of Hydrometeorology*, **12**, 869–884.

Hirji, R. and Davis, R. (2009). Environmental flows in water resources policies, plans, and projects: findings and recommendations. Environment Department Papers, Natural Resource Management Series: 117. The World Bank, Washington, DC.

Hoekstra, A. Y. (2009). Human appropriation of natural capital: a comparison of ecological footprint and water footprint analysis. *Ecological Economics*, **68**, 1963–1974.

Hoekstra, A. Y. and Chapagain, A. K. (2009). Globalization of water: sharing the planet's freshwater resources. *The Geographical Journal*, **175**, 85–86.

Hoff, H. (2011). Understanding the nexus: background paper. Bonn 2011 Nexus Conference. Stockholm Environment Institute, Bonn, Germany.

Hoff, H., Döll, P., Fader, M. *et al.* (2013). Water footprints of cities, indicators for sustainable consumption and production. *Hydrolology and Earth System Sciences*, **10**, 2601–2639.

Hoff, H., Falkenmark, M., Gerten, D. *et al.* (2010). Greening the global water system. *Journal of Hydrology*, **384**, 177–186.

Hoff, H., Gerten, D. and Waha, K. (2012). Green and blue water in Africa: how foreign direct investment can support sustainable intensification. In *Handbook of Land and Water Grabs in Africa: Foreign Direct Investment and Food and Water Security*, ed. Allan, J. A., Keulertz, M., Sojamo, S. and Warner, J. London: Routledge, pp. 359–375.

Instituto Nacional de Pesquisas Espaciais (2012). Taxas anuais do desmatamento – 1988 até 2011. Available at: http://www.obt.inpe.br/prodes/prodes_1988_2011.htm.

Integrated Regional Information Networks (2009). Kenya: selling the cows to feed the children. Available at: http://www.irinnews.org/Report/85806/KENYA-Selling-the-cows-to-feed-the-children.

Intergovernmental Panel on Climate Change (2011). Special report managing the risks of extreme events and disasters to advance climate change adaptation (SREX), summary for policy makers. Intergovernmental Panel on Climate Change, Stanford, CA.

International Energy Agency (2009). *World Energy Outlook 2009.* Paris: International Energy Agency.

International Union for Conservation of Nature (2004). The Lesotho highlands water project: environmental flow allocations in an international river. International Union for Conservation of Nature, Gland, Switzerland. Available at: http://cmsdata.iucn.org/downloads/lesotho.pdf.

Islam, M. S., Oki, T., Kanae, S. *et al.* (2007). A grid-based assessment of global water scarcity including virtual water trading. *Water Resources Management*, **21**, 19–33.

Issar, A. S. (2010). Progressive development by greening the deserts, to mitigate global warming and provide new land and income resources. In *Progressive Development*, ed. Issar, A. S. Heidelberg, Germany: Springer, pp. 37–42.

Janicot, S., Mounier, F., Hall, N. M. J. *et al.* (2009). Dynamics of the West African monsoon. Part IV: analysis of 25–90-day variability of convection and the role of the Indian monsoon. *Journal of Climate*, **22**, 1541–1565.

Jha, S. (2001). Rainwater harvesting in India. Press Information Bureau Government of India. Available at: http://pib.nic.in/feature/feyr2001/fsep2001/f060920011.html (accessed 11 October 2010).

Juusola, H. (2010). The internal dimensions of water security: the drought crisis in Northeastern Syria. In *Managing Blue Gold: New Perspectives on Water Security in the Levantine Middle East FIIA Report*

2010:25, ed. Luomi, M. Helsinki: Ulkopoliittinen instituutti, pp. 21–35.

Kendy, E., Molden, D., Steenhuis, T. S., Liu, C. and Wang, J. (2003). Policies drain the North China Plain: agricultural policy and groundwater depletion in Luancheng County 1949–2000. International Water Management Institute Research Report 71. International Water Management Institute, Battaramulla, Sri Lanka. Available at: http://www.iwmi.cgiar.org/Publications/IWMI_Research_Reports/PDF/pub071/Report71.pdf.

Kendy, E., Wang, J., Molden, D. J. *et al.* (2007). Can urbanization solve inter-sector water conflicts? Insight from a case study in Hebei Province, North China Plain. *Water Policy*, **9**, 75–93.

Kendy, E., Zhang, Y., Liu, C., Wang, J. and Steenhuis, T. (2004). Groundwater recharge from irrigated cropland in the North China Plain: case study of Luancheng County, Hebei Province, 1949–2000. *Hydrological Processes*, **18**, 2289–2302.

Kerr, J. (2007). Watershed management: lessons from common property theory. *International Journal of the Commons*, **1**, 89–110.

Keys, P. W., van der Ent, R. J., Gordon, L. J. *et al.* (2012). Analyzing precipitationsheds to understand the vulnerability of rainfall dependent regions. *Biogeosciences*, **9**, 733–746.

King, J., Brown, C. and Sabet, H. (2003). A scenario-based holistic approach to environmental flow assessments for rivers. *River Research and Applications*, **19**, 619–639.

King, J. and Louw, D. (1998). Instream flow assessments for regulated rivers in South Africa using the building block methodology. *Aquatic Ecosystem Health & Management*, **1**, 109–124.

Kumar, P. (ed.) (2010). *The Economics of Ecosystems and Biodiversity: Ecological and Economic Foundations.* Oxford: Routledge.

Kummu, M., Ward, P. J., de Moel, H. and Varis, O. (2010). Is physical water scarcity a new phenomenon? Global assessment of water shortage over the last two millennia. *Environmental Research Letters*, **5**, 034006.

Lake Victoria Basin Commission and World Wildlife Fund for Nature. (2010). Assessing reserve flows for the Mara River. Lake Victoria Basin Commission of the East African Community and WWF Eastern & Southern Africa Regional Programme Office. Available at: http://wwf.panda.org/who_we_are/wwf_offices/eastern_southern_africa/publications/?193036/Assessing-Reserve-Flows-for-the-Mara-River.

Le Quesne, T., Kendy, E. and Weston, D. (2010). The implementation challenge: taking stock of government policies to protect and restore environmental flows. Nature Conservancy and World Wildlife Fund for Nature. Available at: http://conserveonline.org/workspaces/eloha/documents/wwf-tnc-e-flow-policies-report/view.html.

Liu, J., You, L., Amini, M., *et al.* (2010). A high-resolution assessment on global nitrogen flows in cropland. *Proceedings of the National Academy of Sciences*, **107**, 8035–8040.

Liu, J. G. and Yang, H. (2010). Spatially explicit assessment of global consumptive water uses in cropland: green and blue water. *Journal of Hydrology*, **384**, 187–197.

Lohar, D. and Pal, B. (1995). The effect of irrigation on premonsoon season precipitation over South West Bengal, India. *Journal of Climate*, **8**, 2567–2570.

Lundqvist, J. and Gleick, P. (1997). *Sustaining our Waters into the 21st Century: Comprehensive Assessment of the Freshwater Resources of the World.* Stockholm: Stockholm Environment Institute.

Ma, J., Hoekstra, A. Y., Wang, H., Chapagain, A. K. and Wang, D. (2006). Virtual versus real water transfers within China. *Philosophical Transactions of the Royal Society B: Biological Sciences*, **361**, 835–842.

Malhi, Y., Aragão, L. E. O. C., Galbraith, D. *et al.* (2009). Exploring the likelihood and mechanism of a climate-change-induced dieback of the Amazon rainforest. *Proceedings of the National Academy of Sciences*, **106**, 20610–20615.

Marengo, J., Nobre, C. A., Betts, R. A. *et al.* (2009). Global warming and climate change in Amazonia: climate-vegetation feedback and impacts on water resources. *Geophysical Monograph Series*, **186**, 273–292.

Marshall, C. H., Pielke Sr, R. A. and Steyaert, L. T. (2004a). Has the conversion of natural wetlands to agricultural land increased the incidence and severity of damaging freezes in south Florida? *Monthly Weather Review*, **132**, 2243–2258.

Marshall, C. H., Pielke, Sr, R. A., Steyaert, L. T. and Willard, D. A. (2004b). The impact of anthropogenic land-cover change on the Florida peninsula sea breezes and warm season sensible weather. *Monthly Weather Review*, **132**, 28–52.

Mekonnen, M. M. and Hoekstra, A. Y. (2010). The green, blue and grey water footprint of farm animals and animal products. Value of Water Research Report Series: 48. UNESCO-IHE Institute for Water Education. Available at: http://www.unesco-ihe.org/Value-of-Water-Research-Report-Series/Research-Papers.

Millán, M. M. (2010). *Sequía en el Mediterréneo e inundaciones en el reino unido y Centroeuropa.* Almeria, Spain: Fundación Cajamar.

Millennium Ecosystem Assessment. (2005). *Ecosystems and Human Well-being: Synthesis.* Washington DC: Island Press.

Multi-Resolution Land Characterization Consortium. (2007). National land cover database 2001. Multi-Resolution Land Characterization Consortium, Sioux Falls, SD. Available at: http://www.mrlc.gov/nlcd2001.php.

Nelson, A. (2008). Estimated travel time to the nearest city of 50,000 or more people in year 2000. Global Environment Monitoring Unit, Joint Research Centre of the European Commission. Available at: http://bioval.jrc.ec.europa.eu/products/gam/.

Nicholls, N. (2010). Local and remote causes of the southern Australian autumn-winter rainfall decline, 1958–2007. *Climate Dynamics*, **34**, 835–845.

Nicolis, G. and Prigogine, I. (1989). *Exploring Complexity.* New York: Freeman & Co.

Nikoli, R. (2011). Moisture recycling and the effect of land-use change. MSc thesis, Delft University of Technology, The Netherlands.

Nobre, C. A. and Borma, L. D. S. (2009). 'Tipping points' for the Amazon forest. *Current Opinion in Environmental Sustainability*, **1**, 28–36.

Odum, H. T. (1983). *Systems Ecology: An Introduction.* New York: Wiley.

Olsson, P. and Galaz, V. (2009). Transitions to adaptive approaches to water management and governance in Sweden. In *Water Policy Entrepreneurs. A Research Companion to Water Transitions around the Globe*, ed. Huitema, D. & Sander, M. Cheltenham, UK: Edward Elgar Publishing, pp. 304–324.

Ornstein, L., Aleinov, I. and Rind, D. (2009). Irrigated afforestation of the Sahara and Australian Outback to end global warming. *Climatic Change*, **97**, 409–437.

Ozdogan, M., Rodell, M., Beaudoing, H. K. and Toll, D. L. (2010). Simulating the effects of irrigation over the United States in a land surface model based on satellite-derived agricultural data. *Journal of Hydrometeorology*, **11**, 171–184.

Parry, M. L., Canziani, O. F., Palutikof, J. P., van der Linden, P. J. and Hanson, C. E. (2007). *Contribution of Working Group II to the Fourth Assessment Report of the Intergovernmental Panel on Climate Change.* Cambridge: Intergovernmental Panel on Climate Change.

Pfister, S. and Hellweg, S. (2009). The water 'shoesize' vs. footprint of bioenergy. *Proceedings of the National Academy of Sciences*, **106**, E93–E94.

Pielke, R., Niyogi, D., Chase, T. and Eastman, J. (2003). A new perspective on climate change and variability: A focus on India. *Proceedings of the Indian National Science Academy – Part A: Physical Sciences*, **69**, 585–602.

Poff, N. L., Richter, B. D., Arthington, A. H. *et al.* (2009). The ecological limits of hydrologic alteration (ELOHA): a new framework for developing regional environmental flow standards. *Freshwater Biology*, **55**, 147–170.

Ramankutty, N., Evan, A. T., Monfreda, C. and Foley, J. A. (2008). Farming the planet: 1. Geographic distribution of global agricultural lands in the year 2000. *Global Biogeochemical Cycles*, **22**, GB10003.

Ramirez-Vallejo, J. and Rogers, P. (2004). Virtual water flows and trade liberalization. *Water Science and Technology*, **49**, 25.

Reij, C., Tappan, G. and Smale, M. (2009). Agroenvironmental transformation in the Sahel. IFPRI Discussion Paper: 914. Washington, DC: International Food Policy Research Institute.

Richter, B. D., Baumgartner, J. V., Powell, J. and Braun, D. P. (1996). A method for assessing hydrologic alteration within ecosystems. *Conservation Biology*, **10**, 1163–1174.

Ripl, W. (1992). Management of water cycle: an approach to urban ecology. *Water Quality Research Journal of Canada*, **27**, 221–237.

Ripl, W. (1995). Management of water cycle and energy flow for ecosystem control: the energy-transport-reaction (ETR) model. *Ecological Modelling*, **78**, 61–76.

Ripl, W. (2003). Water: the bloodstream of the biosphere. *Philosophical Transactions of the Royal Society of London. Series B: Biological Sciences*, **358**, 1921–1934.

Ripl, W. and Hildmann, C. (2000). Dissolved load transported by rivers as an indicator of landscape sustainability. *Ecological Engineering*, **14**, 373–387.

Rockström, J., Falkenmark, M., Lannerstad, M. and Karlberg, L. (2012). The planetary water drama: dual task of feeding humanity and curbing climate change. *Geophysical Research Letters*, **39**, L15401.

Rockström, J., Gordon, L., Folke, C., Falkenmark, M. and Engwall, M. (1999). Linkages among water vapor flows, food production, and terrestrial ecosystem services. *Conservation Ecology*, **3**, 5.

Rockström, J., Steffen, W., Noone, K. *et al.* (2009). A safe operating space for humanity. *Nature*, **461**, 472–475.

Rosegrant, M. W. (2008). *Biofuels and Grain Prices: Impacts and Policy Responses.* Washington DC: International Food Policy Research Institute.

Rost, S., Gerten, D., Bondeau, A. *et al.* (2008). Agricultural green and blue water consumption and its influence on the global water system. *Water Resources Research*, **44**.

Roy, S. B., Hurtt, G. C., Weaver, C. P. and Pacala, S. W. (2003). Impact of historical land cover change on the July climate of the United States. *Journal of Geophysical Research*, **108**, 4793.

Rulli, M. C., Saviori, A. and D'Odorico, P. (2013). Global land and water grabbing. *Proceedings of the National Academy of Sciences*, **110**, 892–897.

Savenije, H. H. G. (1995). New definitions for moisture recycling and the relationship with land-use changes in the Sahel. *Journal of Hydrology*, **167**, 57–78.

Scanlon, B., Reedy, R., Gates, J. and Gowda, P. (2010). Impact of agroecosystems on groundwater resources in the Central High Plains, USA. *Agriculture, Ecosystems & Environment*, **139**, 700–713.

Scanlon, B. R., Faunt, C. C., Longuevergne, L. *et al.* (2012). Groundwater depletion and sustainability of irrigation in the US High Plains and Central Valley. *Proceedings of the National Academy of Sciences*, **109**, 9320–9325.

Shah, T., Burke, J. and Villholth, K. (2007). Groundwater: a global assessment of scale and significance. In *Water for Food, Water for Life: A Comprehensive Assessment of Water Management in Agriculture*, ed. Molden, D. London: Earthscan, pp. 395–423.

Shah, T., Singh, O. and Mukherji, A. (2006). Some aspects of South Asia's groundwater irrigation economy: analyses from a survey in India, Pakistan, Nepal Terai and Bangladesh. *Hydrogeology Journal*, **14**, 286–309.

Shah, T. M., Molden, D., Sakthivadivel, R. and Seckler, D. (2000). *The Global Groundwater Situation: Overview of Opportunities and Challenges*. Colombo, Sri Lanka: International Water Management Institute.

Shiklomanov, I. A. (2000). Appraisal and assessment of world water resources. *Water International*, **25**, 11–32.

Siebert, S., Burke, J., Faures, J. M. *et al.* (2010). Groundwater use for irrigation: a global inventory. *Hydrology and Earth System Sciences*, **14**, 1863–1880.

Siebert, S. and Döll, P. (2010). Quantifying blue and green virtual water contents in global crop production as well as potential production losses without irrigation. *Journal of Hydrology*, **384**, 198–217.

Skinner, J. and Cotula, L. (2011). Are land deals driving 'water grabs'? IIED Briefing Papers. International Institute for Environment and Development, London. Available at: http://pubs.iied.org/17102IIED.html.

Smakhtin, V., Revenga, C. and Döll, P. (2004). A pilot global assessment of environmental water requirements and scarcity. *Water International*, **29**, 307–317.

Stalnaker, C., Lamb, B. L., Henriksen, J., Bovee, K. and Bartholow, J. (1995). The Instream Flow Incremental Methodology: a primer for IFIM. US Geological Survey Biological Report: 29. Washington, DC: US Geological Survey.

Steffen, W., Love, G. and Whetton, P. (2006). Approaches to defining dangerous climate change: a southern hemisphere perspective. In *Avoiding Dangerous Climate Change*, ed. Schellnhuber, H. J., Cramer, W., Nakicenovic, N., Wigley, T. and Yohe, G. Cambridge: Cambridge University Press, pp. 205–214.

Steffen, W., Sims, J., Walcott, J. and Laughlin, G. (2011). Australian agriculture: coping with dangerous climate change. *Regional Environmental Change*, **11**, 205–214.

Sullivan, C. A. and Meigh, J. (2007). Integration of the biophysical and social sciences using an indicator approach: addressing water problems at different scales. *Water Resources Management*, **21**, 111–128.

Tennant, D. L. (1976). *Instream Flow Regimes for Fish, Wildlife, Recreation and Related Environmental Resources*. Washington DC: US Fish and Wildlife Service.

Tharme, R. E. (2003). A global perspective on environmental flow assessment: emerging trends in the development and application of environmental flow methodologies for rivers. *River Research and Applications*, **19**, 397–441.

Timbal, B., Arblaster, J., Braganza, K. *et al.* (2010). Understanding the anthropogenic nature of the observed rainfall decline across South Eastern Australia. CAWCR Technical Report: 026. Centre for Australian Weather and Climate Research, Melbourne, Australia. Available at: http://www.cawcr.gov.au/publications/technicalreports/CTR_026.pdf.

United Nations Environment Programme (2011). *Livelihood security – climate change, migration and conflict in the Sahel*. United Nations Environment Programme, Nairobi, Kenya. Available at: http://www.unep.org/disastersandconflicts/Introduction/EnvironmentalCooperationforPeacebuilding/EnvironmentalDiplomacy/SahelReport/tabid/55812/Default.aspx.

van der Ent, R. J. and Savenije, H. H. G. (2011). Length and time scales of atmospheric moisture recycling. *Atmospheric Chemistry and Physics*, **11**, 1853–1863.

van der Ent, R. J., Savenije, H. H. G., Schaefli, B. and Steele-Dunne, S. C. (2010). Origin and fate of atmospheric moisture over continents. *Water Resources Research*, **46**, W09525.

Verma, S., Kampman, D. A., van der Zaag, P. and Hoekstra, A. Y. (2009). Going against the flow: a critical analysis of inter-state virtual water trade in the context of India's national river linking program. *Physics and Chemistry of the Earth, Parts A/B/C*, **34**, 261–269.

von Braun, J. (2009). Addressing the food crisis: governance, market functioning, and investment in public goods. *Food Security*, **1**, 9–15.

von Braun, J. and Meinzen-Dick, R. S. (2009). *'Land Grabbing' by Foreign Investors in Developing Countries: Risks and Opportunities*. Washington DC: International Food Policy Research Institute.

Vörösmarty, C. J., Douglas, E. M., Green, P. A. and Revenga, C. (2005). Geospatial indicators of emerging water stress: an application to Africa. *AMBIO*, **34**, 230–236.

Vörösmarty, C. J., Green, P., Salisbury, J. and Lammers, R. B. (2000). Global water resources: vulnerability from climate change and population growth. *Science*, **289**, 284.

Vörösmarty, C. J., McIntyre, P., Gessner, M. O. *et al.* (2010). Global threats to human water security and river biodiversity. *Nature*, **467**, 555–561.

Water Corporation (2012). *Yearly Streamflow for Major Surface Water Sources*. Water Corporation, Leederville, Australia. Available at: http://www.watercorporation.com.au/D/dams_streamflow.cfm (accessed 30 March 2012).

Wichelns, D. (2010). Virtual water and water footprints: policy relevant or simply descriptive? *International Journal of Water Resources Development*, **26**, 689–695.

World Trade Organization (2008). *International Trade Statistics 2008*. World Trade Organization, Geneva. Available at: http://www.wto.org/english/res_e/statis_e/its2008_e/its08_toc_e.htm.

Yang, H., Wang, L. and Zehnder, A. J. B. (2007). Water scarcity and food trade in the Southern and Eastern Mediterranean countries. *Food Policy*, **32**, 585–605.

Zickfeld, K., Knopf, B., Petoukhov, V. and Schellnhuber, H. (2005). Is the Indian summer monsoon stable against global change? *Geophysical Research Letters*, **32**, L15707.

Part

III

Food production globally: in hotspot regions and in the landscape

Hillside with patchwork cultivation in Rwanda, where every square metre of land is under pressure to be exploited for agriculture. To sustain a growing population in the world, we need solutions to manage land and water efficiently.

Chapter

5 Food production: a mega water challenge

This chapter analyses the pressure on the Earth System caused by escalating agricultural production from the perspective of efforts to feed a growing human population. The focus is on the situation by 2050. Particular attention is paid to improvements in water productivity and efforts to close the currently large yield gap in the developing world. Presented estimates reveal what can be achieved on what is currently cropland. What emerges is a carrying capacity overshoot for more than half the world's population, which must be compensated for through virtual water transfers in food traded from water surplus countries. The chapter analyses the ability of the agricultural system and its support systems to cope with shocks and change in the Anthropocene era, and the adaptability and social-ecological resilience required to deal with a more turbulent world.

5.1 Food demand trajectories and water preconditions

5.1.1 Hunger alleviation and population increase: two strong driving forces at work

Until the beginning of the twentieth century, increasing food production to meet the needs of a growing world population was essentially a case of continuing the expansion of the area of cultivated land. As far back as the nineteenth century there was a growing pessimism about the possibility of feeding a constantly growing population, which was put into words by Thomas Malthus (1766–1834). During the twentieth century the global population increased by more than 350%, from 1.65 billion to more than 6 billion (UNDP, 2004).

After World War II, rapid population increases were not matched by an equal increase in food production in many of the newly independent developing countries. As a result, by the mid-1960s many of these states were dependent on massive food aid from the industrialised world. In 1967, a report by the US President's Science Advisory Committee stated that 'the scale, severity and duration of the world food problem are so great that a massive, long-range, innovative effort unprecedented in human history will be required to master it' (IFPRI, 2002).

In response to the hunger challenge, different public and private sector agencies and organisations, such as the Rockefeller and Ford Foundations, began in the 1940s and 1950s to invest in an international agricultural research system in developing countries. Initially, they focused primarily on developing high-yield varieties of rice and wheat. In the late 1960s, during the so-called Green Revolution, the use of agricultural science and modern techniques of food production resulted in impressive increases in yields for rice and wheat in Asia and Latin America (Comprehensive Assessment, 2007; IFPRI, 2002).

Even if many countries, such as India, managed at the national level to move away from recurrent famines and dependence on food aid and towards food self-sufficiency, a large number of undernourished people continued to exist across the developing world. Since the 1970s the absolute number of undernourished people has oscillated around 900 million. Despite a doubling of the global population, however, the absolute number of undernourished people has remained fairly constant. In other words, global food production has been massive and has been able to keep pace with the global population increase since the 1970s. In addition, the global food system has been able to cater for higher per capita levels of demand for food (see Figures 2.1 and 2.3 in Chapter 2).

The food and economic crises of 2007–09 led to economic and social turbulence in many developing

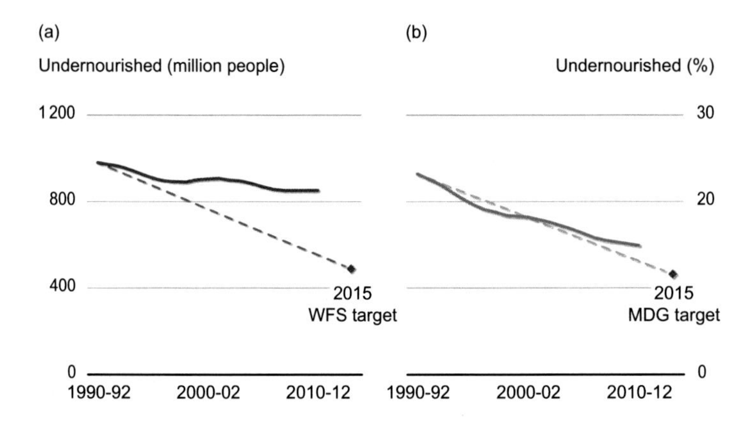

Figure 5.1 Undernourishment in the developing world 1990–92 to 2010–12, and the number in relation to the World Food Summit (1996) target to halve the number between 1990–92 and 2015 (a), and the proportion in relation to the Millennium Development Goal 1, target 1C, to halve the proportion between 1990 and 2015 (b) (FAO, 2012).

countries. Early estimates indicated that the crises had resulted in a dramatic increase in the number of undernourished people, which it was assumed peaked at 1020 million in 2009 (FAO, 2011). However, new data sources and a methodological revision of the indicator for the prevalence of undernourishment mean that new figures for 1990–2012 show a much-improved picture (FAO, 2012). Since 1990, progress with reducing hunger has been more pronounced than was earlier believed, and the price spikes of 2007–08 only slowed down and levelled off earlier progress (Figure 5.1a). Globally, in 2010–12 the number of undernourished people was estimated at 868 million, 852 million of whom were living in developing countries. About 234 million (26.8%) were in sub-Saharan Africa, but the greatest number (304 million) live in South Asia, with almost 217 million in India alone (FAO, 2012).

The global community has set several goals over the years to reduce global hunger. The first World Food Conference, in Rome in 1974, put the global problem of food production and consumption into focus. It declared that 'every man, woman and child has the inalienable right to be free from hunger and malnutrition in order to develop their physical and mental faculties'. The Rome Declaration on World Food Security, from the World Food Summit (WFS) in 1996, reaffirmed the right of everyone to have access to safe and nutritious food, and sets a target to reduce the number of undernourished people by half between 1990–92 and 2015. The United Nations Millennium Declaration of 2000 committed nations to a new global partnership to reduce extreme poverty and set out a series of time-bound targets to be reached by 2015.

The first Millennium Development Goal reformulated the WFS goal, aiming instead to halve the *proportion* of people suffering from hunger between 1990 and 2015 (MDG 1 target 1C). The worsening global food security situation in 2007–09 led to yet another World Summit on Food Security in late 2009. The meeting pledged its renewed commitment to the goals from 1996 and 2000 and stressed the need to accelerate efforts to reverse recent negative trends. The declaration emphasises the need for increased investment in an agricultural sector that has been neglected for the past 20 years, and underlines that agriculture, directly or indirectly, provides livelihoods for 70% of the world's poor. The meeting also adopted sustainability as an aim and agreed to proactively face the challenges posed by climate change to food security.

Figure 5.1a and b show clearly the difference between the WFS target and Target 1C of MDG 1. The number of undernourished people has decreased by 130 million, equivalent to a 13% reduction, but the proportion has fallen by almost 36%, from 23.2% to 14.9%. In other words, continuing population increases in the past decade mean that the MDG target is almost on track, while the number suffering from hunger has remained comparatively static (FAO, 2012).

To be able to improve the food situation for the undernourished it will be necessary for global agricultural production to keep pace with and meet the needs of the expected population increase in the coming decades. In absolute terms, the task of catering for about 2.3 billion extra people (UN DESA, 2011) will be many times larger, in terms of the water for food challenge, than improving the diets of the almost 1 billion undernourished (Rockström *et al.*, 2005). The medium UN population projections forecast a global population increase of more than 2.4 billion, from 6.9 billion in 2010 to 9.3 billion by 2050. In the second half of the century, the increase is projected to be quite modest and the world population is estimated to reach 10.1 billion by 2100. In comparison, the high variant UN projection would result in a population of 10.6 billion by 2050, the low variant is estimated at 8.1 billion and, with a constant fertility rate, the projection is 10.9 billion (UN DESA, 2011). This book uses the medium population projection for its analyses, according to which the population increase in the period 2010–50 will be largest in the early decades and then level out after 2030.

In 2009, for the first time, the majority of the global population was thought to be living in urban areas. Urban growth to 2050 will include the total expected global population increase plus the migration of about 550 million people from rural to urban areas. It is only in the least developed countries, which are mainly in sub-Saharan Africa, that the rural population is expected to increase – by about 30% from 600 million to almost 800 million (UN DESA, 2010). India is expected to be the world's most populous country by 2050, with a population of 1.7 billion – an increase of almost 470 million on 2010. In contrast, China is expected to reduce its population by about 46 million in the coming 40 years. After reaching a peak of around 1.4 billion in 2026, the population is expected to fall to 1.3 billion by 2050 (UN DESA, 2011).

According to FAO estimates, agricultural production will have to increase by 70% by 2050 to cope with the projected increase in the world population as well as projected rises in average per capita food consumption. Nearly 100% of this increase is expected to take place in developing countries. Compared with production in 2005–07, this projection would mean an additional one billion tonnes of cereals and 200 million tonnes of meat being produced annually by 2050 (Bruinsma, 2009). Other estimates indicate an increase in food demand of 40–100% (Foley *et al.*, 2011; Tilman *et al.*, 2011).

Figure 5.2 a and b show that the current prevalence of hunger and the future needs of the expected 2.3 billion extra people are correlated to the same countries, many of which are in the least developed countries group (FAO, 2013b). In many of these countries, agricultural production occurs in the often water-scarce savannah and steppe zones. The fact that water is a limiting factor in many of these countries highlights the importance of understanding the links between water availability, water management and food production. The water used for the cultivation of food crops, feed crops and fodder is thus also directly linked to nutrition. The availability and accessibility of water for agricultural use are therefore also directly linked to **food security**, which according to the 1996 WFS is 'a situation that exists when all people, at all times, have physical, social and economic access to sufficient, safe and nutritious food that meets their dietary needs and food preferences for an active and healthy life'. The concept of food security builds on three dimensions: **availability, access** and **utilisation**. While food **availability** refers to the national or international level of supply, including food production, stock levels and net trade, **access** refers to meeting demand, i.e. inter- and intra-household food distribution. Effective **utilisation** defines the nutritional status of individuals and relates to whether individuals and households make good use of the food to which they have access, e.g. whether food can be prepared in sanitary conditions and if people's health status is such that essential macro- and micronutrients can be metabolised and absorbed.

While water for domestic use is a prerequisite for optimal utilisation, water for agricultural use is a requirement for food availability. For example, in sub-Saharan Africa many smallholders are subsistence farmers who live at the mercy of erratic rainfall and seasonal river flows (see Chapter 6). The volatile and unpredictable availability of green and blue water directly affects the fourth component related to food security – **stability over time**. The low dependability of water for agriculture means that many people are faced with chronic food insecurity and locked into a nutritional poverty trap. It is more often persistent conditions, such as long-term or recurrent water scarcity, that determine the food security situation. Worldwide, in 2004 more than 90% of deaths from hunger were associated with chronic conditions, and only 8% were related to humanitarian emergencies (Barrett, 2010).

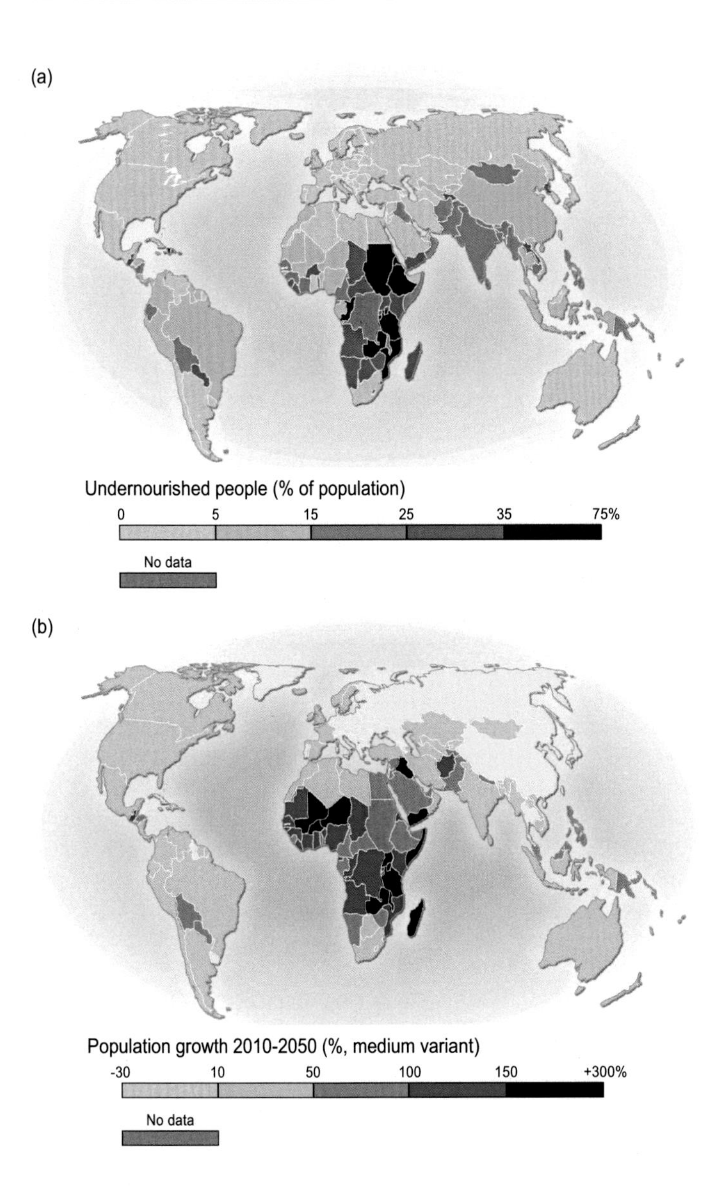

Figure 5.2 Global prevalence of undernourishment 2010–12 (a), and expected global population change 2010–50 (b) (FAO, 2012; UN DESA, 2011).

5.1.2 Dietary improvements increase per capita food requirements

Urbanisation and higher disposable incomes continue to drive the trend for a global convergence of diets, e.g. increasing poultry consumption in most countries, and similar urban eating habits linked to fast and convenience foods. Relatively monotonous diets based on indigenous staple grains or starchy roots, locally grown vegetables, other vegetables and fruits, and limited amounts of foods of animal origin are being replaced by more varied diets that include more pre-processed foods, more foods of animal origin, more added sugar and fat, and often more alcohol (FAO, 2013b; Steinfeld *et al.*, 2006). Increased economic development means that people can afford to move up the food chain. This gradual shift from prevalent undernourishment to richer and more varied diets, often leading to over-nutrition, has been termed the 'nutrition transition' (Popkin *et al.*, 2001). Urbanisation is a key driver as urban migrants lose their connection to traditional, local agricultural production, on their own or their neighbours' fields, and become consumers in the urban, and thus also often the international, food commodity market.

Increased intakes of energy-dense foods, which are high in sugars, fats and salt, and often low in vitamins, minerals and other micronutrients, mean that the number of overweight people worldwide has nearly doubled since 1980. In 2008, more than 1.5 billion adults globally were overweight, a prevalence almost double that of undernourishment. In addition, about 30% of those overweight were obese: 300 million women and 200 million men. Being overweight or obese increases the risk of a number of non-communicable diseases, such as cardiovascular diseases, diabetes and certain cancers. Overweight or obesity leads to at least 2.8 million adult deaths globally each year. About 65% of the world's population lives in countries where being overweight or obese kills more people than undernourishment, including all high-income and most middle-income countries (GHO, 2013; WHO, 2011).

The prevalence of being overweight is highest in upper-middle-income countries, but obesity is now on the rise in low- and middle-income countries, particularly in urban settings. Of the 43 million children under the age of 5 years estimated to be overweight in 2010, more than 80% lived in developing countries, and the fastest rise is in the lower-middle-income group. Thus, many low- and middle-income countries now face a 'double burden' of dealing with diseases linked to undernourishment and those linked to being overweight (GHO, 2013; WHO, 2011).

Global trends in changing food habits contain two major components related to food intake. The first factor is the consumption of more calories per person. From 1961 to 2009, the global average food calorie supply per person per day increased by 29% to more than 2830 kcal. There are clear differences between regions, with Latin America, the Middle East and North Africa region, South East Asia and northeast Asia showing steep increases, while sub-Saharan Africa and South Asia remain almost constant (Figure 2.3 in Chapter 2). National data show much more dramatic changes: an increase of more than 200%, from 1400 to more than 3000 kcal/capita per day

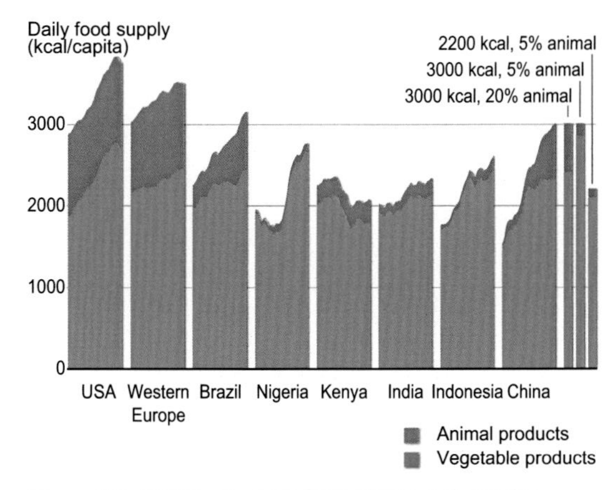

Figure 5.3 Average per capita food calorie supply per day, 1961–2009, divided into vegetal and animal calories for eight countries and regions, and with three standard food supply levels for comparison (FAO, 2013b).

makes China one of the most extreme examples of the past 40 years (Figure 5.3) (FAO, 2013b).

Following the global trend for increases in per capita food supply, many developing countries, such as Brazil and China, are now approaching the average calorie supply level of many developed countries, of almost 3700 kcal/capita per day in the United States and more than 3500 kcal/capita per day in Western Europe. Countries such as India and Kenya, which have had higher population increases and less successful agricultural production improvements than e.g. China, still lag behind at around 2100–2300 kcal/capita per day (Figure 5.3).

The second crucial component relates to the increase in the demand for and consumption of animal products. Between 1961 and 2009, the average global calorie supply of food originating from animal sources increased by 48% to more than 500 kcal/capita per day. While about 30% of the average food supply of calories in the United States and Western Europe consists of animal products (in Denmark and France it is close to 40%), the corresponding percentage is about 20% in China and Brazil, and less than 10% in e.g. India, Indonesia and Nigeria (Figure 5.3).

The dramatic increase in China can be explained by an increase in the food supply of pig meat between 1961 and 2009 from 2 kg/person per year to 37 kg, poultry from 1 kg/person per year to 13 kg and eggs from 2 kg/person per year to 19 kg – a total

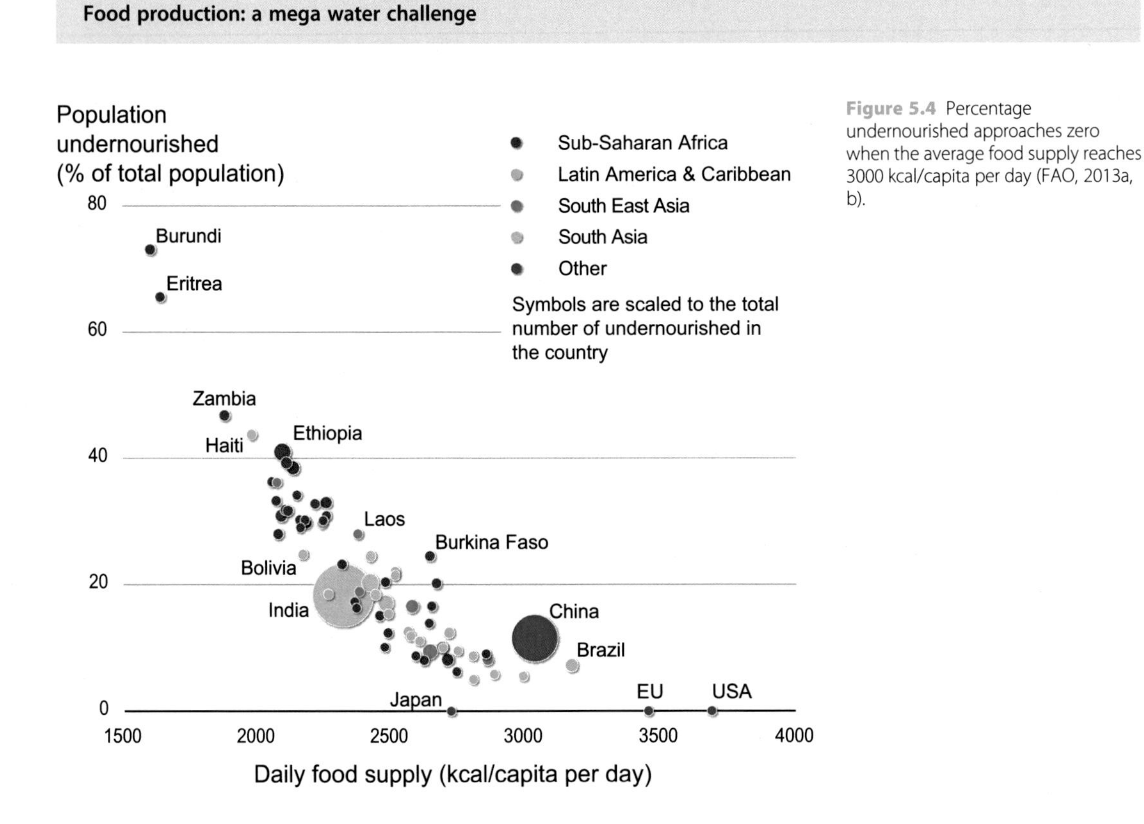

Figure 5.4 Percentage undernourished approaches zero when the average food supply reaches 3000 kcal/capita per day (FAO, 2013a, b).

increase from 5 to 69 kg/person per year for these three food items. The more modest increase in calories of animal origin over the same period in India, where much of the population is vegetarian, can mainly be explained by an almost 50% increase in per capita milk supply (excluding butter) from 38 to 72 kg/person per year. India's production of whole milk increased from 20 million tonnes per year in the 1960s to more than 110 million tonnes in 2009. India is now the world's leading milk producer – ahead of the United States.

However, although per capita animal-sourced food supply has increased in many developing countries, it is still far behind the leading meat eating and milk consuming countries in the world. In 2009 the highest levels were found in the United States, with an annual per capita supply of beef of 40 kg and of poultry of 49 kg. In Western Europe, the annual per capita supply of pig meat in 2010 was 43 kg and of milk (excluding butter) was 266 kg (FAO, 2013b). A major explanation for increased meat consumption is increased incomes (World Bank, 2009). Some of the differences can also be explained by cultural preferences playing out at the regional level, with no beef consumed by Hindus and no pork by Muslims, and unrestrained consumption patterns in many regions.

In contrast to this increasing and improved diet for many in the developed and developing world, 850 million people are still undernourished. Undernourishment exists when calorie intake is below the minimum dietary energy requirement (MDER). The MDER is the amount of energy needed for light activity and a minimum acceptable weight for attained height, and it varies by country and from year to year depending on the gender and age structure of the population (FAO, 2009). The average MDER for all developing countries was estimated to be around 1950 kcal/capita per day in the period 1990–92 to 2010–12. In comparison, the MDER level in the least developed countries is less than 1800 kcal/capita per day (FAO, 2013a). In many developing countries the national per capita calorie supply level is still as low as or less than 2100 kcal/capita per day. This can be explained by a combination of the generally low level, and the high prevalence of undernourishment, which lowers the average. Figure 5.4 shows that the prevalence of undernourishment tends to decrease towards zero only when the national daily

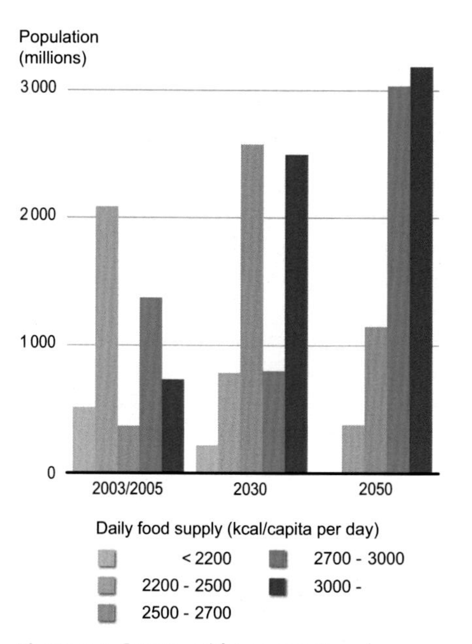

Figure 5.5 Current and future projections for average per capita food supply for the population (million) living in developing countries, 2003/05 to 2050 (Alexandratos, 2009).

per capita food supply approaches 3000 kcal. In the figure a couple of other interesting things can be seen. In contrast to all other countries, China and Brazil are outliers to the general trajectory having both high levels of undernourishment and high average food supply levels (compare Figure 5.3). Japan is another interesting case with, in contrast to the EU and the United States, full nourishment and still quite low average food supply.

Globally, the major future challenge for food production will be to meet the needs and demands of developing countries. Both the expected entire global population increase and the projected diet increases will take place in these countries. Projections for the average per capita demand for food by the population living in developing countries by 2050 indicate that more than 3000 million people will have a food supply greater than 3000 kcal/capita per day, and a further 3000 million an average of 2700–3000 kcal/capita per day (Figure 5.5). By 2030, it is estimated that the average for developing countries will have reached 2864 kcal/capita per day and in 2050 2966 kcal/capita per day. These projections mean, however, that many millions are expected to remain undernourished – about 556 million by 2030 and 370 million by 2050.

The global average food supply, including the limitations above, is expected to reach 3050 kcal/capita per day in 2050 (Alexandratos, 2009), or about the current level in Brazil and China (see Figure 5.3).

5.1.3 Consumptive water use in agricultural production

The expansion of the area of agricultural land has been the major driver behind large losses of ecosystem services from natural ecosystems (Millennium Ecosystem Assessment, 2005; WRI, 2005). Blue water withdrawals for irrigated agriculture directly affect aquatic ecosystems, but the expansion of rainfed agriculture on to non-agricultural land is equivalent to an escalating appropriation of the green water resource earlier used by the replaced terrestrial ecosystems, and thus ecosystem services. Deforestation for agriculture is expanding agricultural land and water use, and results in changes to the hydrological cycle, from the local to the global levels, as is described in Chapters 2, 3 and 7.

The consumptive water use of different food commodities varies according to the total calorie level, the proportion and combination of different vegetal components, and the share and mix of animal food items. The water use by, and water productivity of, different food supply combinations are governed by the water productivity of plant growth for vegetarian foods, and the growth of feed and fodder used for livestock as well as the conversion efficiency from vegetal feed to animal-source foods.

Water for plant production

Evapotranspiration is an inevitable part of all plant growth, and the amount depends on the energy supply, the vapour pressure gradient and the wind, which are determined by meteorological parameters – radiation, air temperature, air humidity and wind. Transpiration is also governed by crop characteristics, environmental aspects and agronomic practices (Allen *et al.*, 1998). For a given crop and climate there is in principle a linear relationship between transpiration (T) and the yield of total crop biomass, i.e. the dry matter in the roots, stems, leaves and fruits/grains. The main variable in total evapotranspiration (ET) is the evaporation (E) (Keller and Seckler, 2004; Molden *et al.*, 2010; Tanner and Sinclair, 1983). While transpiration contributes to productive crop growth, evaporation represents 'collateral' unproductive water losses (Molden *et al.*, 2010).

Food crops fall into three plant categories with regard to how efficiently they can assimilate carbon dioxide (CO_2) in relation to transpiration during the photosynthesis process: C3, C4 and CAM-plants. The gas exchange between a plant and the surrounding atmosphere takes place through the biologically regulated stomata openings on the leaf surface. When the stomata are open to assimilate CO_2, water vapour is lost to the atmosphere at the same time. Plants like wheat, barley, rice, potatoes, lucerne, soybean and pea belong to the less efficient C3 category and are therefore often grown in the temperate climate zone. C4 plants such as maize, sugar cane, sorghum and several other grasses are adapted to hotter climates and use a more efficient CO_2 assimilation cycle. CAM-plants, such as pineapple, are the most effective. Under the same theoretical hydro-climatic conditions, transpiration in C4 plants is about half and in CAM-plants only a quarter of that of C3 plants (Taiz and Zeiger, 2010). Data show water productivity for wheat in the range of 0.6 to 1.7 m^3/kg (mean 0.9 m^3/kg) and for maize of 0.4 to 0.9 m^3/kg (mean 0.6 m^3/kg) (Zwart and Bastiaanssen, 2004).

Water for animal foods production

The feed conversion efficiency rate denotes the amount of feed necessary to produce one unit of meat or other animal product. Monogastric species, such as poultry and pigs, generally have the highest conversion efficiency and consume as a rule only 2–4 kg of grain per kg of meat produced, while ruminants, i.e. cattle, sheep and goats, consume around 7 kg of feed per kg of meat produced (de Haan *et al.*, 2010; Rosegrant *et al.*, 1999). All conversion efficiencies depend on how and where production takes place, i.e. on different livestock systems, using modern or traditional breeds, and managed under various environmental conditions (Herrero *et al.*, 2009; Robinson *et al.*, 2011). One advantage of ruminants is their ability to digest grass, making it possible to base large parts of their feed demand on grazing on pasture land that is unsuitable for cultivation and on crop residues such as straw. The feed conversion ratio in aquaculture fish production is competitive, at about 1.8 kg (similar to chicken), which in some countries might make fish a good protein alternative in the future (Steinfeld *et al.*, 2006).

Production of animal-source foods and vegetable foods compete for natural resources and agricultural production capacity in areas where the cultivation of feed and fodder competes with the cultivation of vegetable foods. On pasture that is only suitable for grazing it is only ruminants such as cattle and sheep that compete with other terrestrial ecosystems. Different animal production systems therefore have different impacts on blue and green water resources. Intensive systems, such as large-scale pig and poultry production, which are entirely or largely based on cultivated feeds, rely on a combination of green and blue water resources, depending on whether the feed has been cultivated on rainfed or irrigated lands, just like for vegetal food produce. In contrast, extensive systems of livestock rearing solely dependent on grazing are only dependent on rainfed green water on pastures, which in most cases cannot be used for crop cultivation. Dairy production is an example that in most cases uses feed from both cropland and pasture (Falkenmark and Lannerstad, 2005; Heinke *et al.*, forthcoming).

5.1.4 Large yield and water productivity gaps to close

In the past 50 years the increase in global agricultural production has been astonishing: an almost tripling of cereal production and a quadrupling of meat production (FAO, 2013b). The appropriation of land for crop cultivation has only about doubled the total area in use, meaning that productivity has increased by even more. Development has generally gone furthest in highly developed countries in North America and Europe, with average cereal yields of up to 10 tonnes per hectare. In many developing countries, however, the large potential for productivity gains remains to be developed. This is especially true for rainfed crop cultivation in developing countries, where a doubling or tripling of production is an important goal in order to meet future food demand and lift many small-scale farmers out of poverty, as is discussed in Chapter 6.

The potential to increase water productivity is particularly high at low yield levels, and improved water resource utilisation corresponds to improved yields for many smallholder farmers in semi-arid and arid areas (Rockström *et al.*, 2007a; see Chapter 6). The need to explore this potential is reflected in the spatial prevalence of undernutrition and projected population increases shown in Figure 5.2a and b, which, as is noted above, largely coincides with the water-constrained savannah and steppe climate zones in sub-Saharan Africa and South Asia.

In addition, the water productivity of animal-source foods can be dramatically improved in many regions and systems. However, livestock production is much more complex than crop cultivation. In principle, animal water productivity is the result of a combination of three components. The first is related to 'direct water use' in livestock production, i.e. evapotranspiration from crops for feed or fodder or from grazing land. The second is the level of conversion efficiency of feed and fodder to animal products such as meat, eggs or milk. This component relates to animal management, such as animal health, shelter or drinking water supply, on the one hand, and fixed factors such as the animal species and breeds used, on the other. The third factor is 'coupled' feed–livestock water productivity, which includes the choice of production system, and the strategic choice of less water-intensive feed or optimisation of grazing pressure to reduce the daily distance animals have to walk to find grass and to reduce evaporation from bare degraded soils (e.g. Heinke *et al.*, forthcoming; Peden *et al.*, 2003; van Breugel *et al.*, 2010).

5.1.5 Consumptive water use depends on diet, food losses and water productivity

A number of the parameters described above govern the total amounts of green and blue consumptive water use in the production of different food items. Due to conversion losses, animal-source foods generally demand an significantly more consumptive water per food unit than vegetal foods. This makes the ratio of vegetal to animal-source foods, in addition to the per capita supply of food, very important (Figure 5.3) in calculating the consumptive water use of any human diet.

Generic food water requirement

The average amount of evapotranspiration needed to produce the daily diet is in the range of 2500–5000 l of water/capita per day (Renault, 2003). The global 'generic human food water requirement' has been estimated at 3600 l of water/capita per day, or 1300 m^3/capita per year. This estimate is based on the assumption that the average evapotranspiration required to produce the equivalent of 1000 kcal of vegetal foods is 0.5 m^3, while 4 m^3 per 1000 kcal is required for animal products, based on a daily per capita energy food supply demand of 3000 kcal with 20% from animal-source foods (Falkenmark and Rockström, 2004; Rockström, 2003; Rockström *et al.*, 1999).

As is noted above, 3000 kcal/capita per day is expected to be the global average by 2050. The 20% proportion of animal-source food is in line with the currently rising global average of 18% and reflects the fact that many less developed countries experiencing rapid economic development have already passed 20% (e.g. China was at almost 23% in 2009) (Figure 5.3).

The generic annual per capita agricultural water requirement of 1300 m^3 is based on previous global assessments at current water productivity levels. In this approach, the differences in evaporative demand in different climatic regions are balanced by assumed differences in transpiration efficiency between C3 and C4 plants. As is described above, more water is required to produce animal-source foods due to conversion losses, because only part of the vegetal energy consumed by animals is transformed into meat, milk or eggs. The values used are equal to or lower than earlier estimates (Chapagain and Hoekstra, 2003; Pimentel *et al.*, 1997).

Losses along the food production chain

Considerable benefits could be reaped if food losses could be minimised. Figure 5.6 shows that the character of these losses differs considerably between developing and developed countries. In developing countries, such as those in sub-Saharan Africa, and South and South East Asia, the major part of the losses and spoilage takes place before the produce reaches the market – in the field or during transportation or processing. These losses could be overcome with modern harvesting, transportation and storage techniques, like those seen in Europe and North America, where the major losses and waste take place in the latter half of the food supply chain (Godfray *et al.*, 2010; Gustavsson *et al.*, 2011). This is also true in relation to meat production. Figure 5.6 shows that sub-Saharan Africa stands out as an extreme case, with massive losses at the animal production stage partly due to a lack of proper animal health management.

Figure 5.6 shows that industrialised Asia, North Africa and Latin America are now approaching the same patterns as the OECD countries. Reduced agricultural and post-harvest losses in the first half of the food chain represent a positive trend. However, if rapidly developing societies like China, Brazil and Mexico follow the same market and consumption paths established in the developed world, they risk replacing one wasteful food system with another (Lannerstad, 2009). In order to reduce the food challenge and the pressure on water resources in the future it will be crucial to reduce food losses in both

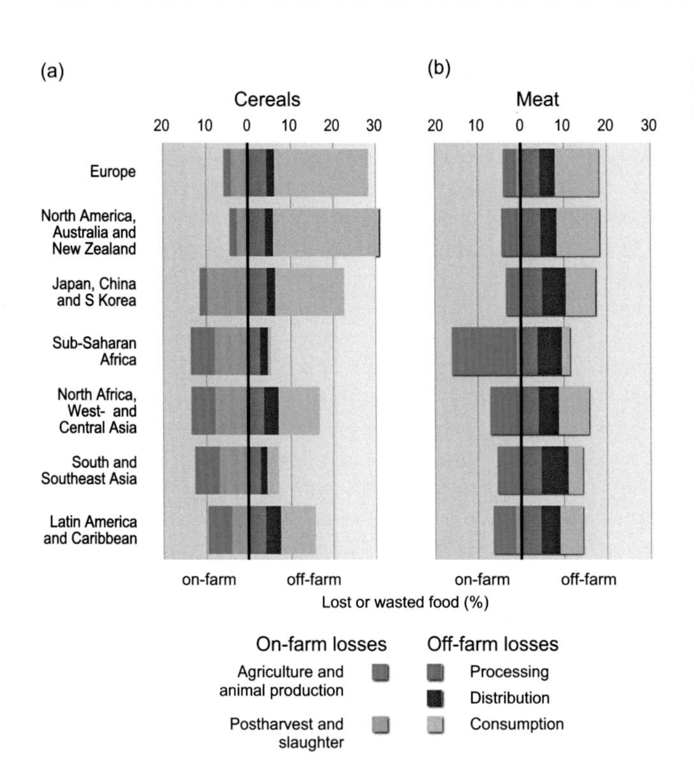

Figure 5.6 Percentage food losses for cereals (a) and meat (b) in seven world regions (data from Gustavsson *et al.*, 2011).

developed and developing countries. A food supply of 2200 kcal/capita per day is considered the minimum necessary food intake level in a 'loss-free' food consumption system (Smil, 2000). This level is also often used as the hunger break-off level by FAO, and is just above the poverty line for the food energy intake of 2100 kcal/capita per day used by the World Bank.

Gains from choosing less water-intensive foods

Another option to reduce the magnitude of water resources in order to address the global food challenge would be to steer away from current dietary trends, in terms of the increasing demand for water-consuming food items, to a consumption level considered satisfactory for human health. This would be particularly effective for animal-source foods produced from animals largely dependent on feed produced on cropland, i.e. those competing for water resources that can be used for the cultivation of edible crops.

Thus, in the simplified estimate of a global 'generic human food water requirement' mentioned above, animal-source foods appropriate eight times as much water per produced calorie, compared to vegetal source foods. Although this assumption depends in reality on a wide range of combinations of animal production systems (grazing, cultivated fodder and cultivated feed) and feed conversion efficiencies (different species and breeds, animal management, animal health, etc.), the principal scale highlights the importance of the food mix. When applied to a diet of 3000 kcal with 20% from animal-sourced foods it means that two-thirds of the consumptive water use behind the daily diet relates to livestock production. However, out of the 4 m^3 per 1000 kcal of consumptive water use also green water from pastures is included in the estimate, and a large part of current consumptive water use is therefore not available for food production. Thus, any water figure for animal-source foods must be viewed with extreme caution and be analysed in its local context. If there is sufficient diversity in the diet, containing all key micronutrients etc., the supply of animal-source foods can be reduced considerably, possibly to 5% of the daily per capita calorie food supply (Smil, 2000).

Four food supply and loss cases

Figure 5.7 visualises food water requirements for different food supply levels, different animal product fractions, and with and without food-loss reductions

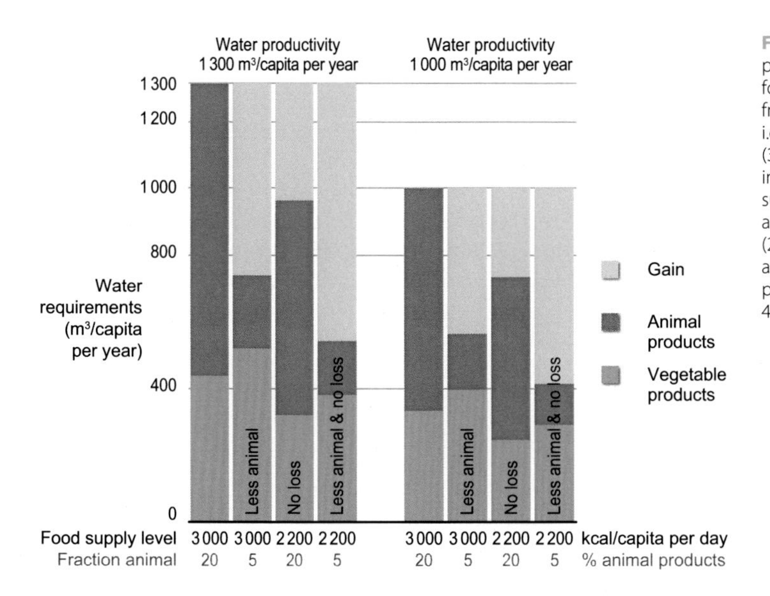

Figure 5.7 Water gains at two different water productivity levels considering different total food supply per capita per day, and different fractions of vegetal and animal-source foods, i.e. an average daily per capita food supply (3000 kcal, 20% animal source, a less animal intensive food supply of 3000 kcal, 5%), food supply with no food losses (2200 kcal, 20%), and a combination of both the lower options (2200 kcal, 5%). Animal-source foods are assumed to demand eight times more water per calorie than vegetal foods, i.e. 0.5 m^3 vs. 4 m^3 per 1000 kcal.

under two alternative water productivity levels. The figure reveals the importance of animal products and food losses to overall food water requirements, when applied to the global 'generic human food water requirement' mentioned above. If the animal foods fraction is reduced from 20% to only 5% (3000 kcal and 5%) the gain is as large as 45%, and if instead food losses can be eliminated (2200 kcal and 20%) the annual per capita water gain is more than 25%. With both a reduction in animal products to 5% and a loss-free food production chain (2200 kcal and 5%) the total gain at the initial water productivity level is almost 60%. A water productivity gap closure that reduces annual per capita demand from 1300 to 1000 m^3/capita per year equals a gain of 25%, and thus less than if all losses are eliminated. The last pillar in Figure 5.7 summarises how annual food water requirements can be theoretically reduced by almost 70%, by combining improved water productivity, less animal products and no losses, i.e. from 1300 to about 400 m^3/capita per year. As is noted above, while reducing food losses is indisputably a gain, there are a multitude of factors to consider in livestock production, which ranges from two extremes, i.e. from entirely grain-dependent animal products competing directly with human edible food production to grass-dependent production of dairy and meat products that do not compete for water resources with crop cultivation for human consumption.

5.2 Consumptive food water requirements by 2050

As is mentioned above, feeding the human population in 2050 presents two challenges from a food production perspective: to cater for the expected population increase of 2.3 billion and to provide an acceptable diet for the currently undernourished 1 billion. Although hunger alleviation in all countries is more a question of sharing the available food, the population increase and shifting diets will inevitably put enormous pressure on global food production. This section assesses the water requirements for feeding the world by 2050 and how different management and demand options can lessen the projected pressure on already scarce water resources.

5.2.1 What can be achieved on current cropland?

The water implications of feeding the world by 2050 presented in this section rest on a thoroughgoing assessment of the potential for food self-sufficiency in each country using currently available agricultural land. The assessment goes beyond earlier studies based on the global generic per capita food water requirement, i.e. the 1300 m^3/capita per year mentioned above. Instead, country-level water values from process-based vegetation and hydrology modelling for different crops are combined with statistics to generate estimates. Country-level food production for self-sufficiency on current croplands is analysed for different per capita food supply combinations at the national level, considering population change, water availability, climate change, water productivity improvements and modelled potential irrigation increases. The assumptions and calculations are described in Box 5.1.

This assessment contains a number of strengths and novelties for analysing the future food challenge in relation to water resource use and linkages to the food system. A primary strength is its detailed country-level estimates, with estimates of existing water availability, alterations in water availability due to climate change and resource constrained irrigation scenarios. Water use for crop production on current croplands is estimated for the water productivity under current management and for potential future water productivity following management improvements, expressed as yield gap closures. For every country, dietary water requirement estimates are made for the current food supply composition, taking into consideration the characteristics of each country's food system such as food system efficiency and livestock production composition and efficiency (see Box 5.1).

Although a great deal of water and food system complexity is captured in the study, some shortcomings remain. The agricultural sector is treated as static in several regards for both crops and livestock. In addition, future production is assumed to operate on current fixed land-use patterns. In crop cultivation, no crop changes or interplay between management improvements and horizontal expansion are captured. The livestock sector in each country is assumed to be static with regard to both the composition of animal type in production and animal water productivity. In addition, the study does not consider the grazing component of feed consumption in livestock production. Thus, it is assumed that the share of feed from grazing in future livestock production in each country will be the same as it was in 2000. In addition, the outcome of the analysis, as is the case for all global studies, is affected by data and modelling uncertainties (see Box 5.1).

The analyses apply an equal per capita food supply level for all countries of 3000 kcal/capita per day, with 20% of the calories derived from animal-source foods, i.e. 600 kcal/capita per day (as is discussed above). The base year is 2000 and scenarios are developed for 2050. By applying a general per capita food supply level, it is assumed that in many developing countries the daily per capita calorie supply will increase from its current low level and that food supply levels in many industrialised countries will have to fall to the common global level. Thus, in many industrialised countries both the calorie level and the animal-sourced food fraction will have to be lowered. In Denmark and France this would mean halving animal-source food consumption, from the present 1200 kcal/capita per day (FAOSTAT; FAO, 2013b).

5.2.2 The water gap by 2050 and potential steps to close it

Water productivity for any food crop differs between countries. Thus, estimates of water deficits or surpluses related to crop cultivation for food self-sufficiency are not easily weighted against each other. Therefore, to enable a comparison between global water food surpluses and deficits, deficit values have been recalculated as a 'deficit as surplus' (Box 5.1). This makes it possible to tell whether summarised potential exports of food from water surplus countries can match the summarised needs of water deficit countries. It is also possible to compare virtual water quantities and gains in exporting countries with those in importing countries. In Figure 5.8 a number of interesting comparisons for different scenarios compare surplus quantities to deficit volumes recalculated as deficit as surplus.

Actual versus uniform food supply levels by 2000

In the first step the baseline country-specific food supply levels for 2000 were compared with a situation in which all countries have a uniform food supply level of 3000 kcal/capita per day with 20% of the calories from animal sources. With current baseline water productivity, global food water requirements would increase from the current 5100 km^3/year to 6900 km^3/year, and the global deficit would be almost twice as large as the surplus, pointing to a large overshoot (bars 1 and 2 in Figure 5.8). The net deficit of almost 800 km^3/year shows that under current productivity green and blue water quantities would not be sufficient to produce a standard food supply for all.

Population and diet changes multiply demand by 2050

By 2050 the world population is expected to have increased by an additional 2.3 billion. Assuming

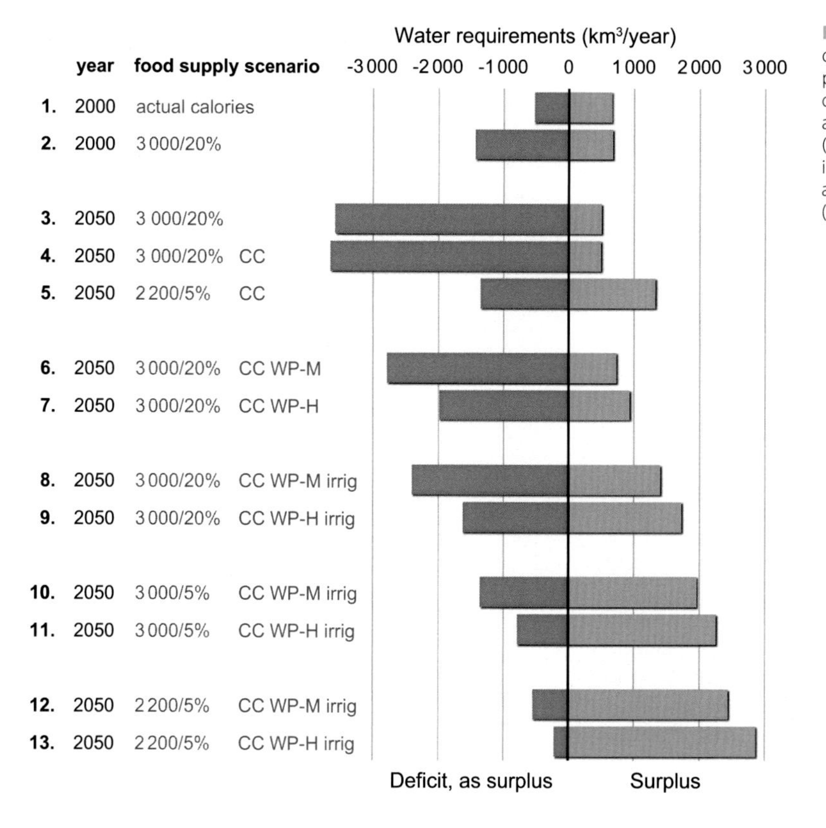

Figure 5.8 Global water surplus and deficits as surplus for the global population in 2000 and by 2050, under different global average food supply assumptions, under climate change (cc), two different water productivity improvement levels (WP-H and WP-M) and potential maximum irrigation (irrig) (see Box 5.1).

current water productivity and a uniform per capita food supply of 3000 kcal/capita per day with 20% from animal foods, global food water requirements would be as high as 11 400 km^3/year. Compared to the baseline of 2000, this would mean an increased deficit of more than 6000 km^3/year in water deficit countries. Recalculated as a 'deficit as surplus', the net deficit of more than 3000 km^3/year is seven times larger than the surplus (bar 3 in Figure 5.8). This impossible situation highlights the urgent and immense need to find different ways to increase water productivity, increase water availability for food production or reduce demand.

Only minor climate change impacts by 2050

In the light of the ongoing trend of climate change with its major impact on water resources and agriculture, these impacts have to be included in any food production scenario. Thus, all the steps in this analysis include the impact of climate change on crop production (Box 5.1). Combined with the challenge of feeding a future global population in 2050 at a uniform food supply level, climate change only increases shortages in water deficit countries by 380 km^3/year. Recalculated to a 'deficit as surplus' the summarised global shortfall is only increased by 80 km^3/year, or less than 3% (compare bars 3 and 4 in Figure 5.8). Thus, population increase combined with increased per capita food supply levels and higher animal-sourced food demand totally dominate the future challenge at the global level – not climate change impacts on water availability.

Standard versus minimum food supply levels

Bar 4 in Figure 5.8 shows that under current water productivity and with existing water resource limitations on current croplands, it is impossible to balance surpluses and deficits through trade, since not enough food can be produced. However, if the food supply level is drastically reduced to a global calorie level of 2200 kcal/capita per day with only 5% of the calories derived from animal foods, available water resources on agricultural lands, at current levels of water productivity, would be sufficient to feed the population in

Box 5.1 Analysis and modelling

Jens Heinke, Potsdam Institute for Climate Impact Research

For the analysis in this chapter, a model-data fusion approach was applied in which climate model projections, LPJmL simulations, FAO statistics and UN population data were combined to produce estimates of current and future food water requirements and availabilities.

Simulation setup

For the period 1901–2000 LPJmL was forced with observed monthly values of temperature, precipitation, number of rain days and cloud cover taken from the CRU TS3.0 climate data set (Mitchell and Jones, 2005). For the subsequent period to 2099, a climate scenario derived from projections by the ECHAM5 model for the SRES A2 scenario was used (for details see Gerten *et al.*, 2011). Soil parameters and historical CO_2 concentrations were as in Rost *et al.* (2008). For the period after 2000, atmospheric CO_2 concentrations were held constant at the level in 2000. Land-use patterns are based on the data set described by Fader *et al.* (2010) and were held constant at the year 2000 level for the entire simulation period. Management intensities for crops, impacting both on yields and water fluxes, were calibrated to match average 1998–2002 yields reported by FAO (FAOSTAT; FAO, 2013b) following the method described by Fader *et al.* (2010).

In addition to the reference simulation two additional sets of simulations were performed to evaluate the sensitivity of water productivity to improvements in agricultural management and rising atmospheric CO_2 concentrations. Two yield gap closing scenarios were used to assess the impact of improved agricultural management. Yield gaps are defined as the difference between actual and potential yields due to non-optimal management. Improved management can increase yields, better exploiting the potential yield and lessening the gap. That yield increases relative to the actual yield are higher for larger yield gaps is essential to this study. It is intended to reflect the large potential to improve management, yields and water productivity in low management systems (Rockström *et al.*, 2007b). A moderate WP-M and a high WP-H management improvement scenario were chosen in which yield gaps were closed by 25% and 50%, respectively. The two management scenarios of water productivity improvement with 19–22% WP increase (WP-M) for the moderate and 40–45% WP increase (WP-H) for the high management improvement scenario, depending on the climate.

Processing the results

The periods 1971–2000 and 2035–2064 were analysed for this study to represent current and future (mid-twenty-first century) conditions. For each the full range of combinations of CO_2 and management scenarios, dietary water requirements and water availability for food production were estimated.

Estimation of water requirements to produce vegetal and animal calories

The water required to produce vegetal calories was calculated as the ratio of total consumptive water use (CWU) of crops (estimated from LPJmL) and total calories potentially supplied from cropland (estimated from FAO's Food Balance Sheets). The total agricultural CWU of a country was calculated as the sum of ET during the growing season of each crop averaged over the time period 1971–2000 or 2035–2064. Because LPJmL does not account for multiple cropping cycles, the modelled values of CWU for each crop and grid cell were corrected using data on actual harvested area from the MIRCA 2000 data set (Portmann *et al.*, 2010). For annual crops not represented by one of the crop functional types (CFTs) in LPJmL, CWU was estimated based on grid cell averages of CWU for current CFTs. For perennial crops, year-round ET from pastures was used.

Total potentially produced calories produced within a country were estimated from FAO's Food Balance Sheets for the period 1998–2002, assuming no imports or exports and accounting for the characteristics of each country's food sector. The food sector of each country was generalised by calculating fractions of waste, and processing of primary crops to secondary products, as well as calories supplied to people per kilogramme of commodity. These values were then applied to the domestically produced primary crops in the Food Balance Sheets to arrive at potential calorie supply. Note that this potential fully accounts for the current country-specific efficiency of the food system to supply calories from the produced crop. For the baseline period, CWU was directly related to the estimated total calories. For all other scenarios, the relative change in agricultural production was approximated by changes to modelled total production from LPJmL's CFTs. Total potential calorie supply was then adjusted accordingly before relating it to total agricultural CWU for the respective scenario/time period.

The water required to produce animal-sourced calories was estimated based on the efficiency of each country's livestock sector to produce animal

Box 5.1 (*cont.*)

calories from calories produced on cropland and previously calculated water requirements for vegetal calories on cropland. The water required to produce the biomass grazed and browsed by ruminants is not accounted for due to lack of data.

Estimation of blue water availability

The total blue water resource was first estimated at the river basin scale and then aggregated to the country level. The total blue water resource for each basin was essentially taken as the sum of all runoff generated from precipitation (excluding return flows from irrigation). Because wetlands are not explicitly included in LPJmL, a correction for evaporation from rivers, lakes, reservoirs and wetlands was applied at the grid cell level before aggregation to the basin scale. The extent of open water surfaces and wetlands for each grid cell was estimated from the Global Lakes and Wetlands Database (Lehner and Döll, 2004). Average grid cell runoff was then calculated as an area weighted average of the modelled runoff for the terrestrial fraction and the difference between incoming precipitation and potential evapotranspiration for the open water and wetland fraction. Calculated grid cell runoff was deliberately allowed to be negative, assuming that the local gap in the water balance is met by lateral influx from upstream areas.

Blue water resource per country was computed as the sum of all river basins within a country. For transboundary basins the blue water resource was allocated to the neighbouring countries as follows: for each country, the sum of discharge values in all cells belonging to the country was calculated and divided by the sum of all discharge values of the respective basin; the total blue water resource of that basin was then allocated to the countries according to these weights (see also Gerten *et al.*, 2011).

Of the estimated total blue water resource not everything can or should be used for human needs. Spatial and temporal mismatches between water requirements and availability as well as the need for environmental flow requirements restrict the amount of water that can be utilised. In the literature, a water demand to supply ratio exceeding 0.4 is often used as a classification for 'severe water stress' (Rijsberman, 2006). It is therefore assumed that only 40% of the total blue water resource is actually available. It should be noted that this represents an optimistic estimate as no other water uses than agriculture are considered here.

Calculation of water surpluses and deficits

From the estimated water requirements to produce vegetal and animal calories, total dietary water requirements were calculated for a range of different diets, populations and water productivity scenarios. Comparing these requirements to total available blue water could lead to an overly optimistic picture if other factors constrain full exploitation of blue water resources for food production.

In order to estimate how much of additional blue water could potentially be utilised for food production on current cropland, all LPJmL simulations were performed for a second time assuming full irrigation on rainfed crops. Total consumptive blue water use (ET from irrigation water) was calculated applying the same corrections as described above.

To take account of imperfect utilisation of blue water due to losses during transport from the water source to the field, the efficiency of the irrigation systems was included. Based on country-specific irrigation efficiencies compiled from census data (Rohwer *et al.*, 2007), overall conveyance losses were calculated of which 50% was assumed to occur as evaporation and therefore not returned to the river (Rost *et al.*, 2008). Total available blue water was reduced by these consumptive conveyance losses to estimate available blue water at the field level.

For comparison with water requirements, the minimum potential blue water use and the blue water available in the field was used. If blue water consumption for actually irrigated crops already exceeded blue water availability, the additionally required blue water was assumed to be available from non-conventional or non-renewable resources.

World Bank income country group categorisation

To be able to give some indication of the potential purchasing power situation in 2050 all countries are grouped according to the World Bank income country group categorisation from 2011 (World Bank, 2011). Countries are divided according to 2009 gross national income (GNI) per capita, calculated using the World Bank Atlas method, giving four groups: low income (L), USD 975 or less; lower-middle income (ML), USD 996–3945; upper-middle income (MU), USD 3946–12 195; and high income (H), USD 12 196 or more.

2050 (bar 5 in Figure 5.8). This highlights the importance of both the average per capita level, partly driven by losses, and the share of animal-source foods, driven by increased affluence allowing consumers to choose more livestock-intensive diets. However, this much lower per capita food supply level is far from the current situation in both developed and many developing countries.

Two water productivity levels and irrigation

By improving water productivity, 'more crop per drop' can be achieved. Two levels of future water productivity improvements are shown in Figure 5.8 (bars 6 and 7). The first level is a yield closure of 25%, WP-H, and the second a closure of 50%, WP-M (see Box 5.1). The width of the yield gap for the crops analysed is country specific. While the potential is large in many water-scarce developing countries, the gap is limited or non-existent in many developed countries. As is noted above, these two alternatives are far from sufficient to achieve a global food water requirement balance. However, with the addition of a modelled maximum possible increase in blue water use on cropland, i.e. potential irrigation, the situation improves dramatically (bars 8 and 9). For WP-H with irrigation the surplus actually exceeds the deficit and the results show that it would be possible globally to secure a 'standard' per capita supply of 3000 kcal/capita per day with 20% from animal-source foods using this combination of measures.

Less animal-sourced foods and loss-free food supply

The last four bars (bars 10 to 13 in Figure 5.8) show how the deficits as surplus and surpluses by 2050 change for two different per capita food supply compositions for the water productivity levels WP-M and WP-H. The food supply level of 3000 kcal with 5% from animal-sourced foods represents a minimum non-vegetarian diet, and 2200 kcal with 5% from animal-sourced foods represents a food supply composition with minimum livestock products and a food system without losses (compare Figure 5.7).

By shifting from a 20% to a 5% animal calorie fraction for the 3000 kcal level the deficit decreases by as much as 43% at the moderate level of water productivity improvement, WP-M (bar 10). As a result, this alternative is the second alternative where the surplus exceeds the deficit and it is possible globally to secure a food supply for all. For WP-H (bar 11), the surplus is almost three times the deficit. If the currently massive food losses could be abolished (bars 12 and 13) a food supply of only 2200 kcal/capita per day with 5% from animal-sourced foods would be enough to feed humanity with considerable surpluses of from four to as much as 14 times the deficit.

In sum, without water productivity improvements and an expansion of irrigation, food water requirements for a general per capita food supply in water deficit countries would be seven times higher than the potential quantity available for food production for export in surplus countries. Only the implementation of a full expansion of irrigation with strongly improved water productivity, WP-H, will lead to a world in which food production for a standard diet will be in balance from a water deficit and water surplus perspective. At the lower level water productivity improvement, WP-M, a balance is first reached when the animal-source food supply is reduced to 5% of total calorie food supply.

5.2.3 Comparing the challenge and the two positive options

When the surplus and deficit as surplus bars change in Figure 5.8, the number of people living in water-short vs. water-scarce countries also changes. It is particularly interesting to compare both the water balance and the number of people for the challenge to feed the global population by 2050 under current water management (bar 4 in Figure 5.8) with the two first possible alternatives, with water surpluses bigger than the deficits (bars 9 and 10 in Figure 5.8). Under current water management the number of people in countries with water deficits by 2050 would amount to as many as 7.7 billion, and only 1.1 billion would live in countries with a surplus. Under the two positive options, the number of people living in water-scarce countries decreases to 5.6–5.8 billion. In other words, even at a globally balanced situation more than 60% of the global population will live in countries too water scarce to reach food self-sufficiency.

Most scenarios indicate the need for large virtual water food flows from surplus to deficit countries. The analysis thus clarifies the fundamental importance of the food trade in the coming decades and the importance of paying attention to the economic perspective in terms of purchasing power. To get a feeling for the potential purchasing power situation in 2050 all countries are grouped according to the World Bank income country group categorisation (World Bank, 2011). Countries belonging to the income groups ML (lower-middle income), MU (upper-middle income) and H (high income) are assumed to have large enough purchasing power to import food to compensate for water deficits. By assuming that the poor country group L (low income) lacks the option to import food, the link between poverty and water scarcity is highlighted (Box 5.1).

In Figure 5.9a, b and c the initial challenge and the two options for balanced global food production are depicted for comparison. The three figures show country water deficits, actual water surpluses and the number of people in each income category (compare bars 4, 9, 10 in Figure 5.8). The correlation between the two lowest income groups and large water deficits highlights the challenges many of these countries will face in the coming decades. In all three cases, there are very few water surplus countries in the low-income group, and the two groups with higher incomes are clearly associated with surplus situations (Figure 5.9a, b and c).

Both the balanced situations, i.e. Figure 5.9b and c, show very similar patterns for water quantities and the number of people in the different income categories. Actual country deficits are dramatically reduced and a large number of people move from deficit to surplus categories. This positive trend is particularly the case for countries currently in the ML income category. However, the number of people living in the poorest and most water-scarce countries remains constant at around 1.5 billion in all three alternatives. Indifferent to improved water productivity or reduced levels of food demand, the same 33–37 countries fall into this precarious category.

5.2.4 Poor water-short countries

The 1.5 billion people living in water-scarce countries by 2050 and possibly too poor to be able to import food will face a dilemma. One option will be to take an adaptive approach by making the best possible out of the water resources they have. However, reducing the demand to the minimum supply level of 2200 kcal/capita per day with 5% from animal-source foods makes only a few smaller countries self-sufficient in food. The majority of the countries in the water-scarce poor group would still suffer a water deficit of 381 km^3/year – and still host a combined population of about 1.5 billion. Nations in Africa constitute the largest share of the 32 poor and water-scarce countries, while five are in South Asia, one in Central America and one in Central Asia. The four largest countries in this group are Bangladesh, Niger, Uganda and the Democratic Republic of Congo, with a total estimated population of more than 0.65 billion by 2050.

It is important to remember that our analysis is carried out on current agricultural land. It would be possible for some countries, e.g. many sub-Saharan countries, to expand agriculture into other terrestrial biomes. For a country like Bangladesh, with 71% of its land area in use as agricultural land (FAOSTAT; FAO, 2013b) and blue water resources available within the country boundary, tackling the food deficit is primarily constrained by land.

5.2.5 Virtual water flows involved

For the base year, 2000, the virtual water trade was about 550 km^3/year from surplus countries, i.e. deficit as surplus. The corresponding local food water requirement in importing countries was 700 km^3/year, i.e. a theoretical water gain of 150 km^3/year. The scenarios for food self-sufficiency in 2050 initially illustrate large water deficits. However, with improved water productivity, increased irrigation and adjusted calorie food supply combinations, the amount of food produced that can be traded becomes large enough to balance deficits and surpluses.

At the starting level of current water productivity and an average global food supply of 3000 kcal/capita per day with 20% from animal-source foods – and including climate change – the summarised national deficits amount to −7133 km^3/year and account for more than 60% of total global food water requirements of 11 600 km^3/year. When the deficit is recalculated as deficit as surplus, virtual water demand amounts to −3644 km^3/year or about seven times the current virtual water trade. More importantly, it is seven times larger than possible exports from surplus countries, i.e. 496 km^3/year, and thus not possible.

For the three alternatives where global water surpluses balance deficits, i.e. current WP 2200/5% and no losses; WP-M 3000/20%; and WP-H 3000/5% (bars 5, 9 and 10 in Figure 5.8), virtual water flows will about triple compared to the current level of virtual water trade of 530 km^3/year. However, given increased water productivity, the food quantities that need to be traded will have to much more than triple.

5.3 Preconditions at the system level: sustaining the ability to produce food for humanity

Section 5.2 examines how to provide food for the human population in 2050 from a static perspective,

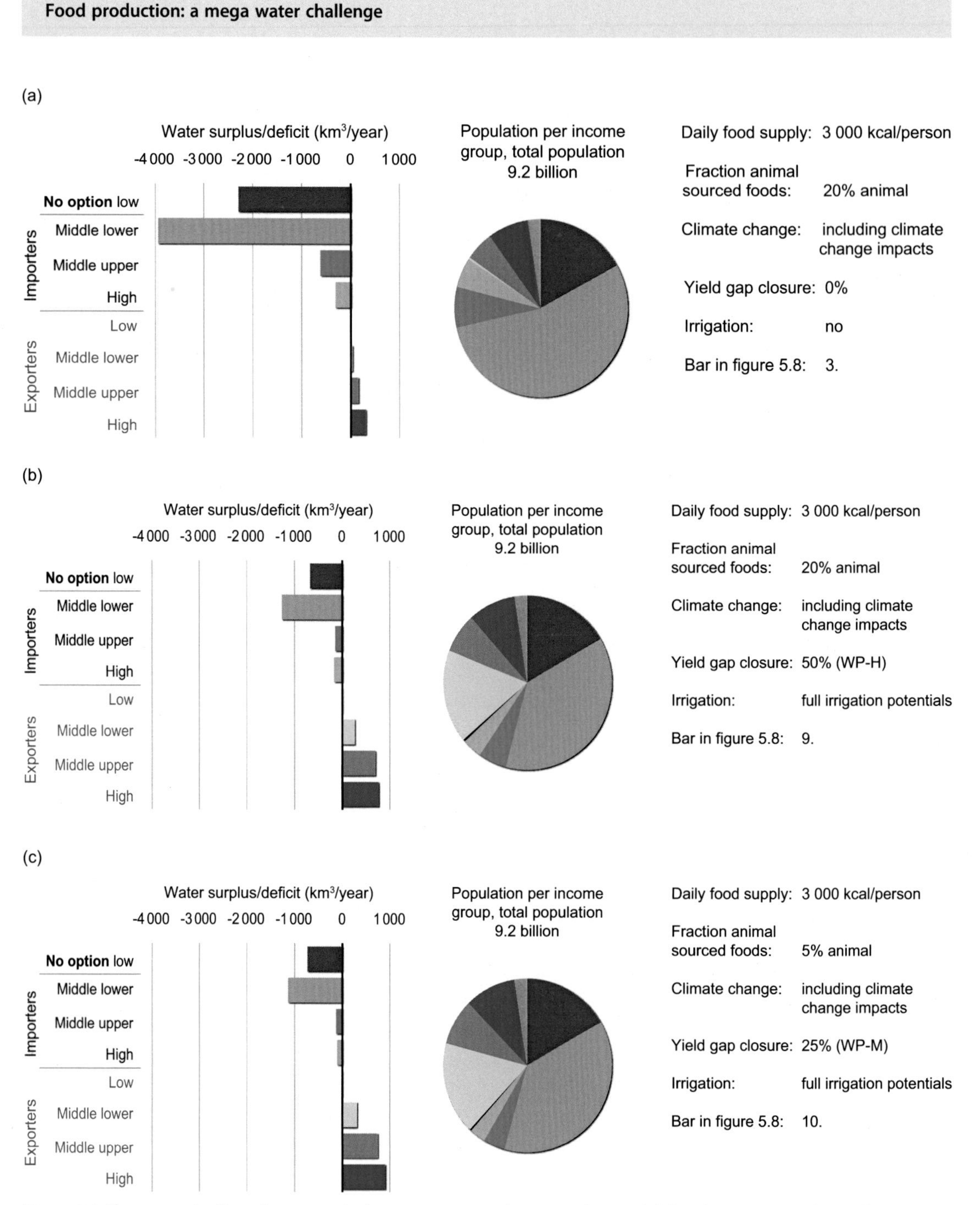

Figure 5.9 Three examples illustrating summarised country estimates of water surpluses and deficits by 2050 grouped according to current economic situation with: (a) the water challenge to feed the world by 2050 with current water productivity and under climate change impacts; (b) a balanced situation with 3000 kcal per person per day and 20% from animal-source foods and a 50% yield gap closure (i.e. WP-H); (c) a balanced situation with 3000 kcal per person per day and 5% from animal-source foods and a 25% yield gap closure (i.e. WP-M). Climate change impacts and modelled irrigation potentials are included in b and c.

comparing water availability with food water requirements. It shows that global food security based on current trends, of 3000 kcal/capita per day and a 20% animal-based component, would imply food water requirements that are too large to be met. A worldwide modernisation of agriculture in the developing world, involving yield gap closure, irrigation expansion or reduced food demand would balance water deficits and surpluses on a global level, assuming commodity trading and the necessary purchasing power. Nevertheless, a set of poor countries are likely to continue to suffer from water scarcity affecting food security.

In reality, however, the food production system is obviously not static. For example, increased precipitation variability due to climate change might shift rainfall seasonally, making crop cultivation impossible, and global or regional economic turbulence might lead to sudden and abrupt food trade impediments, leaving import-dependent countries vulnerable. In other words, the agricultural system is a dynamic system, subject to various types of ecological, social and economic changes, including shocks such as droughts, pests, price spikes and trade disturbances, as well as creeping changes such as the silent processes of soil degradation and demographic change. It is thus important to examine issues related to sustaining the ability to produce food in a dynamic world, i.e. the capacity to deal with change and continue to develop – adding to the resilience perspective on water and food. Two main issues emerge related to biomass production functions and landscape dynamics: first, dealing with disturbances and creeping changes; and, second, securing the functioning of crucial regulatory ecosystem services in the landscape. There are also vital functions to be aware of in the food system, such as the food trade system, competing land uses and possibilities for expansion. Factors that influence moisture feedback, potential thresholds and key disturbances are also important, as well as the potential traps arising from vulnerabilities linked to shocks and slow creeping change.

5.3.1 Supporting system functions

Surplus food production for export

A fundamental assumption is linked to the basic prerequisite of our static analysis – that water surplus countries really can produce enough food for export. Research will be required to assess the realism of this assumption. Even if agricultural water availability does not face any constraints, a large production increase would be a genuine challenge. Even if there is a large enough water surplus beyond food self-sufficiency needs, there might be other constraints on its use to expand agricultural production.

Diet changes

With increasing wealth often comes an increase in the animal-based food component of diets, and related increases in food water requirements. In our analyses we assume that 600 kcal per person per day of the average calorie food supply will come from animal-sourced foods. Viewed from current trends, with rapidly increasing consumption of meat, milk and egg, this assumption appears in many countries to be too optimistic. For many other countries, on the other hand, this level might be far too high. However, as shown above, the level of animal-sourced foods is highly decisive for food water requirement on a country level as well as globally.

Balancing other biomass requirements

The most evident competing biomass requirement is the current emphasis on biofuels, which infringes on the agricultural capacity of countries, requiring a balancing of food production with other bioresource needs (see Chapters 6 and 8). Another rising water requirement of fundamental importance from the resilience perspective is the evaporative demand needed to sustain the carbon stock in living vegetation increasingly set aside for carbon sequestration for climate change mitigation.

A functioning trade system

A further precondition is an upgrading of the global food transportation infrastructure to handle rapidly increasing trade volumes. This will mean investment in land and sea transport as well as other infrastructure such as harbours, ports and food storage, as well as proper global governance of the food trade system to secure reliability in space and time.

The most difficult factor will probably be establishing a fully functioning global food market. Many countries implemented export restrictions during the food crisis of 2007–08 (e.g. over 40% of East Asia and South Asia) (FAO, 2009). Price vulnerability is another threat for the import-dependent, water-deficient countries, which will have a combined population of close to 6 billion people.

Increase food stocks to create redundancy

World cereal stocks oscillate around 500 million, a stock-to-use ratio of around 20%. During the period 1981–2000 world grain stocks on average could balance 15 weeks of consumption. In the last decade the stock-to-use ratio has fallen considerably and now only corresponds to 12 weeks of the world's total grain consumption (FAOSTAT; FAO 2013b). The food crisis 2007–08 highlighted the importance of stocks to buffer turbulences in the global food market. The idea, widespread in other sectors of production, of 'just-in-time' purchasing and management can have fearsome consequences in the food sector, leading to interruptions in food supply to large populations. Expected increased turbulence in the Anthropocene will make it necessary to increase food stocks at global, regional, national and local levels to buffer against unexpected turbulence.

Foreign land acquisition

The food crisis and volatility in the food markets have undermined trust in trade and are major causes of foreign investment in agricultural land, often called land grabbing (von Braun and Meinzen-Dick, 2009). This trend in state-controlled virtual water flows began even before the food crisis of 2007–08, e.g. China began to lease land for food production in Cuba and Mexico and continues to search for new opportunities to feed its large population (von Braun and Meinzen-Dick, 2009). Asian and Middle Eastern companies have already bought or leased some 15% of existing farmland in countries on all continents (Anseeuw *et al.*, 2012).

5.3.2 Land, water and ecosystem dynamics

Vulnerability to rain variability disturbances

There are many kinds of possible disturbances. One obvious example is rain variability, reflected for instance in droughts and floods. One way to protect against sudden interruptions in the food supply system due to droughts is the use of water storage to bridge periods of deficient water availability. Another way to bridge shocks is the use of insurance, such as crop insurance and weather index insurance (Hazell and Hess, 2010).

It will be equally challenging to remain aware of slow disturbances or creeping changes in the production system, linked to altered rainfall patterns, the effects of rising temperatures, a possible decline in land productivity due to poor soil/water management, poor irrigation management generating soil salinity, a rise in the water table, and so on. The production system could also suffer from the possible effects of crop simplification and monocultures, which make the food production system even more vulnerable to droughts and pests.

Poor water-shortage country predicament

Poor countries lacking purchasing power to secure national food security might be forced to choose horizontal agricultural expansion wherever possible to increase domestic food production. The extent to which horizontal expansion is possible by cropping pastures and grassland is therefore an important question. It will be essential to prevent the so-called poor hotspot countries from falling into rigidity or poverty traps (see Chapter 3), especially the most populated hotspot countries such as the Democratic Republic of Congo, Ethiopia and Tanzania.

Securing land productivity

It is important to understand landscape dynamics and key biophysical feedback processes, such as river basin closure as a consequence of excessive consumptive water use, which might introduce a threshold beyond which societal organisation will have to be redefined. Such cases may call for a transformation of water supply management systems to incorporate the reuse of wastewater.

There are many types of silent processes that might interrupt a well-balanced system of crop production: changes related to land productivity, such as slowly advancing erosion, acidification or salinisation, soil crusting, etc.; changes in the hydro-climate linked to climate change or as an effect of land-use change; or the effects of land grabbing, disturbing the national farming system and reducing the potential for food production.

Silent processes might contribute to regime shifts, for instance, due to growing water demand for agricultural production (irrigation) or creeping land-use change. Some regions are more prone to water-related regime shifts than others. Consideration should be given to: management measures, with a focus on the drivers of food demand; early-warning systems, with a focus on slow variables; and resilience building in the landscape. Some locations are more vulnerable in relation to phosphorus, moisture recycling and so on.

Securing vital ecosystem functions in the landscape

There is a more in-depth discussion of ecosystem functions in the landscape in Chapter 6. It will be essential to secure adequate ecosystem diversity, both functional diversity and response diversity (Elmqvist *et al.*, 2003; Kremen and Miles, 2012), and to think about factors of relevance to the health of ecosystems, both terrestrial and aquatic, and so-called environmental flows, to safeguard the essential characteristics of habitats.

5.3.3 Shifting weather patterns

Rainfall variability, shocks and surprises

There has been a lot of debate about the extent to which climate change has caused or will cause changes in the frequency, magnitude and duration of water-related disturbance events such as storms, floods and droughts. Kabat *et al.* (2003) state that a combination of natural and human activities is having a disturbing influence on the water cycle. Dramatically changing weather patterns mean that replenishing water arrives as intense rainfall, overwhelming flood defences and escaping to the ocean before it can be stored to irrigate crops and sustain ecosystems. Large fluctuations in annual river flows bring floods to some and droughts to others, ruining the livelihoods of farmers. The impacts of a variable climate are felt in all parts of the world, but the severity of the impacts and the vulnerability of people and ecosystems differ enormously from place to place. Some hot-spots are obvious: megacities with informal shantytowns in floodplains face huge losses from flooding and storms; and countries already struggling to produce enough food for their growing population are particularly susceptible to seasonal fluctuations in rainfall (Table 5.1).

Countries and regions that have been severely affected by drought in the twentieth century include Afghanistan, Ethiopia, Kenya, Pakistan and India, and the Middle East and northeast Brazil. The African continent, especially the Sahel and Southern and Central Africa, has been exposed more than any other region of the world to recurrent droughts (Kabat *et al.*, 2003). There are observable medium-term trends for annual rainfall and river flows and assumed increases in evaporation losses linked to climate change which raise concerns, and the Sahel is one of the most pronounced examples (see Table 5.2).

Table 5.1 Selected floods in 2002 (from Kabat *et al.*, 2003)

Location	Duration (days)	Affected region (km^2)	Damage (USD/ km^2)
Central and South America			
Brazil (central)	2	780 000	8 974
Brazil (west)	2	2 250	
Chile	12	166 900	190
Ecuador	54	52 930	
Peru	11	333 200	
Uruguay	30	187 500	
Venezuela (west)	11	224 900	13
Other			
Europe (central)	18	252 300	79 270
China (north west)	10	252 000	1 587
Russia (south)	12	224 600	1 945
Trinidad	15	880	3 750

Western countries are also hit by droughts, like that experienced in the United States in 2002 which also affected water-sensitive sectors throughout Canada, with a lack of precipitation in several areas including the prairies, 45% of livestock farms facing water shortages and agriculture yields down by 50–60%.

Shifting anomalies

Ignoring the water challenges of climate change and associated extreme events is no longer an option in the assessment, management and governance of food production. New perspectives are needed to prepare for and adapt to the effects of climate variability and the likely implications of climate change, not only on water and food production but also on water functions and uses for human well-being in general (Kabat *et al.*, 2003).

Hansen *et al.* (2012) found that the distribution of seasonal mean temperature anomalies has shifted towards higher temperatures and the range of anomalies has increased (Figure 5.10). An important

Table 5.2 Average precipitation and river flow changes in Sahelian countries over consecutive 20-year periods. The figures compare average annual precipitation and river flows in the period 1970 to 1989 with those between 1950 and 1969 (from Kabat *et al.*, 2003)

Country		Reduction in precipitation (%)
Cameroon		16
Togo		16
Central African Republic		17
Benin		19
Ghana		19
Nigeria		19
Guinea		20
Chad		20
Ivory Coast		21
Burkina Faso		22
Guinea Bissau		22
Mali		23
Senegal		25
River	**Gauging station**	**Reduction in annual flow (%)**
Comoe	Aniassue	50
Chari	Ndjamena	51
Logone	Lai	39
Niger	Malanville	43
Niger	Niamey	34
Bani	Douna	70
Oueme	Sagon	42
Sassandra	Semien	36
Senegal	Bakel	50
Bakoye	Ouali	66
Black	Volta Dapola	41
Black	Volta Boromo	46
Oubangui	Bangui	30

change is the emergence of a category of extremely hot summertime outliers, more than three standard deviations (3σ) warmer than the climatology of the 1951–80 base period used in the analyses. The hot extreme covered much less than 1% of the Earth's surface during the 1951–80 base period, but in recent years has covered about 10% of the land area. Hansen *et al.* (2012) state that extreme anomalies such as those in Texas and Oklahoma in 2011, and Moscow in 2010 were a consequence of global warming because their likelihood in the absence of global warming would be exceedingly small.

5.3.4 Potential traps and resilience building

Seeing the food supply system from a resilience perspective shows that this social–ecological system should be able to continue to function and deliver food, while experiencing shocks and change. It will be essential to support the development of landscapes for food production that are resilient to change in order to reduce the risk of falling into social–ecological traps. The stewardship of water as the bloodstream of the biosphere becomes critical in this context.

Potential traps

Chapter 3 examined how social–ecological traps can develop in relation to water use and misuse, and gives examples of both rigidity traps, which have their roots in governance or management problems, and poverty traps, which originate from economic phenomena and their interrelations in society. It also gave examples of both kinds of trap.

Getting out of traps or shifting off pathways leading to traps are key challenges in water management and food production. Rigidity traps call for mental shifts and future visions, including water use alternatives and changing the way we think about water use and management. Poverty traps require enhancing the adaptive capacity of people, communities and societies to interact, mobilise alternative resources and develop new pathways, and secure external investment. There is a lot of literature on traps, as well as some in relation to water, food production and resilience (e.g. Cifdaloz *et al.*, 2010; Walker *et al.*, 2009).

Stewardship to sustain rainfall

A less discussed precondition is the resilience of the precipitation system as a whole – the challenge of sustaining rainfall discussed elsewhere in this volume. This calls for a focus on moisture feedback to the atmosphere, providing the moisture needed for the

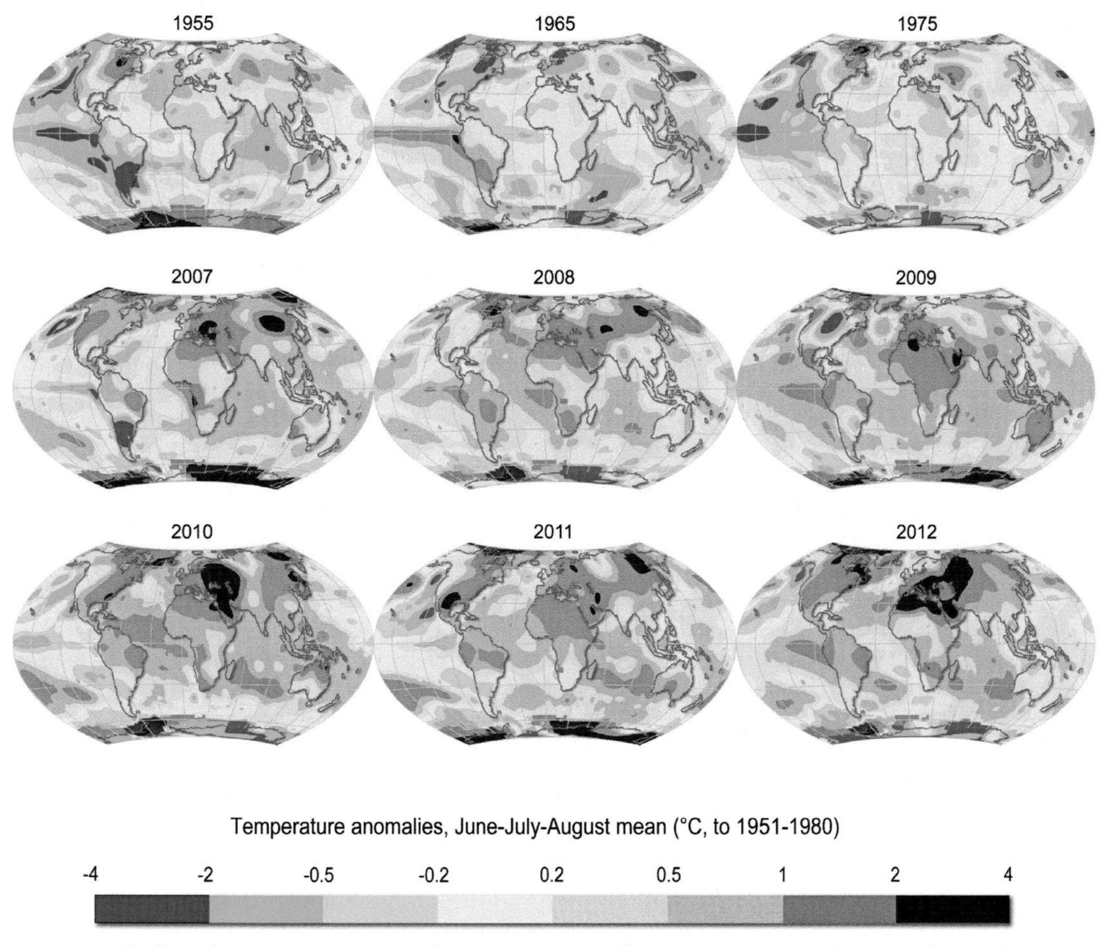

Figure 5.10 The figure shows June–July–August surface temperature anomalies in 1955, 1965, 1975, and 2007–12 relative to the 1951–80 mean temperature in units of the local detrended 1981–2010 standard deviation of temperature. (after Hansen *et al.*, 2012).

production of new rain in critical regions. The balance between green and blue water flows from agricultural land is fundamental from the perspective of the concept of planetary boundaries introduced in Chapter 1. As is noted above, total consumptive blue water use must be limited to 4000 km^3/year, of which 2600 km^3/year is already consumed by current levels of crop production (Rockström *et al.*, 2009). Thus, at the planetary level, additional consumptive blue water use must be limited to some 2400 km^3/year.

The implications of changes in temperature anomalies for water as the bloodstream of the biosphere will be challenging for food production, other ecosystem services and human well-being. Food production and water management may be confronted with less predictability, stronger variability and rainfall patterns and associated disturbance events that are altered in frequency, magnitude and duration, as well as in time and space. In the Anthropocene era, such changes may become more the norm than the exception. Adaptability, diversity and resilience will be required to absorb such events, along with flexible governance systems that can turn social and economic challenges in regions hit by, e.g. prolonged drought or devastated crops, into new opportunities for development. It is in this new, more turbulent context (Folke and Rockström, 2009) that we set out in this book the concept of stewardship to sustain rainfall over landscapes and catchments, and between them.

5.3.5 Future strategic trade-offs

Social–ecological resilience will be required to deal with turbulent times and surprising change (Carpenter *et al.*, 2012). Hence, there will be an important trade-off between food production, aiming for improvements in efficiency and the ratio of water and land used per unit of food produced, and the capacity of the land and the social–ecological system to deal with climate-related changes in rainfall patterns and other forms of unexpected disturbance as a consequence of the Anthropocene era (Chapter 1 and 2). The former approach is often based on static analyses that assume a fairly similar future, with incremental changes as the norm. The latter often takes a complex systems approach, which explicitly accounts for uncertainty and surprise, and includes strategies for insurance and resilience. Food production based on widespread monocultures and a handful of crops may work in the former situation. More diverse food production systems will be required to deal with the latter. New technology and innovation may push agriculture deeper into the current trajectory of efficiency development or help shift and transform it into resilient, social–ecological food production systems that are prepared to deal with a more turbulent world.

Summary

Driving forces increasing food demand

Since the early 1970s, food production has been able to keep pace with the global population largely thanks to agricultural science and modern techniques linked to the Green Revolution. The global population continues to increase, mainly in developing countries, with the medium UN population projection forecasting a total of 9.3 billion people by 2050. With increased economic development, people want to move up the food chain, leading to increased calorie consumption per person and growing consumption of animal-based food products. Many developing countries, such as Brazil and China, are approaching the average calorie level of many developed countries. However, in many developing countries the national per capita calorie supply level remains as low as or less than 2100 kcal/capita per day and globally almost 900 million are still undernourished.

Food water requirements

Production of animal-source foods and vegetable foods compete for natural resources and agricultural production capacity in areas where the cultivation of feed and fodder competes with the cultivation of vegetal foods. On pasture that is only suitable for grazing it is only ruminants such as cattle and sheep that compete with other terrestrial ecosystems. Different animal production systems therefore have different impacts on blue and green water resources.

The consumptive blue water use of different food commodities varies according to the total calorie level, the proportion and combination of different vegetal components, and the share and mix of animal food items. The water use by and water productivity of different food supply combinations are governed by the water productivity of plant growth for vegetal foods, and of the growth of feed and fodder used for livestock as well as the conversion efficiency from vegetal feed to animal-source foods.

The potential to increase water productivity is particularly high at low crop yield levels, often corresponding to smallholder farmers in semi-arid and arid areas. In addition, the water productivity of animal-source foods can be dramatically improved in many regions and systems. Considerable benefits could be reaped if food losses could be minimised. To reduce the pressure on water resources it will be crucial to reduce food losses in both developed and developing countries.

Production on current cropland: a static perspective

Without improved water management in agriculture there will not be enough water on current croplands to produce a global average per capita food supply level by 2050 in line with the global dietary tendency of 3000 kcal/capita per day with 20% from animal-source foods. If current water productivity levels continue till 2050 the summarised water deficits will be many times larger than the surplus. However, with maximum irrigation expansion and a yield gap closure of 50%, global food water surpluses and deficits can be balanced. At a yield gap closure of 25% surplus exceeds the deficit only if the animal calorie fraction is reduced from 20% to 5% for the 3000 kcal level. The analysis demonstrates the importance of water productivity improvements and of dietary composition, and shows that there are different combinations to balance future global water surpluses and water deficits.

Food trade to balance water deficits

Imbalances in food water availability will have to be compensated for by the trade in food. The world is therefore moving towards a future in which around two-thirds of the world population is dependent on the remaining one-third for the supply of part of their food needs. The considerable regional food water deficiencies will have to be met by a well-organised and reliable system of food trade. There is a general correlation between low-income countries and large water deficits. Some 30 poor countries, the large majority of which are in Africa, will most likely have difficulty in balancing food deficits from the global food market. The scale of regional food transfers through trade shows the fundamental importance of economic development in import-dependent, water-scarce countries in order to secure the necessary purchasing power.

A dynamic perspective: to cope with shocks and change

The food production system is not static but dynamic. There are many kinds of possible disturbances and rainfall is already subject to large degrees of variability, which will increase with climate change. The agricultural system has to cope with threats from sudden shocks such as droughts, pests or social phenomena. *Biomass production functions* must be secured to allow the required upgrading of agricultural production to be safely implemented. The functioning of crucial regulatory ecosystem services in the landscape has to be safeguarded. There may be creeping changes in the production system, originating from altered rainfall patterns, the effects of increasing temperatures, or possible declines in land productivity. Vital functions in the *support systems* also need attention: a resilient system of food trading, competing land uses, and the possibility of an expansion of cropland. Disturbances might be related to securing surplus food production for export, a realisation of the desirability of dietary changes, difficulties in balancing food production with other biomass requirements in society, the functioning of the crucial food trade system, protests linked to foreign land acquisition, and so on. Disturbances related to landscape dynamics are discussed in Chapter 7.

Rainfall variability, shocks and surprises

Hot temperature extremes covered much less than 1% of the Earth's surface in the period 1951–80, but they now cover about 10% of the land area. Dramatically changing weather patterns lead to intense rainfall, which overwhelms flood defences and escapes to the ocean before it can be stored to irrigate crops and sustain ecosystems. The impacts of a varying climate are felt in all parts of the world, but the severity of the impacts and the vulnerability of people and ecosystems differ enormously from place to place. Countries already struggling to produce enough food for growing populations are particularly susceptible to seasonal fluctuations in rainfall, although western countries are also being hit by droughts,

Adaptability and social–ecological resilience

The challenge of global food production is moving into much more dynamic terrain than has been assumed in most predictions on food production and associated water demands. The Anthropocene era is a new phase in the life of the Earth, reflecting the sheer pressure of the human dimension combined with climate dynamics such as altered precipitation patterns.

The new situation poses new challenges for sustaining the ability to produce food for humanity. It will require strategies that enhance capacities to deal with change, including sudden and surprising change, and even make use of such events to shift development on to new pathways and towards sustainability. Social–ecological resilience is a precondition for such transformations. More diverse food production systems will be required, and attention must be paid to the important future trade-off between efficiency-oriented food production, and the capacity of the land and the social–ecological system to deal with climate-related changes in rainfall patterns and other unexpected disturbances.

References

Alexandratos, N. (2009). *World Food and Agriculture to 2030/50: Highlights and Views from Mid-2009*. Rome: Food and Agriculture Organization.

Allen, R. G., Pereira, L. S., Raes, D. and Smith, M. (1998). Crop evapotranspiration, guidelines for computing crop water requirements. FAO Irrigation and Drainage Paper 56. Food and Agriculture Organization, Rome.

Anseeuw, W., Boche, M., Breu, T. *et al.* (2012). Transnational land deals for agriculture in the global south: analytical report based on the land matrix database. The Land Matrix Partnership. Available at: http://landportal.info/landmatrix/media/img/analyticalreport.pdf.

Barrett, C. B. (2010). Measuring food insecurity. *Science*, **327**, 825–828.

Bruinsma, J. (2009). *The Resource Outlook to 2050: By How Much do Land, Water and Crop Yields Need to Increase by 2050?* Rome: Food and Agriculture Organization.

Carpenter, S. R., Arrow, K. J., Barrett, S. *et al.* (2012). General resilience to cope with extreme events. *Sustainability*, **4**, 3248–3259.

Chapagain, A. K. and Hoekstra, A. Y. (2003). Virtual water flows between nations in relation to trade in livestock and livestock products. Report No. 13: UNESCO-IHE, Delft, the Netherlands.

Cifdaloz, O., Regmi, A., Anderies, J. M. and Rodriguez, A. A. (2010). Robustness, vulnerability, and adaptive capacity in small-scale social–ecological systems: the Pumpa Irrigation System in Nepal. *Ecology and Society*, **15**, 39.

de Haan, C., Gerber, P. and Opio, C. (2010). Structural change in the livestock sector. In *Livestock in a Changing Landscape. Volume 1. Drivers, Consequences and Responses.*, ed. Steinfeld, H., Mooney, H., Schneider, F. and Neville, L. E. London: Island Press, pp. 35–50.

Elmqvist, T., Folke, C., Nyström, M. *et al.* (2003). Response diversity, ecosystem change, and resilience. *Frontiers in Ecology and the Environment*, **1**, 488–494.

Fader, M., Rost, S., Müller, C., Bondeau, A. and Gerten, D. (2010). Virtual water content of temperate cereals and maize: present and potential future patterns. *Journal of Hydrology*, **384**, 218–231.

Falkenmark, M. and Lannerstad, M. (2005). Consumptive water use to feed humanity: curing a blind spot. *Hydrology and Earth System Sciences*, **9**, 15–28.

Falkenmark, M. and Rockström, J. (2004). *Balancing Water for Humans and Nature: The New Approach in Ecohydrology*. London: Earthscan.

Foley, J. A., Ramankutty, N., Brauman, K. A. *et al.* (2011). Solutions for a cultivated planet. *Nature*, **478**, 337–342.

Folke, C. and Rockström, J. (2009). Turbulent times. *Global Environmental Change*, **19**, 1–3.

Food and Agriculture Organization (2009). *The State of Food Insecurity in the World 2009: Economic Crises – Impacts and Lessons Learned*. Rome: Food and Agriculture Organization.

Food and Agriculture Organization (2011). *The State of Food Insecurity in the World 2011: How does International Price Volatility Affect Domestic Economies and Food Security?* Rome: Food and Agriculture Organization.

Food and Agriculture Organization (2012). *The State of Food Insecurity in the World 2012: Economic Growth is Necessary but not Sufficient to Accelerate Reduction of Hunger and Malnutrition*. Rome: Food and Agriculture Organization of the United Nations (FAO), the International Fund for Agricultural Development (IFAD) and the World Food Programme (WFP).

Food and Agriculture Organization (2013a). FAO: Food security indicators. Rome: Committee on World Food Security (Cfs) Round Table on Hunger Measurement. Available at: http://www.fao.org/publications/sofi/food-security-indicators/en/.

Food and Agriculture Organization (2013b). FAOSTAT online database. Rome: Food and Agriculture Organization. Available at: http://faostat.fao.org/ (accessed multiple dates).

Gerten, D., Heinke, J., Hoff, H. *et al.* (2011). Global water availability and requirements for future food production. *Journal of Hydrometeorology*, **12**, 885–899.

Global Health Observatory (2013). *Overweight and Obesity*. Global Health Observatory, Geneva. Available at: http://www.who.int/gho/en/.

Godfray, H. C. J., Beddington, J. R., Crute, I. R. *et al.* (2010). Food security: the challenge of feeding 9 billion people. *Science*, **327**, 812–818.

Gustavsson, J., Cederberg, C., Sonesson, U., van Otterdijk, R. and Meybeck, A. (2011). *Global Food Losses and Food Waste: Extent, Causes and Prevention*. Rome: Food and Agriculture Organization.

Hansen, J., Sato, M. and Ruedy, R. (2012). Perception of climate change. *Proceedings of the National Academy of Sciences*, **109**, 14726–14727.

Hazell, P. B. and Hess, U. (2010). Drought insurance for agricultural development and food security in dryland areas. *Food Security*, **2**, 395–405.

Heinke, J., Lannerstad, M., Hoff, H. *et al.* (forthcoming). Livestock and water: a blue, green and green continuum. *Proceedings of the National Academy of Sciences*, submitted.

Herrero, M., Thornton, P. K., Gerber, P. and Reid, R. S. (2009). Livestock, livelihoods and the environment:

understanding the trade-offs. *Current Opinion in Environmental Sustainability*, **1**, 111–120.

International Food Policy Research Institute. (2002). Green revolution: curse or blessing? International Food Policy Research Institute, Washington DC. Available at: http://chapters.ewb.ca/pages/member-learning/africa-(culture_-history_-livelihoods)/greenrevreading1.pdf.

Kabat, P., van Schaik, H., Appleton, B. and Veraart, J. (2003). Climate changes the water rules: How water managers can cope with today's climate variability and tomorrow's climate change. Dialogue on Water and Climate, Wageningen, the Netherlands. Available at: http://www.unwater.org/downloads/changes.pdf.

Keller, A. and Seckler, D. (2004). Transpiration: constraints on increasing the productivity of water in crop production. Winrock Water. Paper for Winrock Water Forum, Winrock International, Arlington, VA.

Kremen, C. and Miles, A. (2012). Ecosystem services in biologically diversified versus conventional farming systems: benefits, externalities, and trade-offs. *Ecology and Society*, **17**, 40.

Lannerstad, M. (2009). Water realities and development trajectories: global and local agricultural production dynamics. PhD thesis, Linköping University, Sweden.

Lehner, B. and Döll, P. (2004). Development and validation of a global database of lakes, reservoirs and wetlands. *Journal of Hydrology*, **296**, 1–22.

Millennium Ecosystem Assessment. (2005). *Ecosystems and Human Well-being: Synthesis*. Washington DC: Island Press.

Mitchell, T. D. and Jones, P. D. (2005). An improved method of constructing a database of monthly climate observations and associated high-resolution grids. *International Journal of Climatology*, **25**, 693–712.

Molden, D. (2007). *Water for Food, Water for Life: A Comprehensive Assessment of Water Management in Agriculture*. London: Earthscan.

Molden, D., Oweis, T., Steduto, P. *et al.* (2010). Improving agricultural water productivity: between optimism and caution. *Agricultural Water Management*, **97**, 528–535.

Peden, D., Tadesse, G. and Mammo, M. (2003). Improving the water productivity of livestock: an opportunity for poverty reduction. Paper for the workshop 'Integrated water and land management research and capacity building priorities for Ethiopia'. International Livestock Research Institute, Addis Adaba.

Pimentel, D., Houser, J., Preiss, E. *et al.* (1997). Water resources: agriculture, the environment, and society. *Bioscience*, **47**, 97–106.

Popkin, B. M., Horton, S. and Soowon, K. (2001). The nutrition transition and prevention of diet-related diseases in Asia and the Pacific. Food and Nutrition Bulletin, vol. 22, no. 4 (supplement). United Nations University Press, Tokyo.

Portmann, F. T., Siebert, S. and Döll, P. (2010). MIRCA2000-Global monthly irrigated and rainfed crop areas around the year 2000: a new high-resolution data set for agricultural and hydrological modeling. *Global Biogeochemical Cycles*, **24**, Gb1011.

Renault, D. (2003). Value of virtual water in food: principles and virtues. In *Virtual Water Trade: Proceedings of the International Expert Meeting on Virtual Water Trade*, ed. Hoekstra, A. Y. Delft: IHE Delft, pp. 77–91.

Rijsberman, F. R. (2006). Water scarcity: fact or fiction? *Agricultural Water Management*, **80**, 5–22.

Robinson, T. P., Thornton, P. K., Franceschini, G. *et al.* (2011). *Global Livestock Production Systems*. Rome: Food and Agriculture Organization of the United Nations (FAO) and International Livestock Research Institute (ILRI).

Rockström, J. (2003). Water for food and nature in drought-prone tropics: vapour shift in rain-fed agriculture. *Philosophical Transactions of the Royal Society of London Series B: Biological Sciences*, **358**, 1997–2009.

Rockström, J., Axberg, G. A., Falkenmark, M. *et al.* (2005). *Sustainable Pathways to Attain the Millennium Development Goals: Assessing the Role of Water, Energy and Sanitation*. Stockholm: Stockholm Environment Institute.

Rockström, J., Falkenmark, M., Karlberg, L. *et al.* (2009). Future water availability for global food production: the potential of green water for increasing resilience to global change. *Water Resources Research*, **45**.

Rockström, J., Gordon, L., Folke, C., Falkenmark, M. and Engwall, M. (1999). Linkages among water vapor flows, food production, and terrestrial ecosystem services. *Conservation Ecology*, **3**, 5.

Rockström, J., Hatibu, N., Oweis, T. Y. *et al.* (2007a). Managing water in rainfed agriculture. In *Water for Food, Water for Life: A Comprehensive Assessment of Water Management in Agriculture*, ed. Molden, D. London: Earthscan, pp. 315–351.

Rockström, J., Lannerstad, M. and Falkenmark, M. (2007b). Assessing the water challenge of a new Green Revolution in developing countries. *Proceedings of the National Academy of Sciences of the United States of America*, **104**, 6253–6260.

Rohwer, J., Gerten, D. and Lucht, W. (2007). Development of functional types of irrigation for improved global crop modelling. PIK Report No. 104: Potsdam Institute for Climate Impact Research, Potsdam, Germany. Available at: http://www.pik-potsdam.de/research/publications/pikreports/.files/pr104.pdf.

Rosegrant, M. W., Leach, N. and Gerpacio, R. V. (1999). Alternative futures for world cereal and meat consumption. *Proceedings of the Nutrition Society*, **58**, 219–234.

Rost, S., Gerten, D., Bondeau, A. *et al.* (2008). Agricultural green and blue water consumption and its influence on the global water system. *Water Resources Research*, **44**.

Smil, V. (2000). *Feeding the World: A Challenge for the Twenty-first Century*. Cambridge, MA: MIT Press.

Steinfeld, H., Gerber, P., Wassenaar, T. *et al.* (2006). *Livestock's Long Shadow: Environmental Issues and Options*. Rome: FAO.

Taiz, L. and Zeiger, E. (2010). *Plant Physiology*, 5th edn. Sunderland, MA: Sinauer Associates.

Tanner, C. B. and Sinclair, T. R. (1983). Efficient water use in crop production: research or re-search? In *Limitations in Efficient Water Use in Crop Production*, ed. Taylor, H. M., Jordan, W. A. and Sinclair, T. R. Madison, WI: American Society of Agronomy, p. 538.

Tilman, D., Balzer, C., Hill, J. and Befort, B. L. (2011). Global food demand and the sustainable intensification of agriculture. *Proceedings of the National Academy of Sciences*, **108**, 20260–20264.

United Nations Department of Economic and Social Affairs (2010). *World Urbanization Prospects. The 2009 Revision*. New York: United Nations Population Division. Available at: http://www.un.org/esa/population/unpop.htm.

United Nations Department of Economic and Social Affairs (2011). *World Population Prospects: The 2010 Revision*. New York: United Nations Department of Social Affairs. Available at: http://esa.un.org/unpd/wpp/Excel-Data/population.htm (accessed 21 November 2012).

United Nations Population Division (UNDP) (2004). *World Population to 2300*. New York: United Nations Department of Economic and Social Affairs. Available at: http://www.un.org/esa/population/publications/publications.htm.

van Breugel, P., Herrero, M., van de Steeg, J. and Peden, D. (2010). Livestock water use and productivity in the Nile basin. *Ecosystems*, **13**, 205–221.

von Braun, J. and Meinzen-Dick, R. S. (2009). *'Land Grabbing' by Foreign Investors in Developing Countries: Risks and Opportunities*. Washington DC: International Food Policy Research Institute.

Walker, B., Barrett, S., Polasky, S. *et al.* (2009). Looming global-scale failures and missing institutions. *Science*, **325**, 1345–1346.

World Bank. (2009). Minding the stock: bringing public policy to bear on livestock sector development. Report no. 44010-GLB. World Bank, Washington DC.

World Bank. (2011). Country and lending groups. World Bank, Washington DC. Available at: http://data.worldbank.org/about/country-classifications/country-and-lending-groups.

World Health Organization. (2011). Global status report on noncommunicable diseases 2010. World Health Organization, Geneva.

World Resources Institute. (2005). *Ecosystems and Human Well-being: Wetland and Water Synthesis*. Washington DC: World Resources Institute.

Zwart, S. J. and Bastiaanssen, W. G. M. (2004). Review of measured crop water productivity values for irrigated wheat, rice, cotton and maize. *Agricultural Water Management*, **69**, 115–133.

Mwembe village in the Makanya catchment in northeastern Tanzania - agricultural water management system with supplemental irrigation diverting runoff from a nearby gully. Maize is grown on the terraces, bananas in the ditches and kassava on the soil mounds.

Chapter

6 Closing the yield gap in the savannah zone

This chapter takes a closer look at low-income countries in the savannah zone, where poverty and malnutrition are dominant features. Although high rainfall variability across time and space is a key challenge, this zone has substantial hydrological potential. The chapter analyses how investments in agricultural water management can contribute to closing the yield gap between actual and potential yields in the savannahs, and thereby to building resilience in these regions. The chapter discusses the range of technological, biophysical and social constraints currently limiting the uptake of these promising technologies. Conclusions are finally drawn regarding broad based farming system solutions that target the interacting constraints, and deal with the uncertainty of the future.

6.1 Challenges and opportunities in the savannah zone

This chapter examines the savannah zone, which presents an enormous development challenge but also a unique opportunity to improve food production and build social–ecological resilience. A major constraint on agro-ecological productivity across the world's savannah regions is temporal shortages of water in the crop root zone. This chapter focuses on a set of small-scale practices that improve water management in agriculture to help overcome this constraint. If properly implemented and supported, these practices can unlock the agricultural potential of the savannah zone and provide a window of opportunity for improved livelihoods for millions of smallholder farmers across the world.

Figure 6.1 illustrates the large convergence between regions with high levels of poverty and malnourishment and the world's 'drylands', i.e. regions with an arid, semi-arid or dry sub-humid climate. Excluding the arid regions, which although important are home to a limited number of people on a global scale, these climatic zones cover about 30% of the Earth's land area and are home to about 1.7 billion people (UNCCD, 2011). The main livelihood source is small-scale farming, often practised in combination with keeping livestock. Most farming systems are rainfed and non-mechanised, and external inputs are scarce. Yields fluctuate, but are low, and since the capacity to import food in these agricultural-based economies is low as well, food security plays a central role in the poverty challenge.

Low crop yields in the savannah zone are a complex problem linked to a number of biophysical and socio-economic factors, but the challenging agro-hydro-climate, with high rainfall variability and large unproductive flows in the field water balance in terms of evaporation and runoff, is a fundamental issue (Rockström and Falkenmark, 2000). This already substantial challenge is aggravated by looming global environmental change, and particularly climate change, which is predicted to have some of its largest and most immediate impacts precisely in the semi-arid and dry sub-humid tropics.

There is a strong rationale for focusing on these regions when it comes to strategies to close yield gaps and build resilience because many of them, in contrast to what the term dryland suggests, have substantial and often untapped agro-hydrological potential. This makes the savannah systems of the world global hotspots in terms of water, food and poverty alleviation in the twenty-first century (Falkenmark and Rockström, 2004).

6.1.1 Rethinking drylands

The labelling of semi-arid and dry sub-humid regions as drylands has contributed to a general perception that they are subject to persistent water scarcity, with

(a)

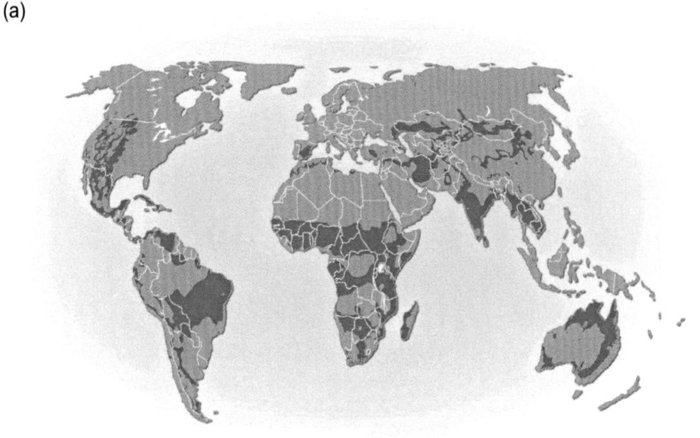

Savannah and steppe climate zones

(b)

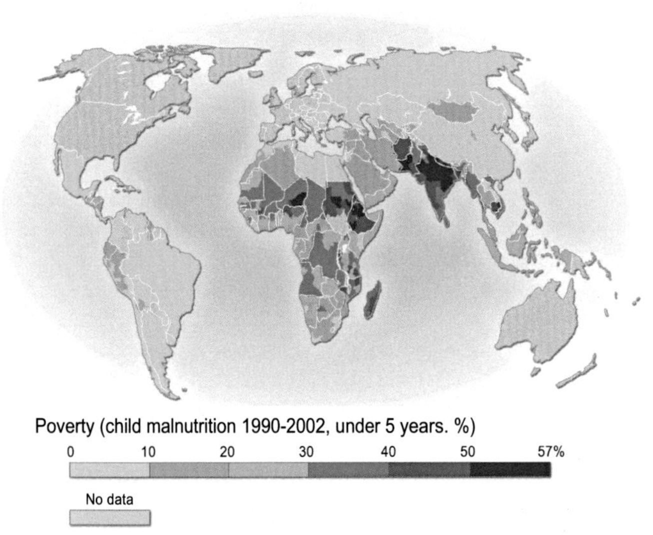

Figure 6.1 Climatic zones and malnourishment. There is a large overlap between the world's savannah regions (a) and the regions of the world hardest hit by poverty (b), measured as the percentage of children under five years of age suffering from malnutrition (CIESIN, 2005; Peel *et al.*, 2007).

sandstorms, rolling dunes, cracking dry soils and thorny shrubs. This misperception has in turn influenced investment in agricultural development in savannah regions. The entry point has been, and still often is, that these areas are notorious for water scarcity and thus have low agricultural potential.

Today, agricultural regions in the savannahs, be it in the Cerrado in Brazil, at the Mossi plateau in Burkina Faso or in Gansu province in northern China, are largely operating way below their agro-hydrological yield potential. Most of these regions have farming systems with productivity levels for their basic staple food crops in the range of 1–2 tonnes/ha (Rockström *et al.*, 2007) when the yield potential, applying current technologies and management practices, could be at least 3–6 tonnes/ha (Rockström *et al.*, 2007; Wani *et al.*, 2009). We believe that the exaggerated perception of dryness in these regions has contributed to the underperformance of their agriculture.

6.1.2 Savannah agro-hydro-climate

Semi-arid and dry sub-humid landscapes are characterised by a sharp divide between the wet and dry periods of the year, with annual rainfall that ranges from 350 mm to 1500 mm. This rainfall can be either mono-modal, as in West Africa with one rainy season per year, or bi-modal, as in East Africa, India and South East Asia, with two rainy seasons per year – one major and one minor. The savannah zone is further characterised by extreme rainfall variability, manifested by coefficients of variation for annual rainfall of between 15 and 50%. This means that average annual rainfall is a concept that has virtually no meaning in this context and that drought years especially, but also flood years are common in these regions.

The atmospheric demand for water is high in the savannah zone, putting constraints on water availability. For example, in the West African Sahel annual potential evapotranspiration (PET) may be in the order of 2000 mm and annual precipitation (P) in the order of 700 mm, giving an aridity index (P/PET) close to 0.3, which indicates a very dry area. However, this gives a false impression, since all rainfall here is concentrated in the rainy season, during which potential evaporation is lower due to increased cloudiness, lower air temperature and higher air humidity. Atmospheric demand for water during this period may amount to 5 mm per day, compared to almost 10 mm during the dry season. This is still a high figure, but much lower than the PET in the dry season – which is a theoretical notion anyway as it is never fed by any moisture, as there is no rainfall and thus no evaporation. For a rainy season of 120 days, which is common in this region, this gives an evaporative demand of 600 mm and an aridity index of 1.2, indicating substantially wetter conditions. Thus, when assessing the agro-hydrological potential of savannah systems it is important to focus on the time of year when crops are actually grown, since during the rest of the year these regions are completely dry.

However, even in the rainy seasons precipitation is highly erratic and often falls during intense events, such as convective storms. Rainfall intensity can reach 75 mm per hour or more, with extreme spatial and temporal variability as a consequence. This means that a large part of the total annual rainfall often falls during a few short showers. The ability to cope with this variability is a key challenge for farmers in the savannah zone. Erratic rainfall also results in large volumes of surface runoff leaving the field before it

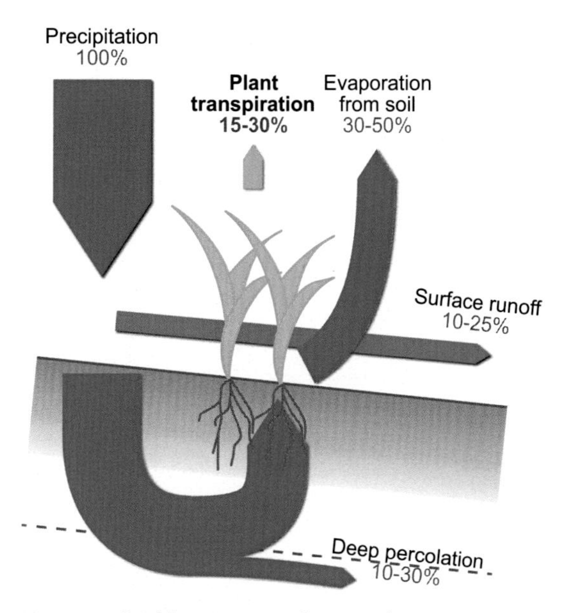

Figure 6.2 Rainfall partitioning in the savannah zone. Losses via runoff, evaporation and drainage are so large that only 15–30% of the rain becomes available for plant growth (transpiration) (Rockström and Falkenmark, 2000).

can percolate. Water availability for agricultural production is thus further reduced. Studies have shown that 40–75% of the rainfall is lost from the farmer's field due to direct evaporation and surface runoff (Rockström and Falkenmark, 2000). Finally, part of the water that infiltrates the soil is also lost via drainage. Thus, only a fraction of the rain that falls on a farmer's field becomes available for biomass production (Figure 6.2).

Another consequence of the variable rainfall is prolonged dry periods of 2–4 weeks between rainfall events, so-called intra-seasonal dry spells. If such sequences of water stress occur during water-sensitive development stages, such as during flowering or yield formation, they seriously decrease crop yields (Rockström and De Rouw, 1997). Bridging these periods is key to improving agricultural productivity in the savannah zone. An important point is that dry spells that cause water scarcity do not show up as reductions in seasonal or annual rainfall figures. Studies have shown that critical dry spells causing agricultural droughts occur in more than 60% of the seasons in semi-arid and dry sub-humid areas in sub-Saharan Africa with annual rainfall below 600 mm (Figure 6.3)

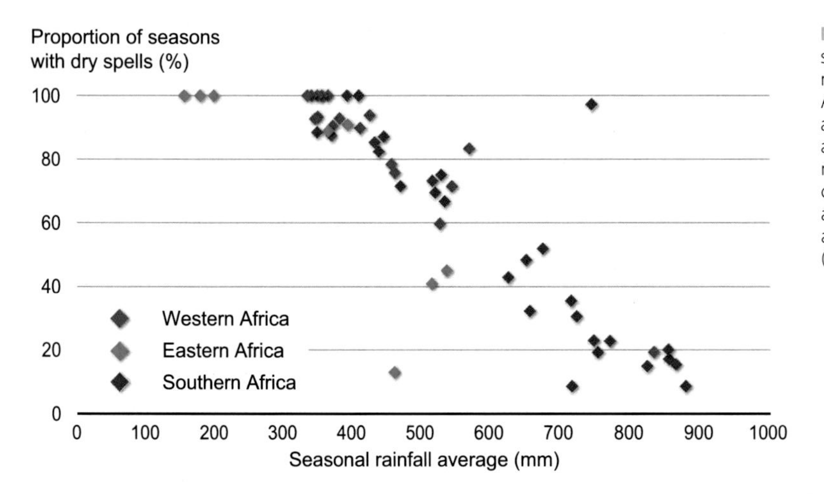

Figure 6.3 Proportion of rainfall seasons with critical dry spells for 64 rainfall stations across sub-Saharan Africa, plotted against seasonal average rainfall. For stations with average rainfall less than 600 mm, more than half the seasons are hit by critical dry spells. For seasons with average rainfall below 400 mm almost all seasons are hit by critical dry spells (Enfors *et al.*, in preparation).

(Enfors *et al.*, in preparation). The frequent occurrence of dry spells in the savannah zone means that they are a more common cause of crop yield reductions and crop failure than meteorological droughts, which on average only occur in one in 10 years and are clearly observable in annual rainfall records (Wani *et al.*, 2009).

Nonetheless, there is frequent confusion around these terms, and yield losses caused by dry spells are often blamed on droughts. This is problematic not only because it contributes to the exaggerated image of drylands discussed above, but also since it directs attention to certain types of remedies rather than others. It is important to distinguish between dry spells, which lead to brief agricultural droughts, and meteorological droughts, which disrupt crop cultivation over an entire season or year, because the management options to handle these challenges are very different. During a meteorological drought, there is not enough water to produce a crop, which means that other types of coping mechanism such as food relief have to take over. In a rainy season subject to dry spells, however, there is enough water but it is not available at the right time. The challenge is to bridge this gap.

Predicted climate change will add to the agro-hydrological challenge in savannah systems. Both rainfall variability and the frequency of extreme events are expected to increase across much of sub-Saharan Africa, for example, and increasing temperatures will increase evaporation (Alcamo *et al.*, 2007; Boko *et al.*, 2007). Interestingly, a synthesis of recent climate change studies from West Africa shows that temperature changes are predicted to have a much larger influence on agricultural production relative to precipitation changes (Roudier *et al.*, 2011). While the warming effect will be mitigated by increasing rainfall in some areas, it will exacerbated by declining rainfall in others. Given the above, Roudier *et al.* (2011) calculate an overall decline in Sahelian crop yields in the order of 20% in the coming decades, excluding the potential for adaptation measures.

In conclusion, the hydrological challenge in savannah systems across the world is not related to an absolute lack of water, but rather to huge fluctuations in rainfall over time and space, and to high water losses from the farming system through evaporation, runoff and drainage. While the current situation is already difficult for smallholder farmers, predictions of the effects of climate change point to worsening conditions in some areas.

6.1.3 Farming and livelihoods in the savannah zone

The challenging hydro-climate is a major reason for the continually low crop yields experienced by many smallholders in the savannah zone, but there are other important factors too. These include other biophysical constraints, such as poor soils and a lack of fertilisers, but also socio-economic constraints linked to the marginalisation and multidimensional poverty often experienced by rural dryland dwellers (Mortimore, 2005; Reynolds *et al.*, 2007).

Although islands of intensive agriculture with substantially higher yield levels exist (see e.g. Tiffen *et al.*, 1994), especially in densely populated areas with good market connections, smallholder farming enterprises are generally the most extensive. This is particularly the case in sub-Saharan Africa. Repeated attempts have been made to intensify these systems, but unlike in South Asia most of these have been short-lived and the Green Revolution has not been sustained. Maize yields, for example, still average only 1.5 tonnes/ha, and sorghum just below 1 tonne/ha. The increase in cereal production over the past 50 years has been achieved through an expansion of farmland rather than intensification and higher yields. It is interesting to compare the development of cereal production in South Asia and sub-Saharan Africa. In both cases, total production of cereals increased markedly in 1961–2001 but while per capita output grew by 24% in Asia, it decreased by 13% in sub-Saharan Africa (Djurfeldt *et al.*, 2005). It will be necessary to intensify current farming systems to be able to meet current and future food needs in this region.

A quick look at the numbers shows that there is no biophysical limitation on doubling or even tripling current yields in the savannah zone. Figure 6.4a provides a global estimate of yield gaps, expressed as the ratio between current and potential yields, for a number of crops, including cereals, roots and tubers, pulses, oil crops, sugar crops and vegetables (Fischer *et al.*, 2011). It shows that yield gaps in the savannah zone are particularly high. Figure 6.4b shows the yield gaps for major grains in a number of countries in semi-arid Africa, Asia and the Middle East. Under similar hydro-climatic conditions, on-station yield levels reach 5–6 tonnes of grain per hectare and commercial farmers can produce as much as 7–8 tonnes/ha. In contrast, on-farm yields reported in government statistics generally oscillate around only 1–2 tonnes/ha. In other words, the yield gap is dramatic, reaching 50–70% (Rockström *et al.*, 2007; Wani *et al.*, 2009).

From a purely biophysical point of view, higher yields are definitely achievable. Closing yield gaps in the savannah zone will require large investments in several areas, such as soil and water management techniques and capacity building among smallholder farmers and extension workers, as well as massive infrastructure development. Considering the rapid social changes currently under way, which include rural to urban migration and economic diversification among smallholders, it should be acknowledged that investment in areas other than agriculture might be equally – or sometimes more – important to alleviate poverty and achieve sustainable development in the savannah zone. Thus, it is important to assess the extent to which, and under what conditions, different development strategies will be able to deliver lasting change.

Nonetheless, millions of people base their livelihoods on small-scale agriculture in the savannah zone and will continue to do so for the foreseeable future. Moreover, the population in the world's least developed countries is projected to continue to increase rapidly to 2050. This is particularly the case in urban regions. Ethiopia and Tanzania are two clear examples from the savannah zone where the urban population is expected to increase by as much as 50% in the coming four decades (UNPD, 2010). This means that successful agricultural development will be necessary to deal with the overall challenge of feeding both the urban and the rural populations in these countries. As is discussed in Chapters 5 and 7 there is an urgent need to halt the conversion of other types of ecosystems into agriculture, as the associated loss of biodiversity and other ecosystem services is approaching critical thresholds, which if crossed may dramatically change the local, regional and global life-support systems on Earth (Rockström *et al.*, 2009b). This means that we need to find innovative ways to improve productivity in existing farming systems in the savannah zone.

6.2 Agricultural water management interventions to increase productivity

Investments to upgrade small-scale farming systems need to be made in a way that increases their overall capacity to deal with both the imminent and future perturbations and stresses that they will face. In other words, efforts to improve productivity in semi-arid and dry sub-humid agro-ecosystems should be made in a way that increases their resilience. This section focuses on agricultural water management (AWM) interventions as a potential way to achieve this goal.

6.2.1 What are AWM interventions?

At the field scale, incoming water is partitioned between runoff and infiltration. Having infiltrated the soil, the water is either taken up by plant roots and used for biomass production, or lost from that particular farming system via evaporation or drainage.

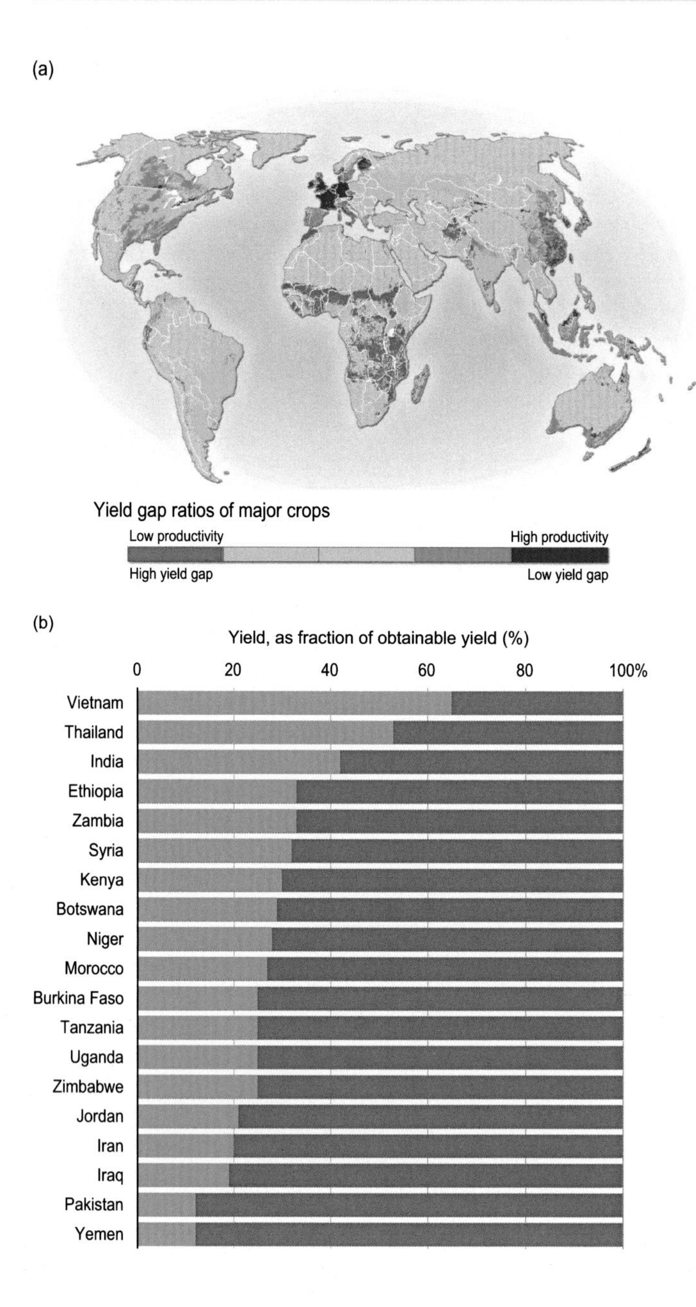

Figure 6.4 (a) An estimate of yield gaps, expressed as the ratio between current and potential yields, for cereals, roots and tubers, pulses, oil crops, sugar crops and vegetables, around the world (Fischer *et al.*, 2011), (b) Observed yield gaps for major grains in selected countries in semi-arid Africa, Asia and the Middle East (Rockström *et al.*, 2007).

AWM interventions include a range of relatively low-tech and low-cost technologies and innovations that aim to modify the water flows and partitioning points in this field water balance so that the amount of water available for crop production increases. Different literatures use different terminology to describe these practices, such as rainwater management and rainwater harvesting (see Douxchamps *et al.* (2012) for a discussion on how these concepts are related). In aiming for a broad view that spans the green to blue and the rainfed to irrigated continuums, the term AWM interventions is used in this chapter.

AWM interventions are often developed from indigenous knowledge, but they can also constitute

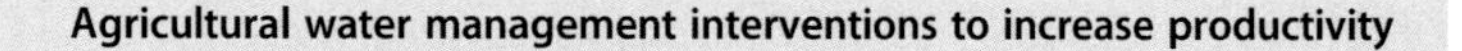

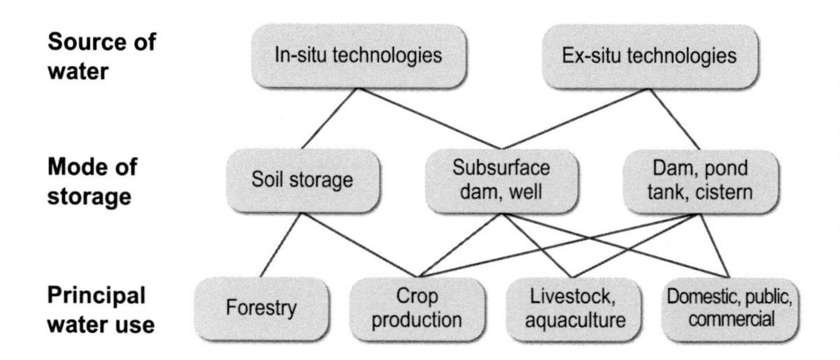

Figure 6.5 Classification of agricultural water management interventions based on the source of water, mode of storage and principal use. *In-situ* technologies make better use of the water that falls on the field by improving infiltration and the capacity of the soil to retain moisture. *Ex-situ* technologies divert runoff from external areas and channel it to the field (adapted after Falkenmark *et al.*, 2001).

entirely new innovations (Critchley *et al.*, 2008). Some of the oldest examples come from India and the Middle East, where rainwater harvesting has been practised for 4000–5000 years (Agrawal and Narain, 1997). There are many different types of AWM intervention. Most make use of green water, but some technologies combine green and blue water sources. Defined by where the water they use is harvested, these technologies can be divided into two major groups: *in-situ* and the *ex-situ* systems (Figure 6.5).

The *in-situ* systems include technologies such as conservation tillage, half moons and *zai* pitting. These technologies aim to increase the amount of soil moisture available for root uptake through specific forms of land management. Increased infiltration and enhanced soil water holding capacity are the two principal approaches used in *in-situ* systems. The systems make better use of the rain that falls on the field, without adding any additional water. In conservation tillage, ploughing is abandoned in favour of tillage practices such as ripping and sub-soiling, which reduce soil inversion and increase rainfall infiltration and root development (Figure 6.6a). In *in-situ* systems the soil is the storage medium for the water. Although this has limitations in terms of how long the water can be stored, due to drainage and evaporation, it also leads to recharge of groundwater, wetlands and shallow wells, thus creating ecohydrological opportunities downstream.

The *ex-situ* systems use a range of runoff-farming technologies, such as sheet, rill or gully water harvesting, to divert runoff to the field from external areas, such as a road or a degraded patch of land with limited infiltration capacity, after heavy rainfall events (Figure 6.6b). Like the *in-situ* technologies, these methods rely on the soil profile for water storage, but in this case additional water is being diverted to the field and made available for storage. Some *ex-situ* systems have built storage structures, which – like runoff-farming systems – capture water from external catchment areas, but the water is stored behind a small dam or in a tank and fed by gravity or simple water pumping techniques, such as treadle pumps, to the field (Figure 6.6c). This means that the water can be stored for longer periods and used as a bridge during more severe dry spells, for post-season irrigation or for other purposes such as a source of domestic water or for cattle to drink (Falkenmark *et al.*, 2001; Malesu *et al.*, 2007).

6.2.2 Effects on crop yields

Research has shown that, if correctly implemented under the right conditions, AWM interventions can contribute effectively to closing the yield gap described above. It has also been shown that these technologies help to reduce soil erosion and the incidence of flash floods, to increase the recharge of shallow groundwater, springs and stream flows, and to sometimes also increased species diversity among flora and fauna (Barron, 2009; Pretty *et al.*, 2006). Examples from around the world of different types of successful AWM interventions in semi-arid and dry sub-humid settings are summarised in Table 6.1.

In the semi-arid Gansu province in China, which has an annual rainfall of 400 mm, large-scale experiments have yielded positive results and demonstrated that *ex-situ* water harvesting systems for supplemental irrigation can be successful even when they only have a limited capacity for storing water (Liu *et al.*, 2005; Box 6.1). Farmers in the Gansu province are poor, with annual incomes of less than USD 1 per day, and staple food crops yield only around 1 tonne/ha. In dry years, the low yields do not even compensate for the purchase

(a)

(b)

(c)

Figure 6.6 (a) Conservation tillage, where a ripper has been used to improve infiltration while avoiding soil inversion. (b) An *ex-situ* rainwater harvesting system, where runoff water from an external catchment is diverted to the field via a simple mud and stone canal after large rainfall event. (c) An *ex-situ* rainwater harvesting system, where runoff water is collected from the roof of the house and collected in a concrete tank and used for supplemental irrigation of a vegetable garden, as well as for household water needs.

of seed, much less create a surplus for sale. After a major drought in 1995, small bottle-shaped sub-surface tanks with a storage volume of 10–60 m^3 were promoted to over 1 million smallholder farmers in this region. The results have been significant. The enhanced capacity to bridge dry spells has increased wheat and maize yields by 40–50%, and many farmers have diversified their farming systems (Liu *et al.*, 2005).

Large-scale experiments with *ex-situ* rainwater harvesting have also taken place among smallholders in semi-arid India (Wani *et al.*, 2009). One widespread practice is to divert surface runoff into a percolation tank, from which the water seeps out, recharging the shallow groundwater table. The water is pumped on to the fields when needed. Although efficient from a productivity perspective, unless water withdrawal is regulated this type of technology can lead to problems of unequal distribution, as groundwater is both a limited and a shared resource. To address the equity and environmental issues related to small-scale water management, the International Crops Research Institute for the Semi-Arid Tropics (ICRISAT) has developed a model for community-based integrated catchment management. This approach provides technological options for runoff water harvesting, groundwater recharge and supplemental irrigation, while at the same time facilitating income-generating micro-enterprises and emphasising collective action and capacity building (Sreedevi and Wani, 2009). The model was piloted in the Adarsha catchment, Andhra Pradesh, in the late 1990s, and has since then been scaled up in about 500 catchments in the region. The results clearly show that the approach has benefitted farmers. Yields for maize increased on average by 72% over the base yield of 2980 kg/ha, for castor by 60% over the base yield of 470 kg/ha, and for groundnut by 28% over the base yield of 1430 kg/ha (Sreedevi and Wani, 2009).

On-farm experiments with improved AWM practices in Burkina Faso have also led to dramatic yield responses. Sorghum harvests increased by up to 300% (above the average 0.5 tonne/ha) in plots where *ex-situ* rainwater harvesting for supplemental irrigation was combined with the application of fertilisers (Fox and Rockström, 2003). Similar experiments in Kenya and Tanzania showed yield increases for maize of 70–100% (Barron and Okwach, 2005; Makurira *et al.*, 2011). These improvements were achieved with only a limited amount of additional water added to the system over the season – 70 mm in the Kenyan and Burkinan cases and 150 mm in the Tanzanian case.

Table 6.1 The effects of AWMs on grain yields in experiments from semi-arid/dry sub-humid regions of China, India, Burkina Faso, Kenya, Tanzania and Zimbabwe

Area	Crop	Yield increase, due to AWM	Reference
China, Gansu	Wheat and maize	40–50%	Liu *et al.* (2005)
India, semi-arid	Maize Castor Groundnut	72% 60% 28%	Sreedevi and Wani (2009)
Burkina Faso	Sorghum	300%	Fox and Rockström (2003)
Kenya	Sorghum	70–100%	Barron and Okwach (2005)
Tanzania	Sorghum	70–100%	Makurira *et al.* (2011)
Zimbabwe, dry season		300–400%	Mazvimavi and Twomlow (2009)

Box 6.1 Rainwater harvesting to enhance dry farming in Gansu, China

Qiang Zhu and Yuanhong Li, Gansu Research Institute for Water Conservancy (GRIWAC)

The loess plateau in the northwest of Gansu province is one of the driest and poorest areas in China. The mean annual rainfall is only 400 mm, and 60–70% of it falls in July to September. Its per capita water resources of 260 m^3 per year are at the lower limit for sustaining human existence. Most of the agriculture is rainfed and totally dependent on the unreliable rain. Droughts occur frequently, causing very low crop yields, and soil erosion is another characteristic of the area. Annual erosion amounts to 5000–10 000 tonnes/km^2. Water shortages make subsistence farming the norm. Farmers cannot grow cash crops without water. The result is that more than 4 million people live below the poverty line. In addition, around 3 million people had no access to safe drinking water before 1995.

Rainwater is the only potential water source as both surface water and groundwater are scarce. Since the late 1980s, however, the Gansu Research Institute for Water Conservancy (GRIWAC) has been carrying out highly successful research and demonstration projects on rainwater harvesting (RWH). In 1995, a once-in-60-year drought hit the area. Millions of people had no water to drink and almost all the summer crops withered. Seeing the GRIWAC's successful results, the Gansu Provincial Government decided to launch the '1–2–1 Rainwater Catchment Project', the first large-scale RWH project in China. It aimed to solve the problem of domestic water use for a rural population of 1 million. The government provided each household with 1.5 tonnes of cement, at a cost equivalent to USD 50 per household, to build a rainwater collection surface, made up of a tiled roof and a concrete-lined courtyard, and two underground tanks at the side of the house. Each household also set up a piece of land irrigated by rainwater to grow vegetables and fruit trees. By the end of 1996, there was 37.2 million m^2 of catchment, and 286 000 underground tanks had been built. Although at first the RWH system could only meet the basic needs of the households, improvements in the economic situation of the households and the further government involvement enabled the scheme to be enlarged. Now, many households have enough water to drink and cook with, and also for sanitation. Many have a shower room and some even own a washing machine (Qiang *et al.*, 2012).

Encouraged by the successful 1–2–1 projects, the RWH Irrigation Project was initiated by the government with technical support from GRIWAC. To reduce costs, the less permeable surfaces of existing structures such as paved highways, country roads and threshing yards were used to collect rainfall. Since the rainwater held in tanks is limited in volume it is used very carefully. A low rate irrigation (LORI) approach has been developed, in which irrigation water is applied at critical periods during crop growing. Small amounts of water are applied just at the root zone of the crops. Simple and affordable but efficient innovations in irrigation methods were applied indigenously for field crops, and drip systems were used for high value crops. By the end of 2005, the RWH irrigation system had the capacity to supply water to about 80 000 ha of rainfed land. Field tests and demonstration projects showed that LORI can increase crop yields by 20–88%, an average of 40%. With water in the tank, farmers started to modify their farming patterns to the needs of the market. The number of greenhouses planted with high value crops or fruit

Box 6.1 *(cont.)*

trees increased rapidly, from zero to around 100 000, and became a way for farmers to increase their income. The plastic roof of a greenhouse is a highly efficient catchment. The water collected from it is enough for one or two harvests of vegetables. A simplified greenhouse with an area of 350 m^2 costs about USD 1000, while the annual income from it can be USD 500–1000, depending on management capability. In some villages, greenhouses are used to keep sheep and pigs, to increase the temperature in winter so they grow more quickly.

RWH has also benefitted the environment and conservation. Before the RWH project, low levels of land productivity led farmers to reclaim as much land as possible to ensure food supply. Even land with a slope steeper than 15 degrees was planted with food crops, despite its very low yield. This further intensified soil erosion and accelerated land degradation, in turn worsening the conditions for agricultural production. Now farmers no longer worry about food. They have changed their practices and started to implement the Land Conversion Programme initiated by the State – shifting sloping land from cultivation to trees and grass.

The 20-year RWH project has dramatically changed the living and production conditions of the rural population in the loess plateau of Gansu. It has been replicated in more than half the provinces in China. The number of beneficiaries in rural China has passed 21 million. RWH has been taken up as the strategic measure for integrated rural development in mountainous areas with water scarcity.

Mr Bao and his wife live in the Xiaojiachuan Village of Qinglan Township, which was a very poor village in the past with an average annual household income of USD 100. In 1997, Bao built two water cellars and two greenhouses. In the first winter, when the temperature dropped as low as –20°C, he succeeded in producing watermelons and sold them in the market. It made a great stir among the villagers. Now he owns nine water cellars with a total capacity of 430 m^3. His two greenhouses earn him USD 500–600 per year and even USD 1200 in a good year. In 2009, he irrigated his corn field twice and the yield was around 11.25 tonnes/ha (a). He updated his house (b) in 2000 and bought a mini-tractor, a motor cycle and an electric tricycle. He has a telephone, television, washing machine and a cell phone. He has accumulated USD 24 000.

Mr Ran Xiong lives in the Daping Village of Qinglan Township. Before 1995, the only water source for the village of 23 households was a small spring. People had to stand in line for more than an hour and then spend another hour taking the water home. In the early 1990s, the spring was drying up and RWH systems were installed. Now each family has a catchment area of 300–400 m^2 composed of a roof and a concrete-lined courtyard and two tanks of 50 m^3 each. Water is no longer a problem. There is enough rainwater to wash clothes and run a shower.

In 1996, with the support of the Gansu Water Resources Bureau, the Pengwa Village of Qinan County paved concrete slabs on hilltop wasteland to provide a rainwater catchment with an area of 4800 m^2 and built 40 water cellars each with a capacity of 50 m^3 around the hilltop. The Project can irrigate 8 ha of apple orchards. Yields increased by 6 tonnes/ha with a value of USD 1800/ha. The total annual benefit amounts to USD 14 400 and the investment of USD 27 000 was returned within two years.

Figure 6.b1.1 (a) The Bao couple, outside their home, (b) on the farm.

Box 6.1 (*cont.*)

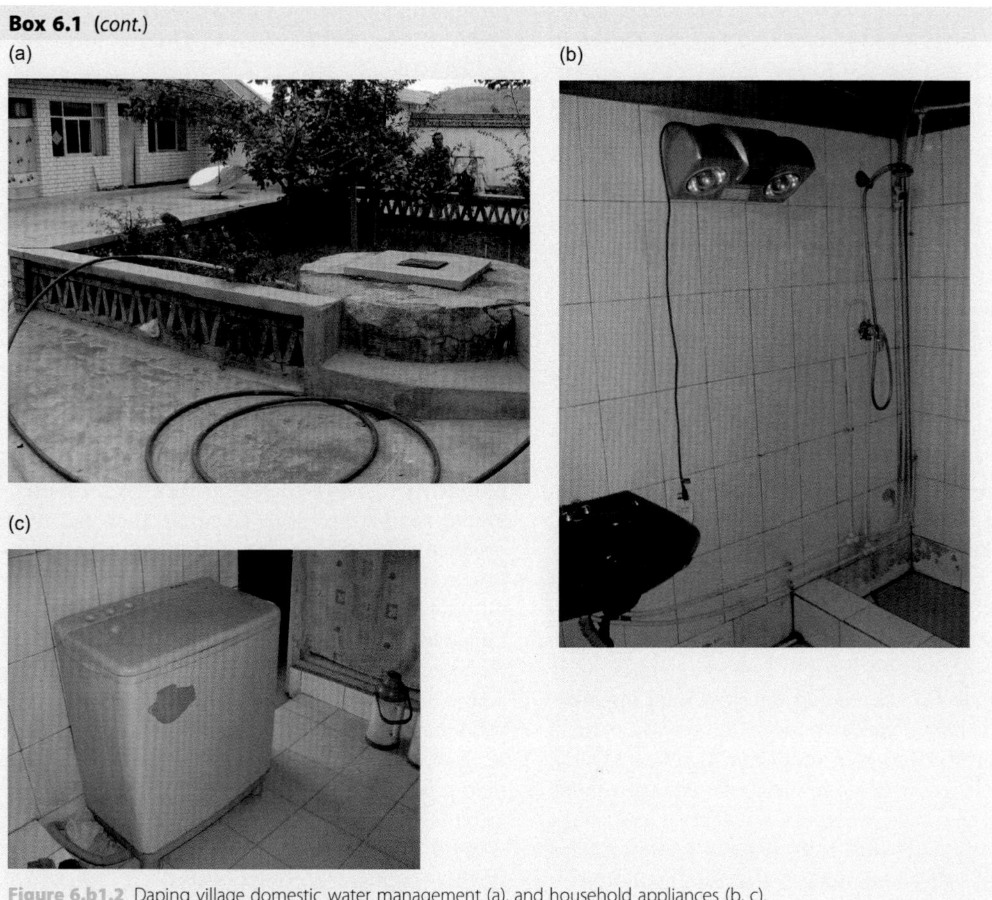

Figure 6.b1.2 Daping village domestic water management (a), and household appliances (b, c).

Figure 6.b1.3 (a) Pengwa village, Qinan County rainwater collection through concrete slabs on hilltop, (b) view over apple orchards irrigated through water management.

The key was that it was added at the right time (Makurira *et al.*, 2011; Rockström *et al.*, 2002). Moreover, in both Burkina Faso and Kenya, in one of the experimental seasons there was complete crop failure on most neighbouring farms due to a long dry spell, while the water harvesting system enabled the harvest of an above-average yield (>1 tonnes/ha). Similar findings have been reported from field trials in Zimbabwe, where *in-situ* water harvesting using deep planting pits increased yields by 300–400% during a drought season (Mazvimavi and Twomlow, 2009). Thus, when rainfall amounts and distribution are close to the threshold for what is needed to produce a cereal crop, improved AWM practices can mean the difference between not being able to harvest anything at all and having a decent yield.

While studies of conservation tillage systems from East and Southern Africa demonstrate an increase in average maize yields of 20–120% (Rockström *et al.*, 2009a), this type of *in-situ* technology does not protect against crop failure to the same degree as the *ex-situ* techniques discussed above (Enfors *et al.*, 2011). In the study from Tanzania referred to above, conservation tillage alone did not provide any significant improvements in maize yields in poor to average rainfall seasons. It did, however, result in a significant (40%) increase in grain yields in an above-average rainfall season. The conclusion drawn was that, at least in the short term, under semi-arid conditions, conservation tillage will help take advantage of already good seasons rather than minimising the risk of crop failure in poorer seasons. In several countries in Latin America, the introduction of conservation tillage led to an agricultural revolution (Derpsch, 1998, 2001). The greater success here is probably due to a combination of somewhat wetter conditions and stronger economic development, making it easier for smallholders to overcome some of the initial challenges associated with conservation tillage, such as the relatively high cost of specialised tillage and planting equipment and of controlling weeds. One of the case studies in Chapter 7 describes soil and water management in the Cerrado ecoregion of Brazil, a 2 million km^2 tropical savannah, that has contributed to closing the yield gap.

6.2.3 Effects on water productivity

While these results are encouraging, examining the effect on yields is not enough. It is important to assess how the introduction of new AWM practices affects the trade-offs between different water uses upstream and downstream in a catchment. One way to do this, as is discussed in Chapter 5, is to calculate water productivity, or the amount of water that is required to produce a unit of crop, under different management regimes. This is critical, as the amount of water used to produce food will determine the amount of water available to sustain ecosystems downstream in the catchment, as is discussed in particular in Chapter 7.

Transpiration increases linearly with increased plant growth and crop yield, but soil evaporation decreases progressively with increased canopy cover as a result of increased shading (Rockström and Falkenmark, 2000). Therefore, increasing crop yields mean that relatively less water is needed to produce each unit of crop (Figure 6.7). The potential for water productivity gains is highest in yield ranges from 0–3 tonnes per hectare, with decreases after that. Water productivity stabilises when yield levels go above 3–4 tonnes/ha. This is because evaporation is then at a minimum and water productivity will only be determined by transpiration, which, as is mentioned above, increases linearly with yield. This clearly demonstrates that the largest untapped potential for rapid yield increases exists in less productive systems, such as those in the savannah zone. Thus, although more food produced always implies that more consumptive water is used, it is relatively easy in these systems to shift to a higher level of productivity with only a smaller increase in total water use (Rockström, 2003). As expected, the experiments cited above report water productivity improvements ranging from 27% in the Burkinan and Kenyan cases to 31% in the Tanzanian case (converted to m^3/tonne) (Makurira *et al.*, 2011; Rockström *et al.*, 2002).

In sum, a growing body of research demonstrates that improved AWM techniques can contribute to closing the yield gap between potential and actual yields in the savannah zone. Improvements in water productivity mean that this can be achieved with only a small increase in total consumptive water use. This is key to creating resilient landscapes that can support the generation of food and other ecosystem services over time.

6.2.4 Evidence of the impact of AWM interventions on savannah systems

The primary reason for investing in AWM is to improve yields, but these systems often fulfil a range of other essential functions, such as domestic and

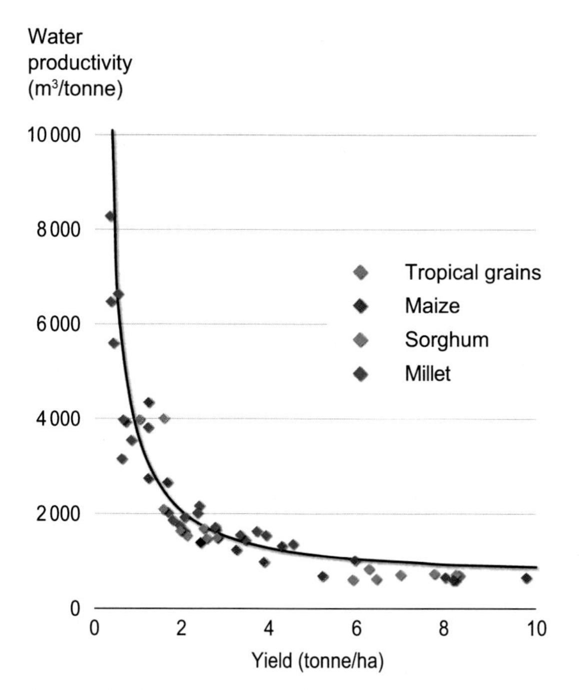

Figure 6.7 Water productivity plotted against yield for different tropical crops. When the crop becomes denser a vapour shift occurs from unproductive evaporation to productive transpiration. This results in a non-linear relationship between water productivity and yields for cereal crops. Large gains in water productivity can be made in low productivity farming systems, such as those in the savannah zone (Rockström, 2003).

public water supply, livestock watering and aquaculture (Barron, 2009). Thus, from a livelihood perspective AWM interventions can provide multiple benefits in terms of improved food security, income generation from selling agricultural produce and labour saving.

In the Andhra Pradesh catchment described above, the community-based catchment model triggered a partial shift to commercial cereal crop production and increased the amount of vegetable and horticultural cultivation (Sreedevi and Wani, 2009). The introduction of new AWM practices provided a whole new set of livelihood opportunities. In Niger and Ethiopia, experiments are taking place in which water harvesting is combined with conservation tillage and productive sanitation, with nutrient recirculation from human urine and faeces. The goal is to achieve what is referred to as a triply green revolution, where more effective green water management produces higher yields in a sustainable way (Rockström and Karlberg, 2010). Preliminary reports cite improved household economy and better health as additional benefits.

AWM interventions may also lead to more fundamental transformations (see Chapter 8) of semi-arid and dry sub-humid agro-ecosystems. In cases where the dominant social–ecological feedbacks trap the system in an undesirable development pathway, the implementation of new water management techniques can be an initial trigger to change system dynamics and begin a transformation. A systems analysis of the Makanya catchment in Tanzania shows that conservation tillage in combination with rainwater harvesting have the potential to disrupt several of the social–ecological interactions that currently lock the system into a land degradation trajectory (Enfors, 2013). Over the past 50 years, increasingly frequent critical dry spells, far-reaching changes in the governance of natural resources and high levels of population growth have interacted with a set of key system variables in the Makanya agro-ecosystem. The result has been a development trajectory where off-farm ecosystem services are being degraded while crop yields remain low and people remain poor. This situation, in which reinforcing feedbacks lock the system into a state of destitution, can be understood as a social–ecological trap (Chapter 3). Low yields and frequent crop failures force farmers to deplete the capital they have accumulated every other year. This reduces the capacity for farm system investments and consequently reinforces land degradation feedback.

Figure 6.8 illustrates land degradation feedback in Makanya, and how conservation tillage in combination with rainwater harvesting-based supplemental irrigation can disrupt the current negative state, opening up opportunities for a transformation to more productive development paths. There are three main mechanisms behind this:

1. Improved yields, which allow farmers to build up buffers and invest in their farming systems, reducing the pressure on surrounding ecosystems and the need for farmland expansion (Figure 6.8, A, B, C, D).
2. Changes in the field water balance, which enable a positive water productivity feedback and reduce erosion (Figure 6.8 E, F).
3. Improved soil health, which enables a better plant response to water availability (Figure 6.8 G).

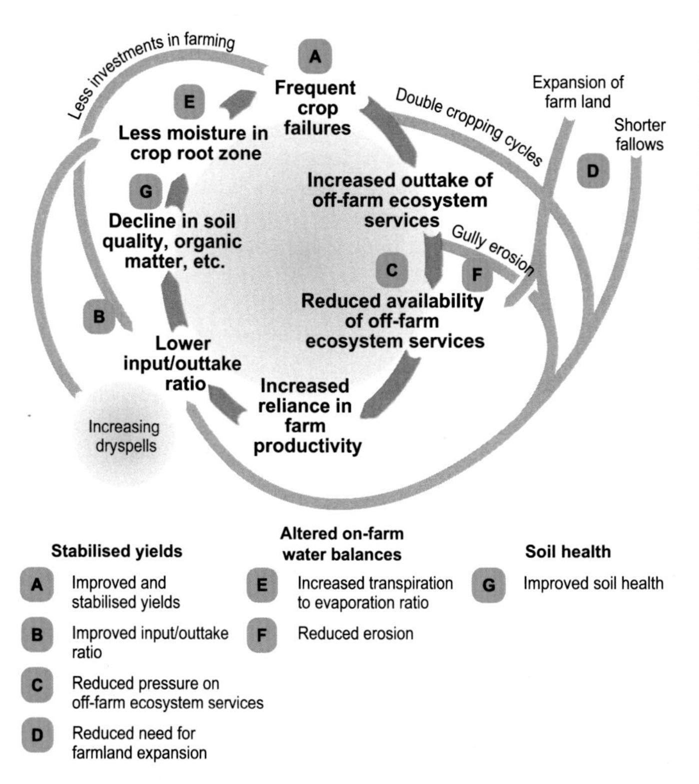

Figure 6.8 The influence of AWM interventions on the dominant feedbacks in the Makanya agro-ecosystem. Over the past 50 years, interactions between external driving forces and key system variables have trapped Makanya on a development trajectory characterised by land degradation. Conservation tillage in combination with rainwater harvesting could disrupt this social–ecological trap through three distinct mechanisms: (1) stabilised yields (A, B, C, D), (2) altered on-farm water balances (E, F), and (3) improved soil health (G) (Enfors, 2013).

6.3 Making small-scale green water management technologies work in a complex, turbulent world

The examples from India, Niger and Tanzania show that AWM technologies could have a substantial impact on both agro-ecosystem dynamics and rural livelihoods in the savannah zone. Implementation of improved AWM techniques can improve yields, increase water productivity, enable farmers to invest in their land and diversify their livelihoods, and thus provide leverage to escape persistent poverty and open up opportunities for a transformation to more multifunctional and resilient landscapes. There have been no comprehensive assessments of the extent to which these technologies are used. In spite of their seemingly great promise, however, they appear to exist mainly in isolated pockets and uptake rates remain low (Barron, 2009). This suggests that certain factors make their implementation a challenge. This section takes a closer look at these factors and discusses the kind of approaches needed to investment to unlock the potential of small-scale green water management technologies.

6.3.1 A range of real-life challenges: technical and biophysical constraints

The positive results obtained from experimental and scientifically monitored trials of improved AWM are often difficult to replicate on a larger scale. Real-world situations often involve technological and biophysical constraints that have to be overcome. It is a major technological challenge to design, dimension and adapt a water harvesting system to be as efficient as possible under the range of conditions, such as soil type, slope, crop system, that exists on a specific site. This often requires professional knowledge and expertise, which are rarely available to rural smallholders, resulting in water harvesting schemes that fail to yield the expected benefits. Thus, the main challenges are not in controlled scientific testing of AWM interventions, which have repeatedly shown positive results, but the practical implementation of the technologies on farmers' fields across a range of possible conditions. This gap must be bridged in order to make AWM interventions viable at a larger scale.

In addition to the technological challenges, there are also critical biophysical issues related to making water management technologies of this kind work in the savannah zone. It may seem logical to expect that yields will increase once water is provided to a desiccated agricultural field. However, it is a sadly common experience among resource-poor smallholders that in parallel with water stress, their crops also suffer from severe nutrient deficiencies (see e.g. Licker *et al.*, 2010). This means that even in conditions of increased water availability, yields will remain low until critical nutrients are added (Barron and Okwach, 2005; Fox and Rockström, 2003).

In semi-arid parts of Ethiopia it is not uncommon for cultivation of *tef* (*Agrostis tef*), the main staple crop, to fail completely if grown without the application of supplemental nitrogen and phosphorus. This highlights the complexity of semi-arid and dry subhumid farming systems. Simultaneous investment in soil fertility management is often necessary to make investment in AWM worthwhile (Rockström *et al.*, 2007; Wani *et al.*, 2009).

Another example of a biophysical dilemma is the competing uses of crop residues and manure when attempting conservation tillage in semi-arid climates. Crop residues are frequently needed to feed animals and manure from cattle, which contains both organic material and nutrients, is often not returned to the agricultural fields but instead used as fuel for cooking. To achieve the net input of organic matter needed to create the longer-term yield benefits in conservation tillage systems related to improved soil structure, crop residues must be left in the field after the harvest and manure added to the field at the beginning of the rainy season. This means very difficult choices for many farmers.

While water, nutrients and soil structure are the three most fundamental challenges in these systems, there are often other biophysical constraints as well. For example, the correct timing of planting, irrigation and weeding is critical in semi-arid farming systems. Late planting might lead to a complete crop failure, even if there is a water harvesting system that can bridge dry spells later in the season, simply because a short season means that farmers cannot afford to miss the first rains. Pest control is another example of a management intervention that if not timed correctly can lead to large yield losses even if water, nutrients and other crop needs are met.

The above discussion highlights the need for an entire *farming systems approach* to achieve the full benefits of AWM interventions. Thus, improved water management needs to be integrated with improved nutrient and pest management, and operations timed according to system- and site-specific requirements. Making general recommendations about the combination of measures to be taken is unfortunately of limited use, since each case and site comes with its own set of challenges. However, an overarching conclusion is that any solution should address multiple problems at the same time in order to have the desired effects and achieve success.

6.3.2 A range of real-life challenges: social and economic constraints

There are also socio-economic aspects to take into account to ensure that water management technologies deliver the intended livelihood benefits (Merrey and Sally, 2008). Studies show that the highest yield impacts generally come from water management technologies that make use of built storage structures. Although this type of more advanced rainwater harvesting is still considered a small-scale and low-cost technology, the initial investment costs of building a dam can be difficult to bear for a smallholder (see e.g. de Graaff *et al.*, 2008). It is important to recognise that new investments in the farming system may also create new vulnerabilities for the farmer. In order to mitigate these risks, smallholders' investments need to be backed up by appropriate fail-safe arrangements. This is especially the case when initial costs are high or when there is a time lag before an investment is likely to yield a return. In this context, both crop insurance and different forms of micro-credit scheme can help farmers make the initial investment needed.

Efforts to promote improved AWM may also need to be coordinated with other relevant investments across scales. The experiment in Makanya, described above, provides a clear example of this. In spite of a very good maize harvest in 2006–07, of up to 3 tonnes/ha, farmers had a shortage of food the following season. Poor access to external markets in Makanya means that maize prices fall very low after a good season. Therefore, instead of selling their surplus, the farmers stored it for later consumption. However, appropriate on-farm storage facilities or cooperative storage infrastructure is lacking in the area, making storage pests a severe problem. In this case, pests destroyed most of the surplus before the next harvest. Thus, attempts to increase yields would have to be combined with improved storage technology or market access opportunities outside the region in order to benefit the farmers in Makanya. Consequently, the fact that improved AWM often relies on small-scale technology should not be confused with 'simple' or 'quick-fix' solutions, since cross-scale investments are often needed in other kinds of infrastructure, in markets and in enabling institutions to ensure that AWM investment pays off (Enfors and Gordon, 2008).

There may also be barriers to the adoption of new technology embedded in the social and cultural context. These barriers are generally difficult to detect from the outside, but it is crucial to take them into account in order to make uptake a genuine option for smallholder farmers. For example, the ability of farmers to operate effectively in emerging free markets is limited in economies where kinship plays a prominent role (Mabogunje, 2007). This means that individual entrepreneurship may be disincentivised, which reduces the likelihood of investments in e.g. AWM if that investment is designed as an individual effort with individual benefits. Under these circumstances, redesigning the project as a collective effort may have a larger effect on uptake. Cultural taboos can also play an important role. In one Tanzanian conservation tillage experiment, for example, many villagers found it unacceptable to use oxen for traction during land preparation, due to their high value, and used donkeys instead (Enfors, 2013). However, the kind of ripper available had been designed for oxen traction, so land preparation was less effective and the conservation tillage system did not work as well as intended, discouraging many farmers from continuing with it after an initial trial run. Awareness of existing social and cultural barriers is required to create appropriate solutions and facilitate adoption.

6.3.3 Navigating change in turbulent times

The above discussion shows that a prerequisite for successful investments in AWM is *farming system solutions* that integrate water management with the management of nutrients, pests and the timing of operations. To make uptake a real option for farmers, efforts to promote AWMs also need to focus on a wider range of issues than the technology in itself, such as fail-safe arrangements, marketing opportunities and social norms. Research on innovation systems further suggests that the successful diffusion of new technologies also requires the existence of entrepreneurs and mechanisms for knowledge diffusion (Hekkert *et al.*, 2007).

Even with such investments, however, it is safe to assume that transforming smallholder agroecosystems in the savannah zone will be a challenge. In recent decades, farmers themselves as well as a range of other stakeholders have made great efforts to upgrade production systems, but yields remain low and many farmers are still poor. The undertaking is tremendously complicated, especially as strong cross-scale feedbacks often seem to reinforce the current state of affairs.

A better understanding of the social–ecological dynamics of savannah systems is required, including the main drivers of change, key variables and feedback between these and thresholds of potential concern (Folke *et al.*, 2010). A thorough analysis of these dynamics would identify leverage points for transformation. In some cases, simple interventions might make a big difference, but most of the time holistic efforts, such as the community-based catchment management model implemented in Andhra Pradesh, will probably be needed.

Given the many parallel processes of social–ecological change under way across the world's savannahs, and thus the large uncertainty about the future there, current investments in small-scale farming also need to be robust (e.g. Enfors *et al.*, 2008). In other words, they have to be made so that they will benefit local communities across a range of potential futures. One way to increase resilience is to increase the diversity of crops cultivated. A farmer from the Makanya experiment used his newly implemented water harvesting system to plant cassava, bananas and beans in addition to the maize that he was already growing. This improved the food security of his family, gave him a small buffer to market price fluctuations and increased the ecological complexity of his field, all of which made his production system better equipped to handle a variety of disturbances.

Sustaining or even enhancing landscape multifunctionality (e.g. Pretty *et al.*, 2006), so that the production of both food and other ecosystem services is sustained over time, will also contribute to increased resilience. First, as ample evidence from around the world shows us, the long-term functioning of the farming system depends on the range of support and regulating services generated in the surroundings. Second, this type of landscape provides smallholders with a range of alternative income sources (see e.g. examples from Zimbabwe in Cavendish, 2000), increasing their livelihood portfolios and thereby their capacity to adapt to changing conditions.

An additional strategy to enhance resilience would be for agricultural intervention programmes to specifically target learning, experimentation and innovation among farmers (see also Röling and Wagemakers, 1998). This would benefit smallholders regardless of future trajectory and in spite of the technological shortcomings of particular AWM technologies.

It is important to acknowledge that we cannot fully foresee what kind of trajectories larger scale adoption of these types of water technologies would open up. For example, while potential water productivity gains offer hope, the aggregated effects at the catchment scale of more intensive green water management on-farm are unknown. Both trade-offs and synergies may arise between water for food and water for other ecosystem services, as is developed further in Chapter 7. Thus, while AWMs are promising in general, not all the transformations based on them will be sustainable. It will be a major challenge to navigate the change process towards desirable outcomes (see e.g. Olsson *et al.*, 2006).

Summary

Semi-arid and dry sub-humid tropical savannahs cover 30% of the land surface of the globe. These regions are home for an almost equal proportion of the population, the majority of whom make their living from small-scale rainfed agriculture and livestock rearing. Their farming enterprises are generally extensive, and yields have remained low for many decades, especially in sub-Saharan Africa. The challenging hydro-climate is a major reason. An intensification of current farming systems will be needed to meet future food needs across the world and halt the expansion of farmland.

Substantial agro-hydrological potential

The good news is that, contrary to popular belief, the savannah zone has substantial agro-hydrological potential. This is because the hydrological challenge is not related to an absolute lack of water, but rather to the huge fluctuations in rainfall over time and space, and the large amount of water lost to the farming system through evaporation, runoff and drainage. Research has shown that by bridging critical dry spells and better managing these flows, there is no biophysical limitation to doubling or even tripling current yield levels.

Modifying water partitioning

AWM interventions include a range of low-tech and low-cost technologies that aim to modify water flows and partitioning points in the field water balance so that the amount of water available for crop production increases. There are many different types of AWM technology, ranging from *in-situ* systems, such as conservation tillage systems in runoff-farming systems, to external catchment rainwater harvesting systems with built storage structures. If correctly implemented under the right conditions, AWM interventions can effectively contribute to closing the yield gap.

Water harvesting for supplementary irrigation

Large-scale experiments with water harvesting for supplemental irrigation have had positive results in the semi-arid Gansu province of China, with 50% increases in wheat and maize yields, largely because these systems have helped to bridge dry spells. AWM experiments conducted on-farm in Burkina Faso have had even better results, with sorghum yield increases of up to 300%. In this case, rainwater harvesting for supplemental irrigation was combined with the application of fertilisers. To tackle equity and environmental issues related to small-scale water management in India, a model centred on improved AWM has been developed for community-based integrated catchment management. The results are convincing, with yield increases of 72% for maize, 60% for castor and 28% for groundnut. These experiments also demonstrate significant water productivity improvements, which implies that it is possible to shift productivity to a higher level in these systems with only a small increase in total water use.

A need to overcome constraints still limiting smallholders' uptake

However, in spite of the promising results of experiments from across the world, the uptake of AWM technologies remains limited, signalling the difficulty of making them work as intended. In real-life situations, a range of technological, biophysical and socio-economic constraints needs to be overcome to make AWM techniques work for smallholders. Whole farm system solutions and broadly based efforts that focus on a wide range of socio-economic issues, such as fail-safe arrangements, marketing and norms, are prerequisites for successful investment in AWM technologies. It is important to remember that interacting constraints often operate at different scales. It will be necessary to target these issues in a coordinated manner in order to make investment in AWM a real option for smallholders.

Finally, given the uncertainty about the future and the difficulty of predicting the kind of trajectories that larger scale adoption of this type of technology would open up, it will be a major challenge to make these investments as robust as possible, and to navigate the process of agricultural transformation towards desirable outcomes.

References

Agrawal, A. and Narain, S. (1997). *Dying Wisdom: State of India's Environment. A Citizen's Report*. New Delhi: Centre for Science and Environment.

Alcamo, J., Flörke, M. and Märker, M. (2007). Future long-term changes in global water resources driven by socio-economic and climatic changes. *Hydrological Sciences Journal*, **52**, 247–275.

Barron, J. (2009). *Rainwater Harvesting: A Lifeline for Human Well-being*. Nairobi: United Nations Environment Programme.

Barron, J. and Okwach, G. (2005). Runoff water harvesting for dry-spell mitigation in maize (*Zea mays* L.): results from on-farm research in semi-arid Kenya. *Agricultural Water Management*, **74**, 1–21.

Boko, M., Niang, I., Nyong, A. *et al.* (2007). Africa. Climate change 2007: Impacts, adaptation and vulnerability. In *Contribution of Working Group II to the Fourth Assessment Report of the Intergovernmental Panel on Climate Change, 2007*, ed. Parry, M. L., Canziani, O. F., Palutikof, J. P., Van Der Linden, P. J. and Hanson, C. E. Cambridge: Cambridge University Press, pp. 433–447.

Cavendish, W. (2000). Empirical regularities in the poverty-environment relationship of rural households: evidence from Zimbabwe. *World Development*, **28**, 1979–2003.

Center for International Earth Science Information Network at Columbia University (CIESIN) (2005). *Poverty mapping project: global subnational prevalence of child malnutrition*. Center for International Earth Science Information Network at Columbia University. Available at: http://sedac.ciesin.columbia.edu/data/set/povmap-global-subnational-prevalence-child-malnutrition (Accessed 29 March 2013).

Critchley, W., Negi, G. and Brommer, M. (2008). Local innovation in 'green water' management. In *Conserving Land, Protecting Water*, ed. Bossio, D. and Geheb, K. Wallingford: CABI, pp. 107–119.

de Graaff, J., Amsalu, A., Bodnár, F. *et al.* (2008). Factors influencing adoption and continued use of long-term soil and water conservation measures in five developing countries. *Journal of Applied Geography*, **28**, 271–280.

Derpsch, R. (1998). Historical review of no-tillage cultivation of crops. In *Conservation Tillage for Sustainable Agriculture. Proceedings from an International Workshop, Harare, 22–27 June. Part II (Annexes)*, ed. Benites, J., Chuma, E., Fowler, R., Kienzle, J., Molapong, K., Manu, J., Nyagumbo, I., Steiner, K. and Van Veenhuizen, R. Eschborn: GTZ, pp. 205–218.

Derpsch, R. (2001). Conservation tillage, no tillage and related tecnologies. In *1st World Congress on Conservation Agriculture, vol. 1: Keynote Contributions*, ed. Garcia-Torres, L., Benites, J. and Martinez-Vilela, A. Rome: Food and Agriculture Organization, pp. 161–170.

Djurfeldt, G., Holmen, H., Jirström, M. and Larsson, R. (2005). Addressing food crisis in Africa, What can sub-Saharan Africa learn from Asian experiences in addressing its food crisis? *SIDA report: SIDA24308en*. Department for Natural Resources and the Environment, SIDA, Stockholm.

Douxchamps, S., Ayantunde, A. and Barron, J. (2012). Evolution of agricultural water management in rainfed crop–livestock systems of the Volta basin. CGIAR Challenge Program for Water and Food Working Papers: 4. International Livestock Research Institute, Nairobi. Available at: http://hdl.handle.net/10568/21721.

Enfors, E. (2013). Social-ecological traps and transformations in dryland agro-ecosystems: Using water system innovations to change the trajectory of development. *Global Environmental Change*, **23**, 51–60.

Enfors, E., Barron, J., Makurira, H., Rockstrom, J. and Tumbo, S. (2011). Yield and soil system changes from conservation tillage in dryland farming: a case study from North Eastern Tanzania. *Agricultural Water Management*, **98**, 1687–1695.

Enfors, E., Keys, P., Barron, J. and Gordon, L. (in preparation). Dryspell frequency and trends over time in semi-arid and dry sub-humid sub-Saharan Africa – implications for smallholder farmers.

Enfors, E. I. and Gordon, L. J. (2008). Dealing with drought: the challenge of using water system technologies to break dryland poverty traps. *Global Environmental Change*, **18**, 607–616.

Enfors, E. I., Gordon, L. J., Peterson, G. D. and Bossio, D. (2008). Making investments in dryland development work: participatory scenario planning in the Makanya catchment, Tanzania. *Ecology and Society*, **13**, 42.

Falkenmark, M., Fox, P., Persson, G. and Rockström, J. (2001). Water harvesting for upgrading of rainfed agriculture: problem analysis and research needs. SIWI Report: 11. Stockholm International Water Institute, Stockholm.

Falkenmark, M. and Rockström, J. (2004). *Balancing Water for Humans and Nature: The New Approach in Ecohydrology*. London: Earthscan.

Fischer, G., Hizsnyik, E., Prieler, S. and Wiberg, D. (2011). Scarcity and abundance of land resources: competing uses and the shrinking land resource base: SOLAW TR02. FAO SOLAW Background Thematic Report-TR02: 2. Food and Agriculture Organization, Rome. Available at: http://www.fao.org/nr/solaw/thematic-reports/en/.

Folke, C., Carpenter, S. R., Walker, B. *et al.* (2010). Resilience thinking: integrating resilience, adaptability and transformability. *Ecology and Society*, **15**, 20.

Fox, P. and Rockström, J. (2003). Supplemental irrigation for dry-spell mitigation of rainfed agriculture in the Sahel. *Agricultural Water Management*, **61**, 29–50.

Hekkert, M. P., Suurs, R. A. A., Negro, S. O., Kuhlmann, S. and Smits, R. (2007). Functions of innovation systems: a new approach for analysing technological change. *Technological Forecasting and Social Change*, **74**, 413–432.

Licker, R., Johnston, M., Foley, J. A. *et al.* (2010). Mind the gap: how do climate and agricultural management explain the 'yield gap' of croplands around the world? *Global Ecology and Biogeography*, **19**, 769–782.

Liu, F. M., Wu, Y. Q., Xiao, H. L. and Gao, Q. Z. (2005). Rainwater-harvesting agriculture and water-use efficiency in semi-arid regions in Gansu province, China. *Outlook on Agriculture*, **34**, 159–165.

Mabogunje, A. L. (2007). Tackling the African 'poverty trap': the Ijebu-Ode experiment. *Proceedings of the National Academy of Sciences*, **104**, 16781–16786.

Makurira, H., Savenije, H. H. G., Uhlenbrook, S., Rockström, J. and Senzanje, A. (2011). The effect of system innovations on water productivity in subsistence rainfed agricultural systems in semi-arid Tanzania. *Agricultural Water Management*, **98**, 1696–1703.

Malesu, M. M., Oduor, A. R. and Odhiambo, O. J. (eds) (2007). *Green Water Management Handbook: Rainwater Harvesting for Agricultural Production and Ecological Sustainability*. Nairobi: World Agroforestry Centre (ICRAF).

Mazvimavi, K. and Twomlow, S. (2009). Socioeconomic and institutional factors influencing adoption of conservation farming by vulnerable households in Zimbabwe. *Agricultural Systems*, **101**, 20–29.

Merrey, D. J. and Sally, H. (2008). Micro-agricultural water management technologies for food security in Southern Africa: part of the solution or a red herring? *Water Policy*, **10**, 515–530.

Mortimore, M. (2005). Dryland development: success stories from West Africa. *Environment: Science and Policy for Sustainable Development*, **47**, 8–21.

Olsson, P., Gunderson, L. H., Carpenter, S. R. *et al.* (2006). Shooting the rapids: navigating transitions to adaptive governance of social-ecological systems. *Ecology and Society*, **11**, 18.

Peel, M. C., Finlayson, B. L. and McMahon, T. A. (2007). Updated world map of the Köppen-Geiger climate classification. *Hydrology and Earth System Sciences*, **11**, 1633–1644.

Pretty, J. N., Noble, A., Bossio, D. *et al.* (2006). Resource-conserving agriculture increases yields in developing countries. *Environmental Science & Technology*, **40**, 1114–1119.

Qiang, Z., Yuanhong, L. and Gould, J. (2012). *Every Last Drop: Rainwater Harvesting and Sustainable Technologies in Rural China*. Rugby, UK: Practical Action.

Reynolds, J. F., Smith, D. M. S., Lambin, E. F. *et al.* (2007). Global desertification: building a science for dryland development. *Science*, **316**, 847–851.

Rockström, J. (2003). Water for food and nature in drought-prone tropics: vapour shift in rain-fed agriculture. *Philosophical Transactions of the Royal Society of London Series B: Biological Sciences*, **358**, 1997–2009.

Rockström, J., Barron, J. and Fox, P. (2002). Rainwater management for increased productivity among small-holder farmers in drought prone environments. *Physics and Chemistry of the Earth*, **27**, 949–959.

Rockström, J. and De Rouw, A. (1997). Water, nutrients and slope position in on-farm pearl millet cultivation in the Sahel. *Plant and Soil*, **195**, 311–327.

Rockström, J. and Falkenmark, M. (2000). Semiarid crop production from a hydrological perspective: gap between potential and actual yields. *Critical Reviews in Plant Sciences*, **19**, 319–346.

Rockström, J., Hatibu, N., Oweis, T. Y. and Wani, S. P. (2007). Managing water in rainfed agriculture. In *Water for Food, Water for Life: A Comprehensive Assessment of Water Management in Agriculture*, ed. Molden, D. London: Earthscan, pp. 315–352.

Rockström, J. and Karlberg, L. (2010). The quadruple squeeze: defining the safe operating space for freshwater use to achieve a triply green revolution in the Anthropocene. *AMBIO*, **39**, 257–265.

Rockström, J., Kaumbutho, P., Mwalley, J. *et al.* (2009a). Conservation farming strategies in east and southern Africa: yields and rain water productivity from on-farm action research. *Soil and Tillage Research*, **103**, 23–32.

Rockström, J., Steffen, W., Noone, K. *et al.* (2009b). A safe operating space for humanity. *Nature*, **461**, 472–475.

Röling, N. G. and Wagemakers, M. A. E. (eds) (1998). *Facilitating Sustainable Agriculture: Participatory Learning and Adaptive Management in Times of Environmental Uncertainty*. Cambridge and New York: Cambridge University Press.

Roudier, P., Sultan, B., Quirion, P. and Berg, A. (2011). The impact of future climate change on West African crop yields: what does the recent literature say? *Global Environmental Change*, **21**, 1073–1083.

Sreedevi, T. K. and Wani, S. P. (2009). Integrated farm management practices and upscaling the impact for increased productivity of rainfed systems. In *Rainfed Agriculture: Unlocking the Potential*, ed. Wani, S. P., Rockström, J. & Oweis, T. Wallingford: CABI, pp. 222–257.

Tiffen, M., Mortimore, M. and Gichuki, F. (1994). *More People, Less Erosion: Environmental Recovery in Kenya*. New York: John Wiley & Sons.

United Nations Convention to Combat Desertification. (2011). *Desertification: A Visual Synthesis*. United Nations Convention to Combat Desertification.

United Nations Department of Economic and Social Affairs (UNDESA) (2010). *World Urbanization Prospects. The 2009 Revision*. New York: United Nations Department of Economic and Social Affairs. Available at: http://www.un.org/esa/population/unpop.htm.

Wani, S. P., Rockström, J. and Oweis, T. (eds) (2009). *Rainfed Agriculture: Unlocking the Potential*. Wallingford, UK: CABI.

Farmers picking cotton in the Kothapally catchment, Andrah Pradesh, India. This catchment, with high population pressure and difficult soil and water conditions has improved from a degraded state through a range of soil and water strategies and structures.

Water resources and functions for agro-ecological systems at the landscape scale

This chapter analyses the implications of agricultural expansion on landscape multifunctionality in terms of water, and the related functions of agro-ecosystem services. Through manipulation of water stocks and flows, and landscape characteristics it is feasible to transform agricultural land from degraded to intermediate levels of multiple ecosystem services or beyond. The focus is on water balance alteration and within-stream and downstream effects for other users. The chapter highlights the landscape-scale resilience perspective, based on a set of landscape-scale indicators. It analyses three landscapes with rainfed cultivation, all in the tropical, semi-arid and sub-humid zone and under rapid transformation, and concludes that there is still potential to better utilise water functions and ecosystem services in most agricultural systems.

7.1 Social–hydrological–ecological systems at the landscape scale

Healthy water flows are key to sustaining the multiple ecosystem services that underpin the sustainability of social–ecological systems at the landscape scale (meso-scale, here defined as 1–10 000 km^2 after e.g. Blöschl, 1996; Montanari and Uhlenbrook, 2004). Water shapes a range of provisioning, supporting, regulating and cultural ecosystem services, directly and indirectly through the presence or absence of water in time and space. This chapter discusses fundamental aspects of water and ecosystem services, and the management of these, with a particular emphasis on tropical, semi-arid and sub-humid zones experiencing rapid transformation and development for human benefit.

This chapter assumes that so-called pristine landscapes are rare globally, and can be considered an extreme form of land use in the Anthropocene era – at least in the areas of human settlement (Ellis, 2011). Despite this, we tend to compare water flows and quality, as well as the provision of various ecosystem services, with an ideal type of pristine-state landscape when evaluating positive and negative consequences. That said, ecosystem services are only services as long as someone or something benefits from them.

The focus is on the landscape scale as this is where land-use changes can aggregate and affect ecosystem services, and consequently livelihoods and development opportunities as well as ecosystem sustainability. Agricultural landscapes are the main interest, as most people live in this zone, and this is the single greatest human activity affecting water flows in landscapes. Of all human activities, agricultural development has the most far-reaching positive and negative impacts on ecosystem services and landscapes. Changes and intensifications in agricultural production are also the main challenges to the sustainability of water resources and ecosystem services for the near future. Changes in agricultural landscapes will undoubtedly take place on a substantial scale over the next 20–50 years because of increasing demand for food, fodder and fibre both locally and globally. Thus, addressing water resources for multiple demands as part of complex systems is urgently needed to explore routes for the provision of multiple benefits with the limited water resources available.

7.1.1 Water and social–ecological systems at the landscape scale

The starting point for this chapter is that a range of social and biophysical processes merge into complex social–ecological systems, with systems features and behaviours, at the landscape scale. Water – in its various states of vapour, liquid and solid –provides multiple social and ecological functions in landscapes. Humans depend on, directly or indirectly, and have an impact on these processes locally and globally (Figure 7.1).

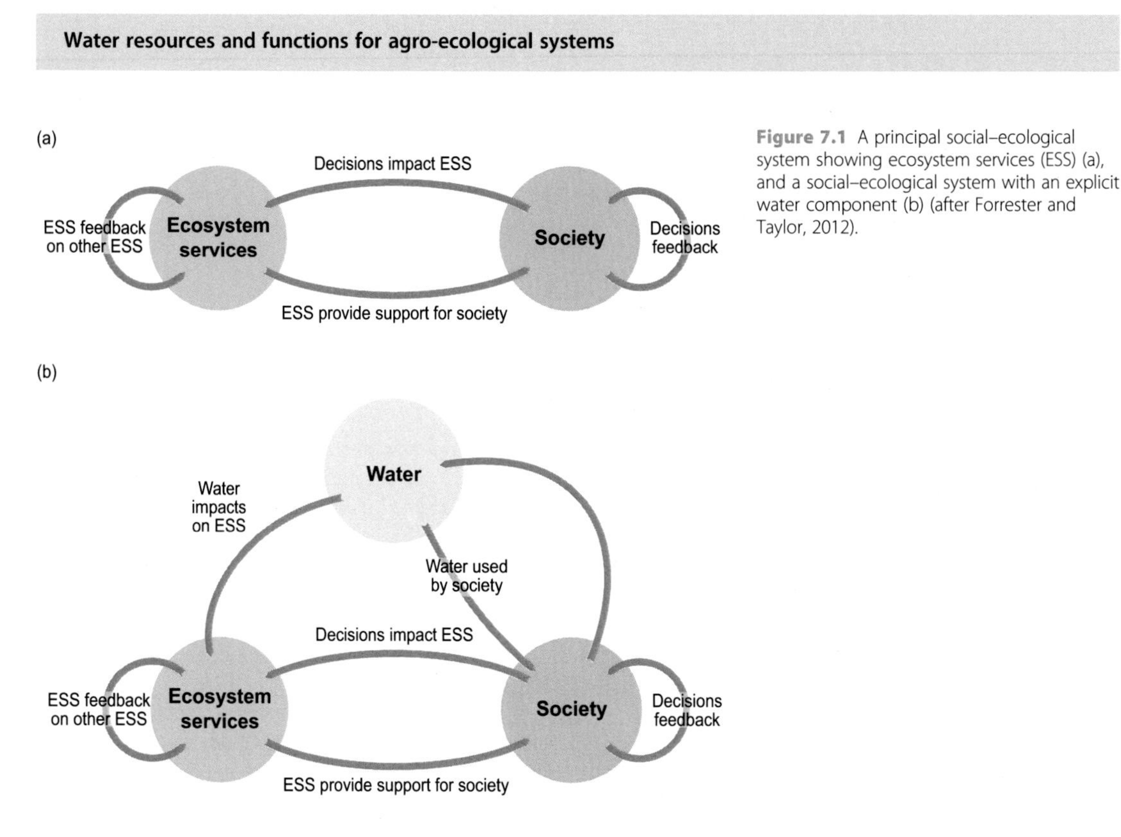

Figure 7.1 A principal social–ecological system showing ecosystem services (ESS) (a), and a social–ecological system with an explicit water component (b) (after Forrester and Taylor, 2012).

Water resources in social–ecological systems face increasing challenges both within and beyond the landscape scale, when water appropriation and the partitioning of water flows change due to human-induced land-use change. This chapter discusses the multidimensional impacts of transformed or transforming landscapes, and distinguishes between the potential benefits and disbenefits of various trajectories of development.

The chapter discusses systems definition in time and space and the potential implications for hydrological processes. A set of landscape case studies in agro-hydrological transition are explored to determine the changes in variables that control flows at the landscape scale and the consequences for various indicators of sustainability. Finally, section 7.4 discusses the possibilities of and potential consequences for alternative water functions to ensure development, possibly including the transformation of social–ecological systems at the landscape level.

There is no single framework for linking ecosystem services and water that is relevant at all temporal and spatial scales, or for all social–ecological systems. A simple model of a social–ecological system can be visualised, using water flow as the link in the biosphere (Figure 1.13, Chapter 1). Others have tried to break down the features of water in landscapes into key attributes that determine the fundamental essence of ecohydrological processes (e.g. Brauman *et al.*, 2007), which in turn form the basis of the ecosystem services in a given spatial and temporal setting. However, as all the attributes and all the ecohydrological processes are fundamental to shaping the availability and function of various ecosystem services, it is not evident how such linkages can be made (see Figure 7.9).

7.1.2 Complex feedbacks between water, ecosystem services and society

A generic framework between ecosystem services and water functions needs further work to provide a meaningful bridge between various sciences related to water, land and ecosystem services. Figure 7.2 illustrates the water–ecosystem–human nexus with principal flows of supply and demand, and the impacts in between. Even in this simple illustration, it is clear that the linkages provide complex feedbacks between the three entities of water resources, the ecosystem services that water sustains and society with its human

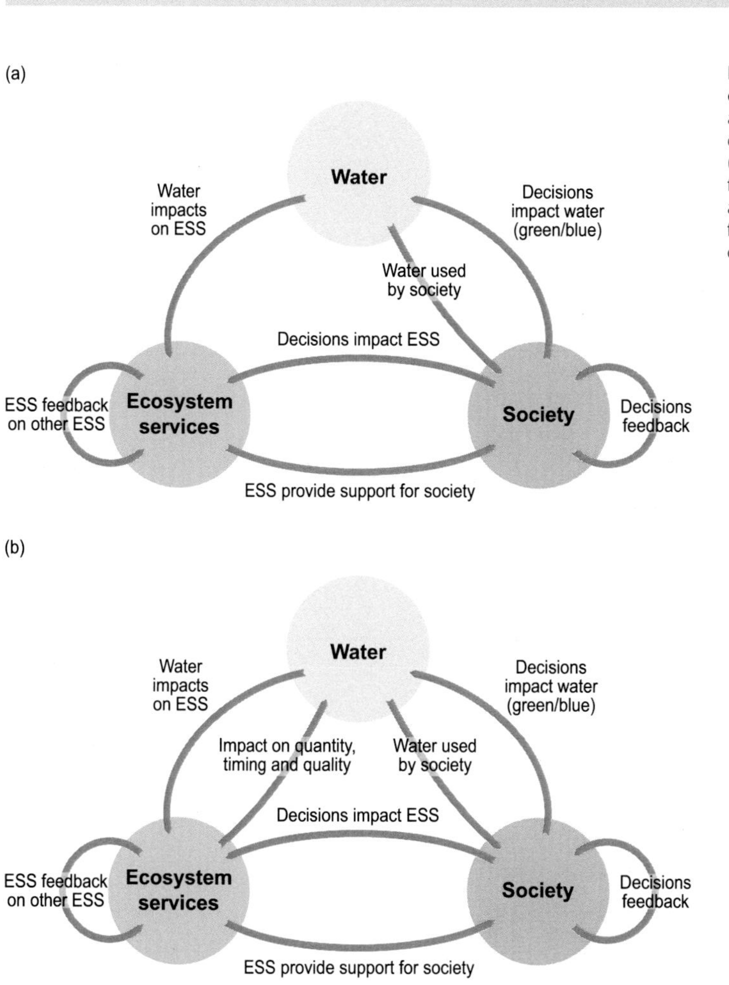

Figure 7.2 A schematic social–hydrological–ecological system for assessing causalities in agro-ecological landscapes. Often assessments on water availability for ecosystem services (ESS) is done (a) without accounting for the feedback of ESS on water resources (b). Even in an extremely simplified scheme, the mode of feedbacks and the potential impacts are challenging to address.

dependencies for water and ecosystems services. Untangling these relations and determining the resilience of a specific social–ecological system will therefore be case-specific in time and space, as the social–ecological system will be defined in a time/space-specific state.

At the landscape scale, the challenge is to translate knowledge and decisions into action on the sustainable use of land-water and vegetation resources (Figure 7.3). Through physical changes in soil and water resources use at the field to landscape scale, in particular through the use and alteration of vegetation in agricultural practices (crops, other biomass produce), we deliberately or inadvertently affect a range of ecosystems services, as well as the water functions that sustain these services, both locally and beyond the landscape scale. Typically affected hydrological attributes were summarised by Kiersch (2002) as part of an open consultation by FAO and van Noordwijk *et al.* (2004) (Figure 7.4). According to this document, based on measurements of water changes at the landscape scale, the impacts on quantity attributes rarely traverse scales beyond 10^2 km^2, whereas water quality attributes can be detected much further away from the source at up to 10^5 km^2. At larger scales, e.g. $>10^2$ km^2, water quantity attributes are more dominated by natural processes rather than changes in local land use. These limits are only intended as broad guidelines. Various case studies and reviews support the ranges of impacts in Figure 7.4 – experimental evidence is scale- and site-dependent as a minimum.

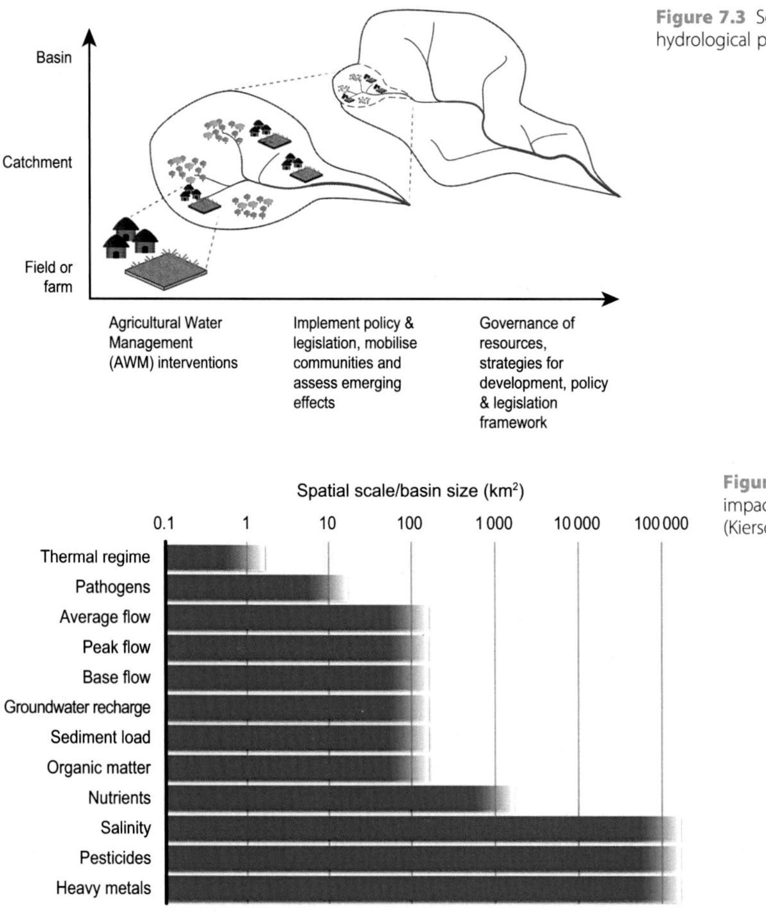

Figure 7.3 Schematic illustration of typical scales for hydrological processes.

Figure 7.4 Proposed spatial scales of the impact of various hydrological attributes (Kiersch, 2002).

7.1.3 Poor generic understanding of the impacts of land-use change

As an example of impact we can look at reviews of the hydrological impacts of afforestation and deforestation at the catchment scale. They show a range of emerging trends and inconclusive evidence on impacts, while cautioning against over interpretation due to biased or small sample sizes. Ilstedt *et al.* (2007) show a three-fold increase in water infiltration from afforestation using plot-scale measurements. An increase in infiltration could easily be interpreted as a reduction in runoff, and it is often assumed, sometimes by default, that more trees (afforestation) result in reduced stream flow. This is, however, not necessarily always so straightforward. Andréassian (2004) largely concludes that afforestation has extremely unpredictable outcomes, as evidenced by 134 paired catchments. Only one impact of increased forest cover could be clearly stated – increased base flow after deforestation – but even this impact was uncertain over time, as revegetation and other land-use changes take place in catchments. The study by Andréassian (2004) was biased towards temperate catchments. Locatelli and Vignola (2009) focused on humid tropics afforestation, and found lower total flow and lower base flow for catchments under afforestation. Brown *et al.* (2005) added significant dry forest systems to their review of 166 case paired catchments, and highlight the importance of the time dimension on impacts. In their review, afforestation takes longer to stabilise flow regimes compared to deforested land. Thus large knowledge gaps still remain, in part due to underrepresentation of particularly dry forest hydrology and land-use change, and interactions with water flows in landscapes.

Systematic reviews and/or meta-syntheses of heterogeneous landscapes and the impacts of use change

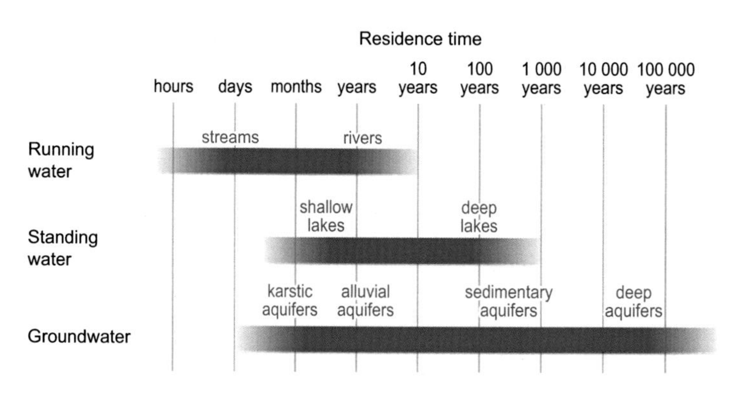

Figure 7.5 Typical order of residence time for freshwater storage in landscapes (after Chilton and Seiler, 2006).

on water flows are providing even less generic understanding. A review of water quality aspects in Australian streams and rivers for various land uses shows the scale effect of Figure 7.4. Suspended solids, total nitrogen and total phosphorous compounds in water decreased significantly as catchment size increased (Bartley *et al.*, 2012). However, determining the dominant hydrological process and the consequences of land-use change at various scales remains a challenge. For example, Oudin *et al.* (2008) used a number of experimentally based rainfall-runoff relations to test whether accounting for land use matters. They tested original equations without a land-use component against equations with land-use components for more than 1500 catchments in various climatic conditions, and suggest that incorporating land cover makes only a marginal difference but still leads to significantly improved predictions, most notably in the smaller catchments and in wetter conditions. However, these models can only give annual mean, long-term predictions, and do not necessarily help predict the short-term impacts of land use on water resource availability in social–ecological systems.

7.1.4 Assessing water flows and functions

Humans significantly modify soils and vegetation, thereby affecting water flows. Models of 'undisturbed landscapes' may not necessarily function well in the highly managed soil and vegetation systems of agriculture and afforestation–deforestation landscapes. For example, hydrologic connectivity is altered by various man-made structures to collect and infiltrate water. A study of disrupted hydrological connectivity in agricultural landscapes through the development of banks and small reservoirs in Western Australia showed a reduction of up to 87% in area connectivity with basin outlets (Callow and Smettem, 2009). The shaping of the landscape clearly reduced hydrological predictability because it affected hillslope dynamics. Modifying landscape descriptors such as the digital elevation model (DEM) and correcting for actual shapes and structures was shown to improve hydrological predictability.

Similarly, water processes and associated functions in landscapes have temporal dimensions, often varying in different spatial scales. For example, water storage, or the residence time of water in landscapes, can range from minutes to days to thousands of years (Figure 7.5). Accounting for the time lag of water storage in a landscape is often as challenging as describing the dominant spatial processes. A key feature in temporal scales is the distribution of water, such as rainfall, which is often the principal resource input into social–ecological systems at the landscape scale.

Schulze (2000) showed that fundamental mistakes can be made on water resource availability by using inappropriate samples or inappropriate averages in time and space to estimate flows and storage in hydrological processes (Figure 7.6). These errors will propagate through estimates of ecosystem services generated by the rain and water flows. This can lead to significant over- or under-valuations of resource impacts, simply by not adequately accounting for natural variations in rainfall and other water processes at the landscape level.

7.1.5 Revising our understanding of water in landscapes

In the past 10–15 years, developments in hydrological signalling have represented a paradigm shift in our understanding of how hydrologic processes operate in the time–space continuum (e.g. McDonnell *et al.*,

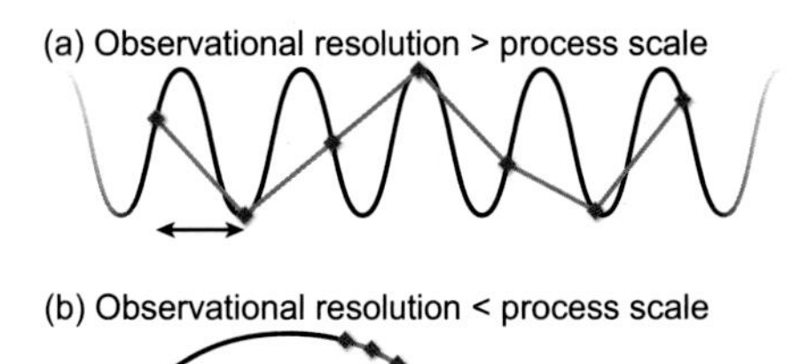

Figure 7.6 Examples of errors in estimates of hydrological attributes due to flawed sampling and poor analysis of common hydrological and climatological processes (after Schulze, 2000).

2007; Sivapalan, 2005; Tetzlaff *et al.*, 2008; Wagener *et al.*, 2007). This paradigm shift in hydrological science will affect the way the water–ecosystem services–human systems are linked, and in particular, how the impacts of land and water use at the landscape scale are assessed, predicted and ultimately managed.

In all systems, the management of system feedbacks is essential to deal with the sustainability of system functions. In the social–ecological system, the focus is often on ensuring or increasing the provision of ecosystem services (benefits), while coping with or reducing the disbenefits for certain local and/or global users. However, because of the intrinsic feedbacks between people, landscapes and ecosystem services, any action that changes land, water and vegetation has impacts directly or indirectly in time or space on other resource beneficiaries (Figure 7.4).

We can hypothesise that a landscape unit with a limited amount of water sustains a given set of ecosystem services within a given timeframe. In this system definition, resources are finite. However, this type of system rarely represents the reality of landscapes: there are always flows in and out in time and space, and stocks that complicate resources accounts. Social dimensions such as skills, knowledge, economics, organisation and adaptability in management shape the accounts of water and ecosystem services, but cannot easily be defined in time or space, or estimated in additive or subtractive terms. The management of stocks and flows will determine the resilience of the social–ecological system to change. Referring to our generic model of linkages between water, ecosystems and society as a representation at the landscape scale (Figure 7.7), it becomes evident that an undesirable link between any of the three components can quickly put the system into a set of negative changes to the flow and stock of water, ecosystems and benefits (Figure 7.7a). Managing even well-defined systems of water resources, ecosystems and society for multiple benefits is clearly even more challenging. We focus below on the management of stocks of water, nutrients and biomass as indicators of social–ecological system resilience in particular landscapes.

7.2 Agricultural expansion and intensification affects landscape multifunctionality

Humans manage landscapes in order to enhance or suppress certain ecosystem services. Agriculture is the greatest man-made activity affecting landscapes and ecosystem services. Most estimates of water use for agriculture, however, disregard the benefits generated by the flows of water, which in itself generate multiple regulating and supporting services not well captured in water accounts. Fundamentally, agricultural production aims to maximise particular provisioning services such as economic yields from biomass, livestock or other biological systems. The change in water appropriation for agriculture affects the surrounding landscape and the ecosystem services it supplies. However, these impacts can be marginal, acceptable or not beneficial to other water users, or for the ecosystem services reliant on the same water attributes that have been changed.

A highly generic set of impacts on water processes by agricultural water appropriation and practices is

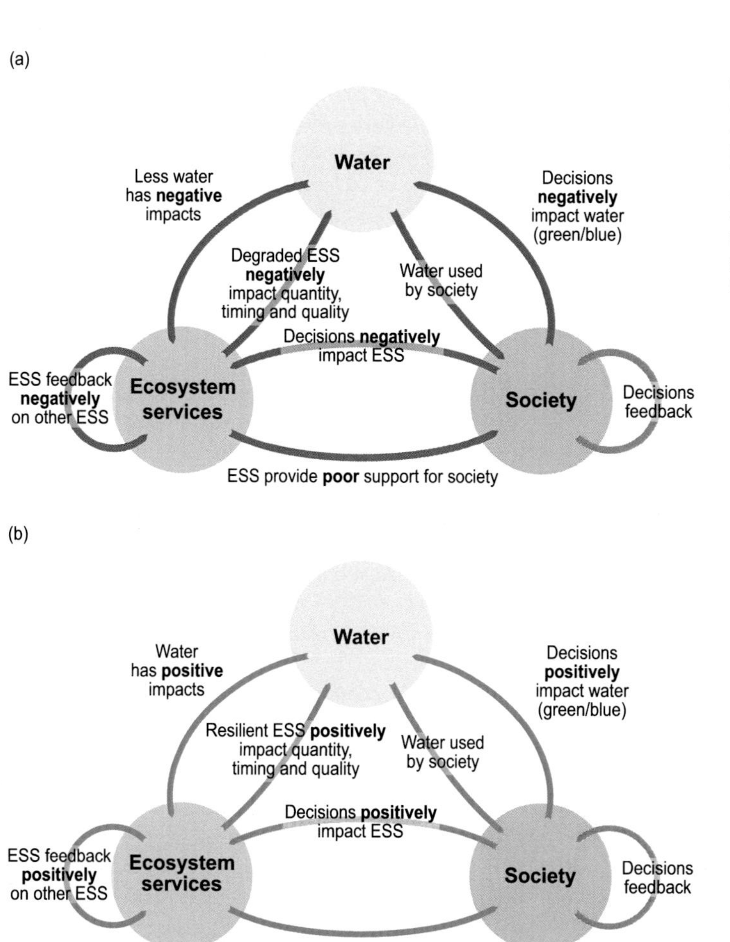

Figure 7.7 Example of degraded and resilient social–ecological systems with an explicit water component, indicating tentative feedbacks. A tentative negative state is shown (a), and a possible positive state of systems (b) indicates the various processes that need to be maintained to attain landscape resilience. The complexity to attain the positive trajectories of states and processes is a challenge for governance and management to maintain sustainable trajectory. ESS, ecosystem services.

listed below. Although these are often cited as generic impacts of agriculture, there is a wealth of knowledge and experience of how to manage these impacts from a biophysical perspective at multiple scales. There is also growing evidence that these impacts are less well evidenced and/or not as generic as is sometimes claimed.

Changes to water quantity in streams, lakes and groundwater for irrigation

According to current estimates, approximately 70% of terrestrial freshwater is appropriated for irrigation purposes annually (Molden, 2007). Not all of what is withdrawn is consumed. In the order of 30–70% is return flow and usable downstream for additional irrigation or other purposes. So at every unit of land irrigated, a substantial amount of withdrawal actually flows back to support downstream uses by agriculture or ecosystem services of the very same withdrawn water. It is very unclear how this return flow is accounted for in the global numbers of irrigation withdrawal. Future trends suggest that the availability of irrigation water will be a critical constraint on food production due to growing demands for domestic and industrial needs as well as requirements for environmental flows and the effects of climate change (e.g. Strzepek *et al.*, 2010).

The development of small to large storage reservoirs for irrigation and other water supply purposes

Globally, there is an estimated 6197 km^3 of storage capacity in dams and reservoirs that are larger than 0.1 km^3 (Lehner *et al.*, 2011). In addition, there is a large amount of water storage capacity in reservoirs, tanks and dams that are smaller than 0.1 km^3, as well as groundwater sources. There are no data on current storage capacity for irrigation purposes alone.

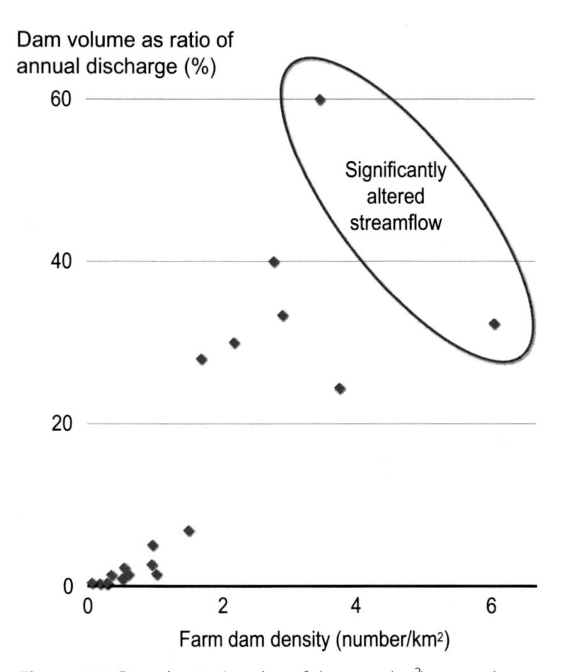

Figure 7.8 Dam density (number of dams per km^2) versus dam volume (% of annual stream flow). Data adapted from Schreider *et al.* (2002).

The impact of dams on stream flow depends on the scale of assessment. Nilsson *et al.* (2005) concluded that more than half the 273 large river basins globally are severely affected by human interventions, and in particular the eight most biogeographically diverse. Yet, there is multi-scale evidence that small dams do not necessarily affect stream flow significantly. The impact depends on the scale of assessment. For example, the Volta basin contains more than 1200 small reservoirs north of the Akosombo Reservoir. According to Andreini *et al.* (2009) and Liebe (2002) these small reservoirs do not affect overall water flow at the basin scale.

Schreider *et al.* (2002) modelled the impacts of small farm dams in catchments of 90–900 km^2 in a similarly semi-arid environment in South Australia. The results showed that not until total storage volumes exceeded 30% of annual stream flow could a significant decrease in stream flow be detected (Figure 7.8) (Schreider *et al.*, 2002). At the scale of the Murray Darling basin, more recent modelling suggests that even with an increase in small storage of 10% from the current 2.2 km^3 storage capacity, the average impact on basin runoff would be less than 1% (0–3% variation for sub-basin regions) (Chiew *et al.*, 2008).

Drainage of wetlands

Historically, the drainage of wetlands to develop agriculture has led to a substantial reduction in the number of biodiversity hotspots. Today, wetlands are among the most protected land uses, through the Convention on Wetlands of International Importance especially as Waterfowl Habitat (the Ramsar Wetland Convention of 1971), which has 163 state parties (FAO, IUCN and UNEP, 2013). According to the Millennium Ecosystem Assessment (2005), Ramsar-designated sites cover approximately 15% of all global wetlands – a total of 1280 million ha globally. Today, further expansion of agriculture in both inland and coastal wetlands is often heavily contested and subject to substantial evaluations compared to potential conversions from other habitats and land-use types. However, the challenge remains especially in less biodiverse local and smaller wetlands, which are being rapidly developed for agricultural purposes with few safeguards for other provisioning and regulating ecosystem services. There is a large knowledge gap on this type of loss of water-related ecosystem services and the long-term implications for local livelihoods and economies.

Changes in soil water infiltration

Poor crop–soil management or changes in vegetation type can reduce soil infiltration capacity and result in unproductive water losses at the field scale, as well as sediment loss at the field to landscape scale. There are suggestions that both globally and regionally, conversion of permanent vegetation (forest) to annual cropping has increased global stream flows (e.g. Piao *et al.*, 2007) but results are scale dependent (e.g. Alkama *et al.*, 2013). Similar hydrological signals have also been detected using landscape-scale modelling and assessments (e.g. Favreau *et al.*, 2009).

Soil water storage (potential water holding capacity) and root depth

Annual crops have a tendency to have a shallower root depth compared to permanent vegetation. In addition, mechanised soil tillage often creates compaction, which reduces the root depth and total potential soil water holding capacity. Studies also suggest that changes in soil water storage capacity can be traced at the basin scale (e.g. Mahe *et al.*, 2005), and may subsequently affect stream flow generation. However, there are a range of methods and practices of crop-soil management using easy management techniques that can substantially enhance infiltration and soil water holding capacity for the benefit of plants or crops.

Changes in water tables and groundwater

Landscape changes and agricultural development affect various dimensions of groundwater. The use of groundwater in dry areas is particularly efficient for the development and intensification of agriculture, enabling critical water supply to boost biomass production. The consequence is rapidly decreasing groundwater tables, in particular in semi-arid and arid regions, both shallow and deep (e.g. Aeschbach-Hertig and Gleeson, 2012). Groundwater resources are also affected by changes in vegetation that may either decrease or increase recharge of groundwater. For example, water policy in South Africa is deliberately designed to regulate invasive vegetation that extracts water beyond the natural vegetation, although some have questioned the scientific evidence (Gorgens and Van Wilgen, 2004). The opposite process can also affect water resources and groundwater levels at the landscape scale. In parts of South and Western Australia, the clearance of longstanding forests by settlers in the nineteenth and twentieth centuries caused a rise in water tables, affecting the quality of agricultural soils and in some parts causing serious and substantial losses due to salinisation as evaporation transported salts to the root zone (e.g. Coram *et al.*, 2000). Reversing this process through reforestation has been possible in a controlled trial (e.g. Bell *et al.*, 1990).

Water quality

Water quality may actually be a more imminent and tangible issue for social–ecological system change at the landscape scale. Water quality is affected by a range of physical (sediment), chemical and biological compounds in both particulate and soluble form. Human settlement and agro-ecological development, including land-use change, often have considerable impacts on local water quality.

The erosion and water transport of sediments is a natural process, which can be accelerated or slowed by human interventions. On a global scale, assessments of sediment loads in large river systems suggest that sediment transport has been reduced by man-made water-course interventions (e.g. Walling, 2006). Large differences exist, depending on hydro-climatic and land-use combinations, and on the scale of assessment (e.g. Syvitski *et al.*, 2003). Water quality has degenerated in the past, and is likely to continue to deteriorate in the near future (e.g. Scanlon *et al.*, 2007; Seitzinger *et al.*, 2010). Lack of data makes it a challenge to estimate local and global changes. The lack of environment-related indicators to complement the human drinking water standards of the World Health Organization is a further challenge. A comprehensive summary based on global water quality data is available from UNEP GEMS/Water, showing both desired trends in water quality and undesirable trends based on human activities in land and water resource management (UN GEMS Water, 2008).

In sum, the above brief description of water stocks and flows suggests that the impacts of social–ecological changes at the landscape level can be detected at a global scale. Data also confirm that there are no generic causes and consequences, or definite pathways of causality across spatial and temporal scales. Most water resources exist in a context where limits and undesirable impacts may not be well represented at the global scale, given the challenges of scaling hydrological processes. In fact, some evidence at the global scale may even challenge our perception of impacts. For example, the estimates of reductions in sediment load to oceans is a very different discourse from the local field-scale message of degraded land and soil erosion, as is the 'general understanding' that the aggregated impact of certain vegetation cover does not necessarily reduce water availability. Clearly, many water resource factors are still poorly understood due to biased evidence in time and space as the examples given above illustrate. Interpretations of water dimensions in multiple functions related to water flows and stocks that generate desired benefits have been undertaken by several researchers (e.g. Brauman *et al.*, 2007; de Groot *et al.*, 2010; Lamarque *et al.*, 2011). Typical water-related functions are associated with the existing or changing attributes shown in Figure 7.9. The development of landscape water functions and ecosystem services, however, is still in its infancy as this represents a scale of increased complexity in hydrology, ecosystem services and social dimensions which has multiple feedback loops in between – as is demonstrated in Figure 7.7a and b.

The desired benefits may have impacts at the local and/or the global scale, which may or may not need management. Section 7.3 reviews three landscapes from an agro-ecological perspective, assessing changes in water and associated water functions. The cases exemplify common development trends,

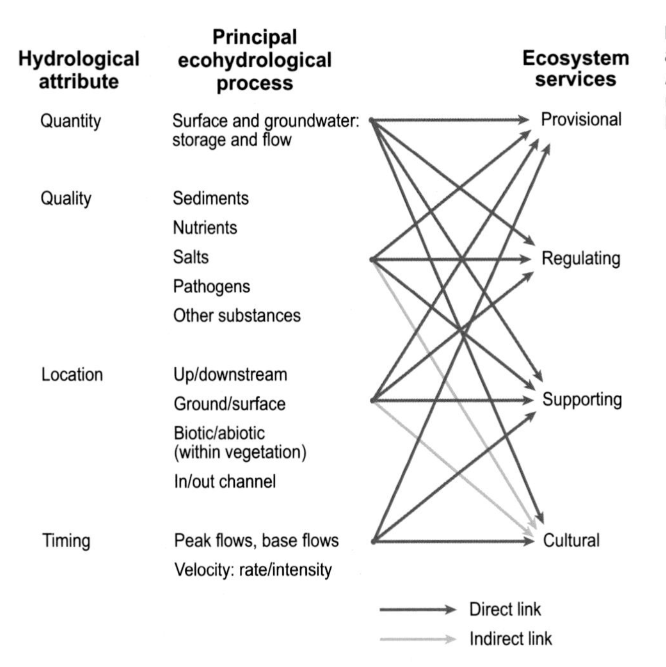

Figure 7.9 The challenge of linking hydrological attributes in a generic framework to ecosystem services. All ecosystem services, both direct and indirect, are the result of time–space-specific hydrological attributes (after Brauman *et al.*, 2007).

and provide various indicators that can be used to assess whether change is acceptable, beneficial or could give rise to internal or external disbenefits, which might be managed through governance and management interventions.

7.2.1 Manipulating water to transform agro-ecological landscapes

Impacts on and changes to water flows in landscapes can have substantial effects on a range of ecosystem services, as is shown above. Agricultural production, which in itself is an ecosystem service, is primarily a way to manage particular benefits arising from ecosystems. Doing so may affect other ecosystem services, depending on the scale of management in time and space. Historically, the management of landscape, including its freshwater resources, has had significantly negative impacts on various local and global ecosystem services. These impacts have been costly and difficult to restore. In view of the current and expected future demand for food, fodder and fibre there is an urgent need to ensure that water and land are efficiently used, while maintaining other ecosystem services of agro-ecosystems beyond these areas, which yield produce. These non-farmed areas in the landscape mosaic also support and enable various ecosystem services to function. This section examines three agricultural landscapes in various stages of degradation or unproductive states, and their transformation to productive states. These cases can help identify potential opportunities for and constraints on building resilience.

Depending on the state of individual, or bundles, of ecosystem services in an agro-ecosystem we can classify landscapes into various states, ranging from degraded to high yielding agro-ecosystems (see Figure 7.10). Although a degraded landscape is undesirable from any perspective, there may be local preferences for an intermediate landscape where agricultural benefits are in balance with other ecosystem service benefits. This can be compared to an agro-ecosystem that has maximised agricultural output at the possible expense of other specific ecosystem services. Through manipulation of water flows or water characteristics in landscapes, it is feasible to transform multiple ecosystem services from degraded to intermediate levels or beyond. The outcome, however, must be negotiated – whether to maximise a few benefits from agro-ecosystems at the expense of reducing the capacity of others.

7.2.2 Indicators of landscape-scale resilience

Landscape-scale resilience is assessed using five landscape-scale indicators, which are related to a range of ecosystem services and their underpinning water flows. These indicators are used to assess the

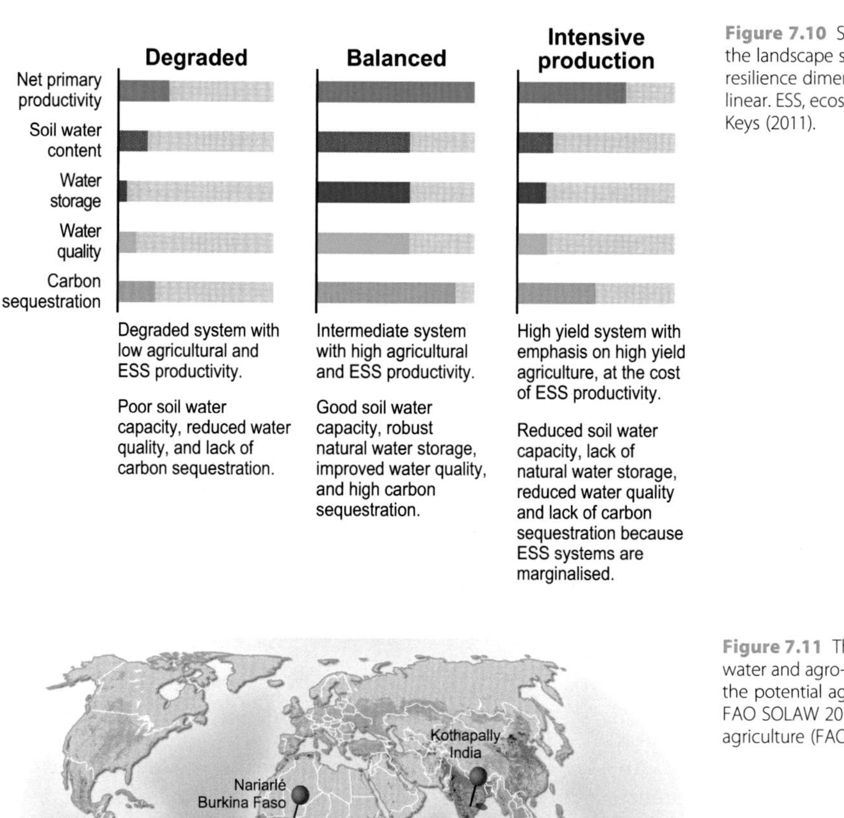

Figure 7.10 States and selected indicators at the landscape scale for water- and land-related resilience dimensions. Note: transitions are not linear. ESS, ecosystem services. After Barron and Keys (2011).

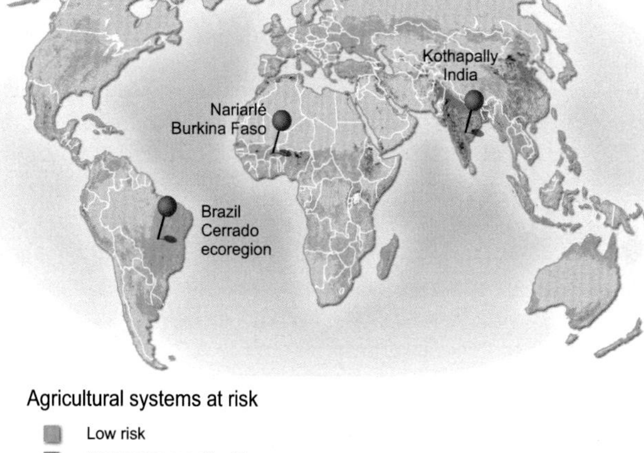

Figure 7.11 Three case studies as assessed for water and agro-ecological changes mapped on the potential agricultural risk assessment by FAO SOLAW 2011: Land and water scarcity for agriculture (FAO, 2011).

relative state of landscapes, ranging from a degraded state with low capacity for generating services that benefit human well-being and societies, to maximised production in agro-ecosystems, with high (direct) economic benefit to human well-being and society. Maximised production is probably unsustainable given the supporting and regulating services that underpin the provisioning capacity. Table 7.1 sets out a set of indicators at the landscape scale selected (see de Bruin *et al.*, 2010) to help assess the different states of landscapes.

7.3 Three examples of landscape transformations and possible transitions

Globally, agricultural development has transformed landscapes and their associated water flows and stocks, as is described in Chapter 3. This section reviews the evidence for landscape transformation through changes in water stocks and flows, and associated ecosystem services. The three cases are located

Table 7.1 Potential indicator themes for assessing the benefits and disbenefits of agricultural transformation related to water resources (modified after de Bruin and Barron, 2012; de Bruin *et al.*, 2010)

Indicator theme	Ecosystem service	Hydrological characteristics	Indicator
Soil	Regulating-supporting	Infiltration Soil water holding capacity	Soil surface infiltration capacity Soil organic matter Soil water holding capacity
Water quantity and quality	Regulating-supporting	Water partitioning Residence times Irrigation availability Sediment transport Pollution and nutrient cycling	Water levels Rainfall amounts and distribution Stream flow Groundwater levels/accessibility Water storage capacity, types of storage Water quality indicators: sediments anthropogenic
Land use	Regulating-supporting	Water partitioning Pollution and nutrient cycling	Agricultural, non-agricultural land use
Habitats (aquatic, terrestrial)	Provisioning	Water partitioning Pollution and nutrient cycling	
Production (yield) related	Provisioning	Water partitioning	Biomass/NPP Food, fodder, fibre outtake/yields

in semi-arid to sub-humid agro-climatic zones, in recent or currently rapidly developing country contexts (see Figure 7.11).

7.3.1 Intensifying rainfed production in Kothapally, India

Our first example is the development of Kothapally, India, from a degraded agro-ecosystem to maximised production, and the impacts this had on the surrounding landscape. In India, the primary targets for rural communities – as well as the local state and the national levels – are self-sufficiency in food supply and fibre supply, as well as viable incomes. Significant investments have been channelled through the public, donor and private sectors to enhance livelihoods through improved agricultural production using catchment management approaches (e.g. Kerr, 2002). At the local level, this has resulted not only in yield increases but also in improved livelihoods and ecosystem service benefits (Joshi *et al.*, 2008).

A densely populated and dry-spell exposed landscape

The Kothapally catchment (Figure 7.12) is a well-researched example (Garg *et al.*, 2012a, b). Before 2000, this 3 km^2 catchment was considered a significantly degraded area, with water and soil condition in a declining state, severely limiting crop production and productivity that severely limited crop productivity. Multiple factors constrained efforts from both a biophysical and a social perspective. Rainfall variability is very high, and is distributed in time and space, with a long-term average of 860 mm and approximately 60 days of rainfall in a 150-day rainy season. The soil is shallow, at less than 1 m, and has a low potential water holding capacity of 0.26% (±0.07%), affecting its capacity to store rainfall to bridge the frequent dry spells linked to the rainfall variability. Most of the catchment was under rainfed cultivation before 2000, and crops were affected by frequent dry spells. High population pressure, with an average farm size of 1.4 ha per family, meant that no crop area expansion could take place to produce more food, and that the only option was intensification on existing land.

Intensive action was required to manage water flows and storage in order to increase development and sustainability in the catchment (e.g. Wani *et al.*, 2011). The roles and partitioning of water in landscapes are fundamental to providing food, fodder and

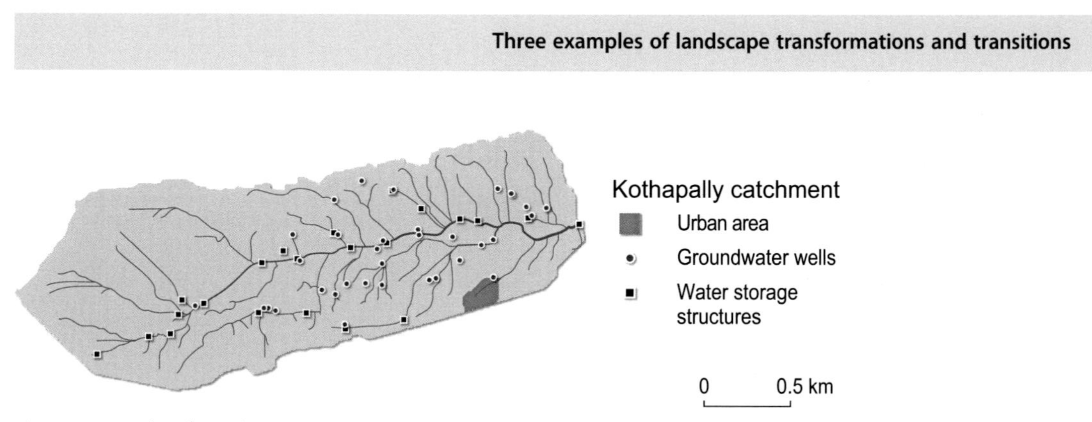

Figure 7.12 Kothapally catchment showing water storage structures and recharge wells (Garg *et al.*, 2012a).

other basic needs, and water is also a joiner of various users, i.e. community and individual beneficiaries of these services. The catchment interventions in Kothapally initiated a range of soil and water strategies and structures to alter the partitioning, storage and flows of water to:

- address soil moisture deficits during the rainy season and dry season;
- recharge groundwater;
- limit the loss of valuable soil depth and soil organic matter.

Enabling two cropping seasons

Efforts to transform agricultural production and land use included a range of measures to enhance rain infiltration using *in-situ* and *ex-situ* structures (see Figure 6.5 in Chapter 6). The *in-situ* structures – typically, infiltration bunds – could only bridge smaller dry spells and, although efficient to some extent, were quite limited for intensification of the current crop system. The shallow groundwater in the catchment was recharged during rainy seasons using *ex-situ* surface water collection and storage (mostly checkdams). It was then abstracted during the post-rainy season for irrigation purposes. This allowed a cropping system change to take place, enabling two cropping seasons instead of one. The introduction of dry season crops was rewarding in terms of cash and food security, because the storage of food decreases after the rainy season, as market prices tend to increase in the dry seasons.

However, additional crop system improvements were necessary to ensure maximisation of the benefits of the water flow and storage changes. First, nutrients were added to the soil, from inorganic and organic sources. These additions maximised the use of water by crops, resulting in improved water productivity. Second, the implementation of these water, crop system and soil fertility changes only occurred with substantial human effort, i.e. manual labour. In order to transform the Kothapally landscape into a more productive state, a number of work hours had to be invested by individuals and communities. Strengthening community cohesion and empowering farmers and other users in rural landscapes was thus a complementary aim of the catchment intervention strategy (e.g. Joshi *et al.*, 2008).

Resilience considerations

Using the potential states depicted in Figure 7.10, the Kothapally catchment can be envisaged as transforming from a degraded state pre-2000 into a highly productive, intensively cultivated landscape today. Several indicators show that the landscape has moved to a healthier state (Table 7.2), but the maximised use of water in the catchment is now affecting the accessibility of water downstream. Garg *et al.* (2012a) modelled the catchment balance for climatic dry, medium and wet years, and concluded that the current implementation of soil and water structures reduces the outflow from Kothapally by 10–15% of the annual water balance, which in dry years is insignificant (Figure 7.13). Managing water partitioning points to benefit crop production and improve human benefits has made the internal catchment status healthier, but it has reduced downstream opportunities to do the same.

Pre-2000, the catchment could be seen as an example of an undesirable resilient state, as substantial investments in labour and resources were required to change the landscape into its current more productive and more desirable state. It is yet to be determined how resilient it will be in its current state to internal and external change. As long as local people maintain the catchment bunds and check dams, the landscape will remain in a productive state. Even in a scenario of reduced rainfall or higher

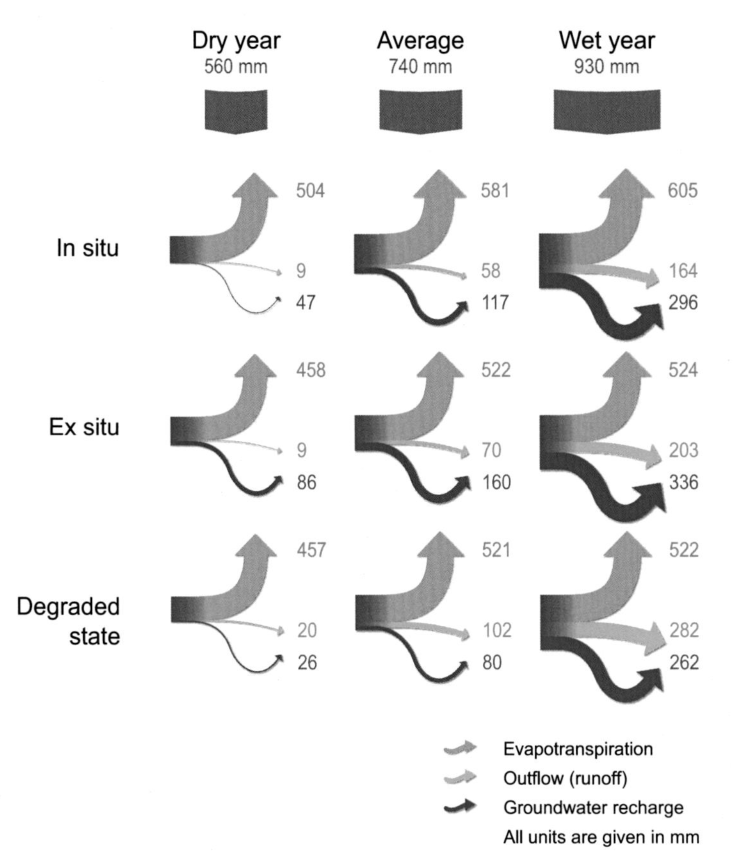

Figure 7.13 Water flow partitioning changes for *in-situ*, *ex-situ* and degraded states in the Kothapally catchment. Long-term average flow partitioning is shown with average rainfall per year for dry, normal and wet years. Partitioned flows are represented by actual evapotranspiration, surface runoff outflow and groundwater recharge (after Garg *et al.*, 2012a).

variability of rainfall, this productive state can probably be maintained. Thus, it could be said that these resilience features are more dependent on social factors, such as the capacity to manage the water structures that keep the catchment productive.

However, if the downstream users increase demand for stream flow, the overall impact of local increases in resilience may reduce landscape-scale resilience. Reduced landscape resilience could occur from downstream users adjacent to the catchment demanding water, as well as the need for adequate reservoir water in the Osman Sagan reservoir to supply the urban area of Hyderabad. There are a range of benefits and trade-offs to consider when deciding whether to use water upstream or downstream in the Osman Sagan sub-basin (Garg *et al.*, 2012b). In light of the demand for water downstream, of the catchment is unlikely to be maintained in its current productive state, as more water will need to be released for other users.

7.3.2 An agricultural transition: Nariarle, Burkina Faso

Transforming savannah to agricultural land

The Nariarle catchment is a typical semi-arid tropical landscape that has changed from a landscape dominated by natural vegetation in the 1950s to an intensely cropped area, adjacent to a peri-urban area. During this time, substantial shifts in land use have occurred in response to various internal and external drivers, including population growth within the catchment and growing consumer demand in Ouagadougou. Historically, rainfall has also shifted from a wetter domain before the 1980s, to a drier rainfall regime, to a possible trend for a wetter regime in recent years.

Rainfall has varied in the past 50 years around a long-term average of 756 mm/year (standard deviation 172 mm/year; Figure 7.14). During this time, there has been land-use change, and population density has increased from 61 inhabitants per km^2 in 1996

Table 7.2 Changes in water- and land-related resilience indicators for Kothapally catchment (after Garg *et al.*, 2012a)

Indicator theme	Indicator	Pre-2000	2009
Soil	Soil surface infiltration capacity	Modified with USDA curve number +6	
	Soil organic matter	SOM approx. 0.4–2.3%	SOM increase from baseline due to integrated nutrient and compost approaches
	Soil water holding capacity	0.75% of current measured soil water holding capacity	
Water quantity and quality	Rainfall amount distribution	Annual average: 840 mm/year, 3–8 times per season of dry spells of 5–7 days	
	Stream flow	19% of annual rainfall = 151mm	10% of annual rainfall = 70 mm
	Groundwater levels/accessibility	Deep groundwater with limited accessibility	Shallow groundwater accessible 32 boreholes and wells
	Water storage capacity, types of storage	Natural storage: no estimate	Small recharge structures: check dams, soil bunds; full soil water holding capacity >14 structures
	Water quality indicators: sediments	Sediment loss 68 mm soil profile loss over 31 years simulated	Reduced to >10 mm soil profile loss simulated over 31 years
	Anthropogenic	N/A	N/A for agricultural chemicals
Land use	Agricultural, non-agricultural land use	90/10%	90/10% + 30% crop irrigated post-monsoon
Habitats (aquatic, terrestrial)		None	None
Production-(yield-) related	Biomass /NPP Food, fodder fibre outtake/yields	Only crops during rainy season, often affected by dry spells	Rainy season crops, plus 1 crop with supplementary irrigation

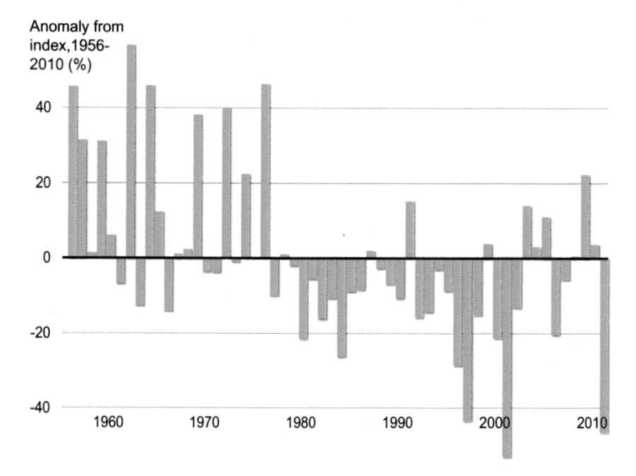

Figure 7.14 Annual rainfall normalised over the period 1956–2010 for Ouagadougou airport, bordering the Nariarle catchment (Royal Netherlands Meteorological Institute, 2013).

to 68 inhabitants per km^2 in 2006 – more than 10% in 10 years.

The earliest land-use maps, from 1956, depict a landscape dominated by savannah vegetation, which covered more than 70% of the catchment. Today, agricultural land takes up more than 70% of the catchment, and savannah vegetation has shrunk to 11% (Figure 7.15). Alongside the expansion of agriculture, more dense forest areas have increased from 3% in 1956 to 12% of the catchment today, which might seem counterintuitive. With these land-use changes, there has also been a significant change in water management in the landscape. The Nariarle catchment has, like many other areas in central-southern Burkina Faso and Ghana, been subject to heavy investment in small reservoirs (e.g. Douxchamps

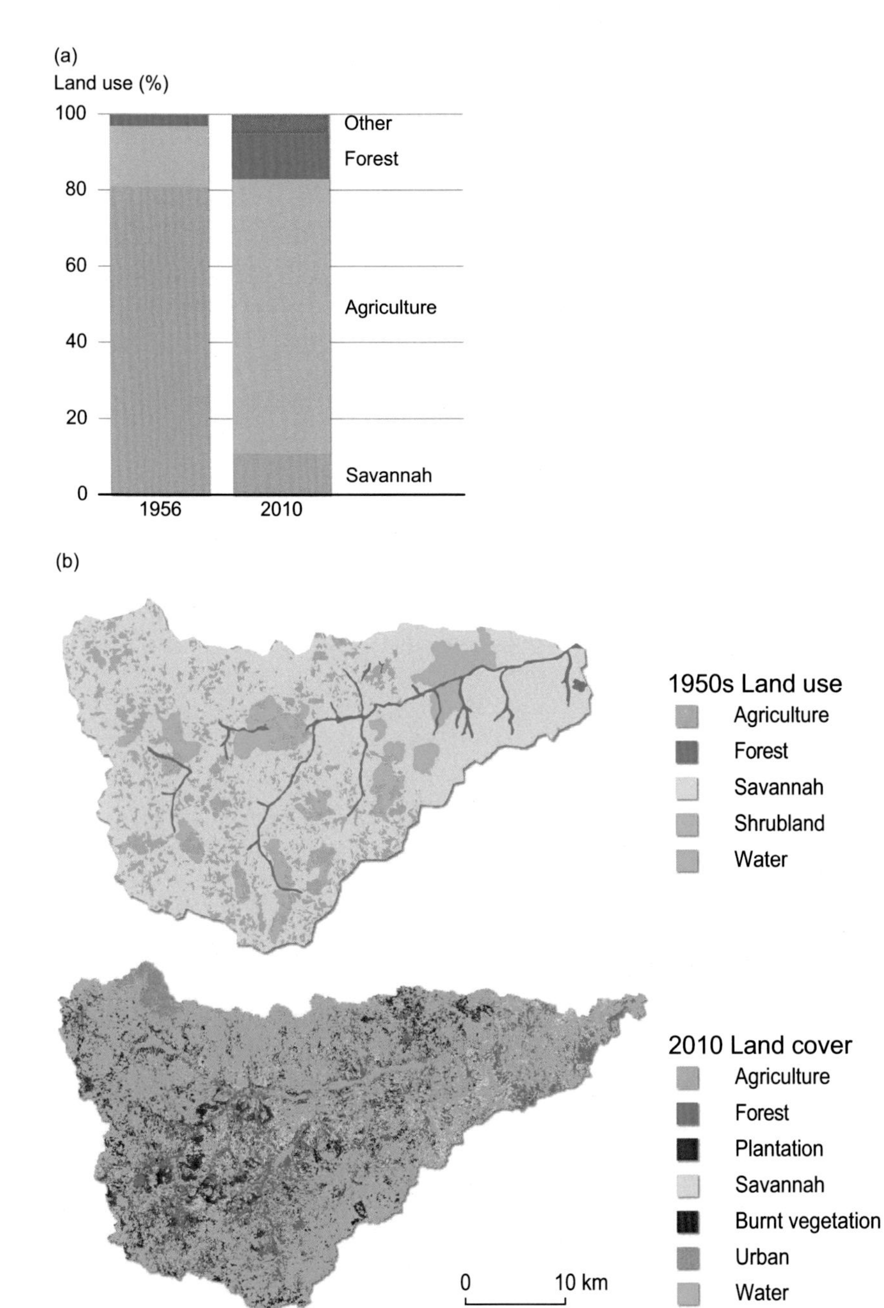

Figure 7.15 Land use in the Nariarle catchment, 1956 and 2010, ratio of agriculture to natural biomes (a) and land-use maps (b) (after Cambridge and Barron, forthcoming).

et al., 2012; Venot and Cecchi, 2011). Within the catchment, the water storage capacity of man-made structures has increased from insignificant storage in the 1950s to a potential storage capacity of more than 10.7 million cubic metres (MCM), or circa 2% of total annual rainfall. Yet, the catchment is near closure, as very little stream flow exits the catchment. Current partitioning of 20-year rainfall

Table 7.3 Changes in water- and land-related resilience indicators for Nariarle catchment (Barron, 2012; Cambridge and Barron, forthcoming)

Indicator theme	Indicator	1956	2010
Soil	Soil surface infiltration capacity Soil organic matter Soil water holding capacity	Soil organic content approx. 0.5–1% in natural soils	SOM approx. 0.1–0.5%
Water quantity and quality	Rainfall amount distribution	Annual average: 840 mm/year, CV=20%	Annual average: 680 mm/year, CV = 19%
	Stream flow Groundwater levels/accessibility	Shallow groundwater accessible in dry streams	Shallow groundwater accessible in dry streams
	Water storage capacity, types of storage	Natural storage: no estimate	Small reservoirs: 2% of annual rain volume
	Water quality indicators: sediments anthropogenic	N/A	Severely deteriorated by agrochemicals and human livestock; toxic algae at times
Land use	Agricultural, non-agricultural land use	16/84%	72/25% + 1% water surface
Habitats (aquatic, terrestrial)		Agricultural land dispersed in natural vegetation allowing for potential habitat connectivity	No natural habitat remains Introduction of aquatic habitats throughout dry season
Production-(yield-) related	Biomass /NPP Food, fodder fibre outtake/ yields	N/A N/A, but no irrigation in dry season	N/A 1% irrigation area for dry season high value income crop

indicates that the Nariarle represents a closed catchment in which most annual rainfall is consumed as evapotranspiration (ET_a = 87%), and outflow as stream flow and groundwater are less than 6% and 7%, respectively. Groundwater is only significantly recharged in the 3 years out of 20 when annual rainfall exceeds 800 mm/year (Barron *et al.*, forthcoming). In 1956, the much more undisturbed land use of natural and permanent vegetation cover meant that the water balance and water partitioning were significantly different (ET_a = 77%, stream outflow = 11% and groundwater recharge = 12%). In absolute terms, this means that water balance partitioning has decreased ET_a by almost 60 mm/year, whereas total water yield (stream flow and groundwater recharge) reduced on average by 100 mm/year. Thus, two fundamental hydrological changes have occurred: a reduction in average rainfall of 160 mm/year (or by 20%); and the closure of the catchment by a combination of reduced stream outflow and reduced groundwater.

Although the agricultural area is dominated by rainfed systems, the 1% of the catchment that is used for informal-formal irrigation (using the reservoirs) is cultivated 2–3 seasons per year and provides substantial support for livelihoods through cash incomes (e.g. Ouattara *et al.*, 2012; Sanou, 2008). The catchment now supports more food and incomes as the population has quadrupled since 1956.

Resilience considerations

As we have no methods for measuring the degradation of landscape-scale productivity, we must speculate based on knowledge of other studies of the region, which have explored soil quality, and the historical development of various other indicators (Table 7.3).

The current landscape configuration of high crop production and a volume of water storage capable of sustaining more people (possibly four times the population of 50 years ago) is considered 'better' or more 'desirable'. Nonetheless, there is substantial scope for further improvement, in particular to close the yield gap in rainfed crop production. There are also several issues with the current cropping system, which could be made more sustainable, including more active sequestration of carbon into agricultural soils and better management of agro-chemical use, which would bring the landscape into a more balanced state with healthier ecosystem services.

However, addressing the reduced stream flow to downstream users appears most challenging. Despite a storage volume of approximately 2% of average rainfall, 88% of water is already 'lost' from the catchment as evapotranspiration. The catchment is closed in terms of water and land resources, and further improvements must be sought to improve the efficiency of the current use of green and blue water, i.e. water productivity within the catchment. Another highly plausible scenario for the region is an increase in demand for water from the urban periphery of Ouagadougou. This would reallocate water away from its use in agriculture to meet the demands of more affluent urban consumers who can pay more per unit of water consumed.

The current catchment of Nariarle has improved its production capacity alongside its population growth. This analysis does not allow an assessment of the income generation associated with these developments, nor of non-economic human well-being effects. We can only speculate that improving water availability and accessibility in time and space has increased the opportunities for livelihood improvements. It is impossible to say how livelihoods and well-being depend on resources and assets generated outside the catchment, through the importation of supply inputs, incomes for investment, knowledge and health services that improve overall lives and well-being. Thus, it is not possible to separate local improvements to build productive and resilient landscapes through water and land management from external drivers. We can speculate that the construction of small reservoirs, currently at 2% of the total catchment rainfall volume, has in principle enabled additional livelihood and income possibilities in a water-deficient area. One impact is that outflow to downstream users has decreased in quantity, but this is essentially an impact of decreased rainfall rather than the construction of small reservoirs.

We conclude that although at first sight the landscape may have lost some of its natural production capacity, particularly in regard to biodiversity, it might not necessarily be considered degraded. In its current state there is still scope for increased production through improved use of land and water resources (i.e. closing the yield gap) as well as through managing resources towards healthier states (increased carbon sequestration, improved water quality, etc.) that will sustain regulating services. There is still some potential for more water storage, which could enable a higher volume of water use to the benefit of the economy and human well-being. Actions taken to improve nutrient management will be critical to achieving the first, whereas knowledge of agro-chemical application as well as good management of water accessibility will be fundamental to the second. A possible third issue will be to recreate some of the lost species richness. Although some afforestation is apparent, it is probably limited to a few select species, as is typical in agricultural landscapes (see similar studies by e.g. Herrmann and Tappan (2013) for Senegal, and Paré (2008) and, Ouedraogo *et al.* (2011) for southern Burkina Faso), compared to the natural vegetation that occurred in the 1950s.

7.3.3 High-intensity production in Brazil

As a third example of how major anthropogenic land-use change transforms water flows and their related functions, some of the vast research on and experiences of conservation tillage and the adoption of best land management practices taken to scale in, for example, Latin America, parts of the United States and Australia is highlighted below. It draws insights from the Brazil context, which has been extensively studied by researchers and practitioners from around the world.

Large-scale conservation tillage and best management practices

It is an established fact that the adoption of conservation agriculture practices and their adaptation to scale has been a significant part of the development of crop production systems in Brazil. Today, more than 30%, or 25.5 million ha (FAO, 2013), of cultivated areas in the country are under conservation agriculture practices, combining a range of soil water and anti-erosion measures with minimum/no tillage and nitrogen fixing and/or cover crops practices. We consider minimum/no tillage systems to be a use of non-inversion soil technologies, but it is still possible to

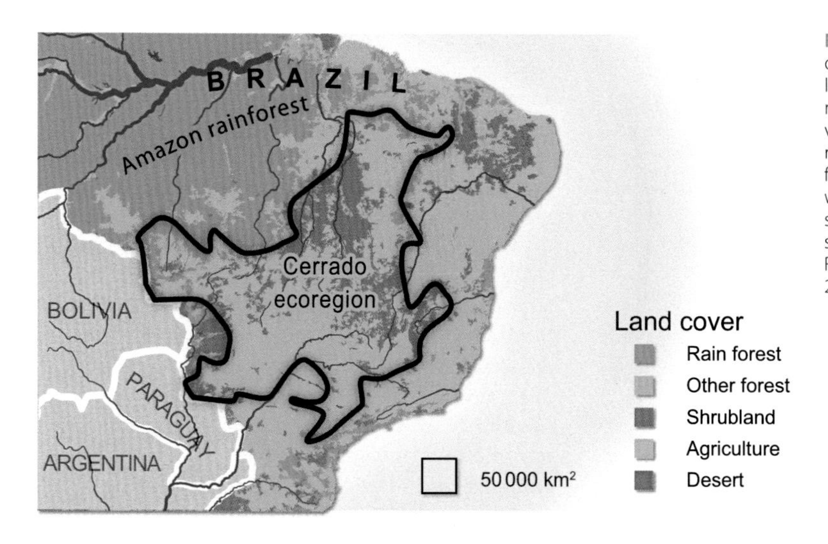

Figure 7.16 The Cerrado ecoregion of Brazil is a vast tropical savannah, located on the plateaus in the central region of the country. The natural vegetation is characterised by a mosaic of shrubs mixed with dry forest. Modern agriculture is widespread and dominated by cattle, soy, maize and rice production systems (European Commission Joint Research Centre, 2003; Olson *et al.*, 2004).

use crust breaking such as harrowing or deeper so-called ripping. According to the current 'school' of conservation agriculture, minimum/no tillage practices need to be combined with cover crops, i.e. crops grown solely to be returned as green manure and build organic matter. For a full, formal definition of conservation agriculture see, for example, FAO (http://www.fao.org/ag/ca/). In addition, the upscaling of minimum/no tillage has been associated with the adoption and adaptation of best management practices at the catchment level. The development of suitable rotations with pasture and nitrogen fixing is also adding to the successful adoption of these conservation practices, and their impacts on soil and sustainable water use.

Agricultural development in Brazil, in particular of cereal, oilseed and sugar crops, is discussed by e.g. Pereira *et al.* (2012). It is a remarkable transition of production and productivity. Since the mid-1970s, the area under cultivation has expanded by 32%, while productivity (yield per hectare) has increased by 240%. Although often blamed for the conversion of rainforest, it is mostly the moist and dry savannah (Cerrado) biome that has been converted to crop production (Figure 7.16). Despite this extensive land conversion, The Brazilian Agricultural Research Corporation (EMBRAPA) estimates that currently only one-sixth of the country's potential agricultural land of 300 million ha is being used.

Although not all yield increases can be attributed to the adoption of conservation agriculture, it should be noted that most conservation agriculture is undertaken by large-scale farmers (e.g. Bolliger *et al.*, 2006) who combine no-till with efficient machinery, improved crop varieties and the use of fertiliser and other agro-chemicals to transform inherently acid and erosive soils to high yielding crop media. According to Landers (2001, 2007) zero-tillage mechanised development began in the early 1980s but did not involve smallholder farmers until 1990, when non-governmental organisations (NGOs) and extension services had the means to develop suitable approaches for such farmers. Nonetheless, it is the adoption of conservation agriculture among large-scale farmers and commercial farms that has taken adoption numbers from scale to scale, not least because 54% of farmland is owned by 2% of farms (Berdegué and Fuentealba, 2011). Various studies on the potential impacts of conservation agriculture and the associated transformation have been undertaken in multiple locations at the landscape and catchment scales.

Resilience considerations

It has been argued that substantial emissions of CO_2 have taken place over the past 60 years due to land conversion from native Cerrado. A review by Zinn *et al.* (2005) involving paired soil sampling of various land uses showed that soil organic matter can decrease in intensively tilled crop systems, but not necessarily, as compared to non-tilled soils in various ecoregions of Brazil. However, in some cases, no difference was detected for the land-use classes intensive annual tillage crop, degraded pasture and non-tilled crop systems. Similar, somewhat inconclusive, evidence

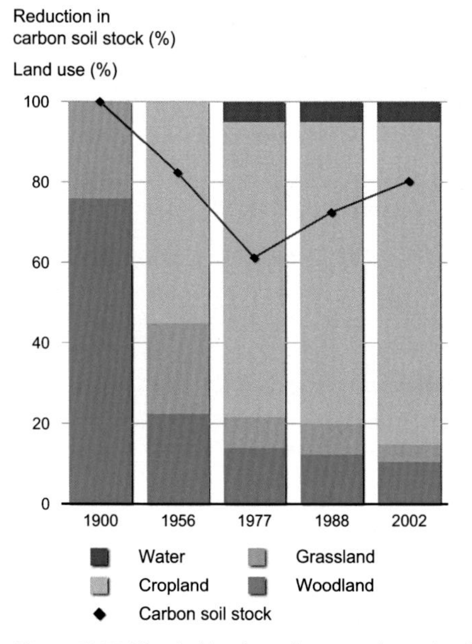

Figure 7.17 Historical land-use change and associated modelled change in soil carbon stock from Ibiriba region, Rio Grande do Sul, Brazil (modified after data from Tornquist *et al.*, 2009).

was discussed in a review of soil organic carbon changes in Cerrado systems by Batlle-Bayer *et al.* (2010). The inconclusive evidence on soil organic matter was from Bolliger *et al.* (2006). The authors added that low/no tillage soil organic matter was of a different quality, usually in more labile fractions, which affects the quality and function of the organic matter in the soil. Looking at temporal development of soil carbon under land-use change, Tornquist *et al.* (2009) investigated various stocks of soil carbon in a simulation study of the Ibiriba region of Rio Grande do Sul. It assessed various land uses since 1900, with a forecast to 2050, and investigated changes in soil organic matter under various conventional and conservation agricultural practices compared to the native vegetation in 1900. It concluded that the adoption of no-tillage soil management and best soil and crop management practices had increased soil carbon stocks from a low value of 3.71 Tg in 1985 to the current 3.97 Tg for the 830 km^2 area (Figure 7.17). Although this is still approximately 20% less than the original soil organic matter under native vegetation, it suggests a remarkable recovery from the lowest values under conventional agricultural tillage and crop systems in 1985, with a projected increase to a potential 4.21 Tg by 2050. A similar study of a lower altitude site with slightly different crop systems (less maize) indicated reductions ranging from −11% to −30% (with an extreme of −54%) compared to naturally vegetated areas, depending on the initial soil texture, the length of time since the conversion from natural vegetation and the amount of recirculated organic matter in the no-till system adopted (Bortolon *et al.*, 2011).

Water flows have been affected by the land-use changes in these transformed landscapes. Several large-scale studies (>10 000 km^2) suggest an increase in stream flow, in particular high flows due to the deforestation of Cerrado vegetation (e.g. Coe *et al.*, 2011; Costa *et al.*, 2003). These changes are attributed two-thirds to reduced vegetation uptake for evapotranspiration, and one-third to changes in rainfall amounts and/or distribution patterns (Coe *et al.*, 2011). At a finer, micro- to meso-scale (0.1–10 km^2; 10–10 000 km^2), various studies with hydrological models show a relative decrease in stream flow as areas are converted to no-tillage combined with best practice crop systems, including crop–pasture rotation and mulch from cover crops. Some examples are shown in Table 7.4, where measured and modelled studies based on best practice soil and crop management were compared to conventional practices in agriculture-dominated landscapes. The various studies of different climatic, land use and soil conditions clearly show the relative change brought about by improving soil and water conservation to scale. First, although macro-scale processes might be increasing stream flow due to land-use change, the meso to micro scale suggests that surface runoff and stream flow have decreased with the implementation of best practices upstream, and sediment loss has reduced relatively more than the reduction in stream flow.

However, it has been proposed that the relative importance of erosion processes has also changed. A study by Castro *et al.* (1999), and confirmed by Wantzen *et al.* (2006), suggests that the reduced overland flow has in part been compensated for by increased stream and channel erosion due to increased flows. This means that the protection of stream channels needs further reinforcing, in the absence of natural gallery forest to further slow flows, stabilise stream banks and reduce overall sediment losses in these converted landscapes.

The Cerrado has been one of the main frontiers for agricultural production and productivity growth. Covering approximately 2 million km^2, approximately 50% has been converted to agricultural land

Table 7.4 Summary of the impact on surface runoff and sediment comparing conservation agriculture and best practices with conventional practices, selected studies in Brazil.

	Rocha *et al.* (2012)	Castro *et al.* (1999)	Minella *et al.* (2009)	Strauch *et al.* (2013)
Area catchment (km^2)	28.5	77	1.2	188
Annual average rainfall (mm)	1200	1700	1600	1350
Climate type	Humid sub-tropical	N/A	N/A	Semi-humid tropical
Relative change with no till and best conservation practice over study period				
Impact in stream flow	−18%	(−84%)	−20 to −60%	−0 to −40%
Impact on sediment loss	−66%	−95%	−40 to −80%	−20 to −40%

use, with massive effects on biodiversity and the fragmentation of biotopes (Figure 7.17). Only 20% remains intact with natural vegetation, and only 2% is protected in bio-reserves. This has obviously affected flora and fauna: species diversity is on the decline, and both flora and fauna, of which up to 40% is endemic, will be permanently lost (e.g. Klink and Machado, 2005). A study analysing trends in soy bean expansion in the northeast Cerrado has concluded that selected large species of mammals will lose 30–50% of their habitat, and birds 6–66% compared to today if the current rate of expansion is maintained (Barreto *et al.*, 2012).

Finally, the land-use change in the Cerrado, its development into no-tillage and conservation agriculture, and further integration with livestock production has led to substantial greenhouse-gas emissions beyond the CO_2 release associated with the conversion of permanent vegetation cover to annual crops. Although remotely linked with water functions, the need to address greenhouse-gas emissions could have co-benefits for water partitioning (Bolliger *et al.*, 2006). Yet, there may still be scope to reduce both nitrous oxide emissions and CO_2 emissions in the refinement and further adoption of reduced tillage in combination with better fertiliser practices (Hillier *et al.*, 2012).

7.4 Discussion

7.4.1 Landscapes emerge as highly complex social–ecological systems. . .

The landscape scale is where land-use changes can aggregate and impact ecosystem services and consequently livelihoods, development opportunities and ecosystem sustainability. This is also the scale where management and governance systems need to operationalise, or put policy and legislation into practice (see Chapter 8). Through physical changes in soil, crop and water resource use at the field to landscape scale, we consciously and inadvertently affect water functions that sustain these ecosystem services. There are no immediate ways of predicting changes to and impacts on connected social–ecological systems due to changes in water storage and flows in time and space, and there are no given solutions for how to obtain desirable, resilient water functions to support human well-being and healthy, diverse ecosystem services. Nonetheless, there are a growing number of examples of agro-ecological landscape trajectories of change and transformation, which can serve as sources of knowledge on short- and long-term impacts.

7.4.2 . . .and provide new challenges for defining causalities and impacts

Humans manage landscapes to enhance or suppress certain ecosystem services. We are no longer dealing with natural or pristine landscapes. Water and ecosystem services related to water functions are particularly affected by the development of agricultural practices. Agriculture, which in itself is an ecosystem service, is primarily a way to manage particular benefits from ecosystems, while other ecosystem services may be affected, depending on the scale (in time and space) of management. Through manipulation of water flows and water storage in landscapes, it is feasible to transform such landscapes from degraded to intermediate levels of multiple ecosystem services or beyond.

Table 7.5 Net primary production (NPP) potential and actual estimated for selected catchments/landscapes, extracted from data in Haberl *et al.* (2007)

NPP data Buffer g carbon/m^2 per year	NPP (potential) 50 km mean	NPP (actual) 50 km mean	Mean difference between potential and actual
Nariarle (Burkina Faso)	560	370	−34%
Mekelle (Ethiopia)	520	210	−59%
Kothapally (India)	520	360	−30%
Machakos (Kenya)	870	670	−23%
Massikaoni (Niger)	480	400	−18%
Warzou (Niger)	470	390	−19%
Makanya_L1 (Tanzania)	750	610	−19%

7.4.3 Landscapes are being transformed by changing water partitioning, affecting hydrological signals

There is still potential to better use water functions to address ecosystem services in most agro-ecological landscapes. In certain areas, it is about enhancing water and land productivity, while in others it is about enhancing water quality and aquatic habitats, reintroducing species diversity, mitigating greenhouse-gas emissions or sequestering organic matter in soils. Many landscapes under rapid agricultural change, including some examples in this chapter, exhibit a gap between potential and actual NPP (Table 7.5). This is an indication that there is room to enhance biomass through better management of various water and land resources and associated ecosystem services. Biomass production and productivity are constrained by management rather than biophysical conditions at the landscape scale.

In the cases presented above, it is clear that improved water management in agricultural production can transform landscapes into more productive systems, but this requires substantial effort and inputs through investment, technical know-how and, not least, labour or energy inputs. Good landscape management is knowledge intensive, and local knowledge is rarely sufficient as multiple external and internal drivers push systems towards desired or undesired change. All three cases explored above can be said to have improved from their degraded states. Although there is scope to improve biomass production and productivity further, the challenge is to do so without undermining water quantity and quality for other users, and thereby undermining well-being at other time and space scales.

There are other examples at the landscape scale of similar impacts from substantial *in-situ* and *ex-situ* soil and water management change (Figure 7.18). Here, the increase in local infiltration for agricultural use can be seen by the decadal data on inflow to the dam from the 840 km^2 catchment. The measured data are plotted against empirically derived catchment scale hydrological functions that link hydro-climatic and compounded surface characteristics, i.e. rainfall, potential evaporation and surface runoff (Oudin *et al.*, 2008). It is clear that in this highly modified landscape, the empirically derived functions are no longer predictors of runoff generation – at least not for this particular Rajasthan catchment. Over time, the Rajasthan catchment increased its deviation from the 'natural functions' expected by various studies.

The three examples of landscapes presented in section 7.3 indicate a need for both governance structures to deal with potential impacts, and structures to negotiate what is desired and what are undesirable or unacceptable losses of water functions in landscapes. It is not necessarily a given that the desired trajectory is restoration of all water-related ecosystem services to a 'pristine level'.

7.4.4 The changes in land and water are multi-scale, affecting the predictability of impacts

Because of this extensive change in land use to man-made landscapes, there is a need to reconsider how to address hydrological principles in methods and models for water flow and stock assessments. Careful and critical approaches to scaling in time and space will be needed to avoid either flawed aggregation of local-scale causalities or superimposing downscaled global-scale findings.

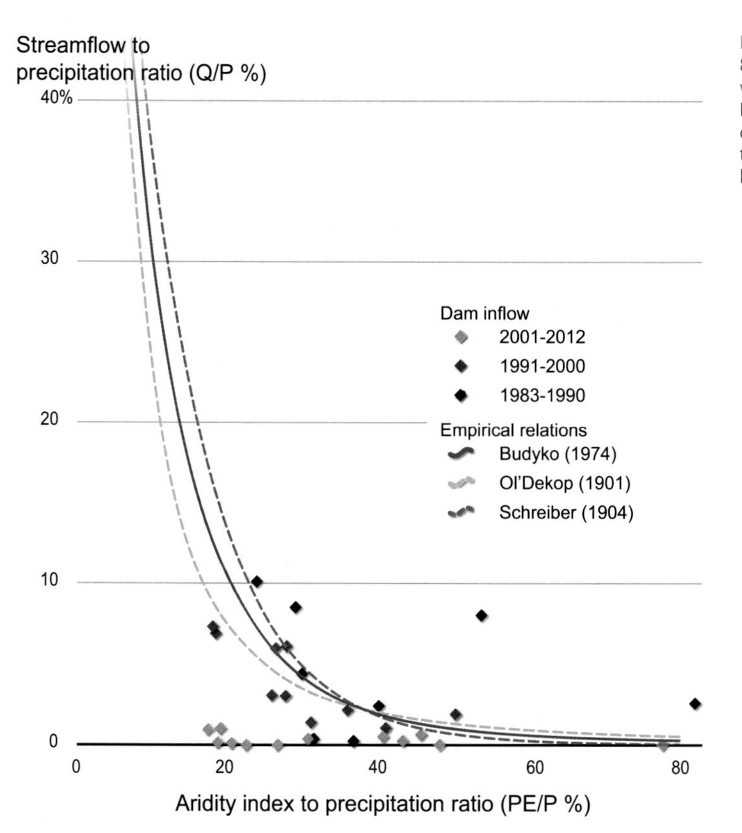

Figure 7.18 Annual inflow to dam from 840 km^2 catchment under increased soil and water management structures for agriculture in Rajasthan (Government of Rajasthan, 2013) compared with empirical relations of more than 1500 catchments in various hydro-climatic locations as presented by Oudin *et al.* (2008).

Addressing consumptive water use alone will distort the water functions in our soil–plant–atmosphere continuum. It is also the water flow that generates ecosystem services, in particular regulating and supporting ecosystem services. Addressing water use for provisioning ecosystem services alone, however, will undermine the long-term sustainability and resilience of agro-ecological landscapes (e.g. Barron and Keys, 2011; Keys *et al.*, 2012).

7.4.5 There are examples of enhanced resilience in developing water resources in agro-ecological landscapes

Three landscapes were analysed from an agro-ecological perspective, assessing changes in water and associated water functions, and exemplifying common development trends. Landscape-scale biophysical resilience was assessed using five indicators to assess the relative state of landscapes, ranging from natural, to a degraded state with low capacity to generate provisioning services that benefit human well-being and societies, to maximised production in agro-ecosystems, of high direct economic benefit to human well-being and society.

The three cases discussed above show that substantial transformation has taken place both towards undesirable, degraded states (see section 7.3 above for Brazil in the 1970s–1980s; Kothapally pre-2000) and towards more sustainable practices and productive current states (Cerrado high-intensive agriculture, and low-mechanics input systems in Kothapally and Nariarle). Nonetheless, all the systems had been highly modified and lost biodiversity in particular and tentatively also increased GHG emissions. There are also indications that water availability downstream may have been reduced (Kothapally, Nariarle). These losses in specific water functions have been as a result of increasing both green and blue water storage as increased soil water infiltration for crop water uptake and to recharge shallow surface or groundwater storage to cope with natural climate variability.

Thus, the active management of water can be said to have strengthened the capacity of the agro-ecological production system to produce, even during short and longer intra- and inter-seasonal water shocks.

7.4.6 Dealing with a mismatch of social and biophysical processes and scales

In the three cases, improved water access and availability have been fundamental to shifting agro-ecological systems into more productive states. Most of the resilience of the three cases is perhaps human determined and embedded in the knowledge and management capacities of these systems, i.e. the social and institutional dimensions, whether internally or externally provided. The challenge will be to persist with and improve the current landscape and embed the knowledge of how to manage biophysical processes in time and space. One fundamental problem is the mismatch between human and social memory and knowledge, typically 20–40 years (Manson, 2008), and the need for sustained management of water, land and vegetation to maintain regulating and supporting services which often operate at longer timescales and sometimes space scales (Dearing *et al.*, 2010). Because short-term, small-scale benefits are often maximised in agro-ecological systems, the water functions of regulating and supporting ecosystem services can be overlooked. Governance must more explicitly include slow and large-scale processes and implications in conjunction with the agro-ecological landscape resilience defined above. For example, the commonly applied IWRM principles are inherently not explicit in addressing water-related supporting and regulating ecosystem services. These will need to be revisited to ensure the inclusion of large-scale and long-term water-related processes that support shorter term regulating and provisioning services (Keys *et al.*, 2012) (see Figure 8.2, Chapter 8).

It is inevitable that demand for water for multiple purposes at the landscape scale will continue to grow, while supply is finite for each defined land and time unit. Building sustainable and productive water functions will require informed choices and equitable dialogues on the trade-offs and impacts of supply and demand. As is shown above, there are still ways and means to improve biophysical productivity, but it will be essential to couple biophysical change with knowledge and social processes, and organisation that can capture and negotiate desired changes. Investment in the development of improved agro-ecological systems should be balanced with similar efforts to enhance the human and institutional capacity to manage and safeguard the desired functions of water resources in local landscapes and beyond. To provide stable trajectories for agro-ecological landscapes in rapid development (Barron and Keys, 2011) is therefore likely to be more a question of well-adapted governance structures for compliance, and the management and distribution of benefits and disbenefits, than of biophysical changes to water functions for agriculture and other ecosystem services at the local scale.

7.5 Conclusion

There is still potential to better utilise water functions and consequently ecosystem services in most agricultural systems. Since many landscapes continue to exhibit gaps between potential and actual NPP, there is room to enhance biomass production and productivity through better management of various water functions and ecosystem services. Limitations are related to knowledge and management rather than biophysical constraints. Good farming and good landscape management are knowledge intensive, and local knowledge is rarely sufficient. Agricultural intensification is necessary and will continue to alter water resource functions at the landscape scale and possibly beyond. It needs to be developed by maintaining a focus on landscape multifunctionality, and aiming for the best possible landscape configuration in relation to water partitioning, while staying away from trajectories that have undesirable impacts. Solutions to sustainability and resilience in productive and highly human-influenced landscapes need to be context specific, and valued internally and externally in order to obtain the desired states.

Agricultural intensification is necessary and will continue to alter water resource functions at the landscape scale and possibly beyond, but it needs to be done with a focus on landscape multifunctionality, and while aiming for the best possible landscape configuration in relation to water partitioning and avoiding critical thresholds. Our science needs to keep pace with our changing landscapes. It is not necessarily best practice to desire 'pristine' flows and functions of water, nor to make assessments using methods and models developed for pristine landscape conditions. There is an urgent need to develop a new hydrological understanding in the Anthropocene era. Finally, in order to maintain stable and desirable development trajectories for highly complex social–ecological systems, governance issues must be addressed with equal vigour as biophysical issues. Social, economic and institutional dimensions will govern the trajectories and maintain or improve water functions for resilience in development.

Summary

Fundamental aspects of water and ecosystem services, and the management of these, are discussed above with particular emphasis on tropical, semi-arid and sub-humid zones under rapid transformation linked to agricultural development to improve human well-being and incomes. The focus is on the landscape scale (1–10 000 km^2) as this is where land-use change aggregates to affect ecosystem services, and consequently livelihoods, development opportunities and ecosystem sustainability at scale. Through physical changes in the use of soil and water resources at the field to landscape scale, we either deliberately or inadvertently affect water functions that sustain these services. There are no easy ways to predict changes and impacts on coupled social–ecological systems due to changes in water storage and flows in time and space, and there are no ready solutions to attaining the desired resilient water functions to support human well-being and healthy, diverse ecosystem services. Nonetheless, there are a growing number of examples of agro-ecological landscape trajectories of change and transformation that can serve as sources of knowledge on the short- and long-term impacts of change.

Agricultural practices affect ecosystem services related to water functions

We are essentially no longer dealing with 'natural' landscapes. Humans manage landscapes to enhance or suppress certain ecosystem services. Water and ecosystem services related to water functions are particularly affected by the development of agricultural practices. Agriculture, which in itself is an ecosystem service, is primarily a way to manage the particular benefits of ecosystems, while other ecosystem services may be affected, depending on the scale (in time and space) of management. Through the manipulation of water flows or water characteristics, it is feasible to transform landscapes from degraded states to providers of an intermediate level of multiple ecosystem services or beyond.

Because of the widespread change of land use to man-made landscapes, there is a need to reconsider how to address hydrological principles in methods and models of water flow and stock assessment. Careful and critical approaches are needed to scaling in time and space to avoid both flawed aggregation of local-scale causalities and superimposing downscaled global-scale findings.

Three landscapes were analysed from an agro-ecological perspective, assessing changes in water and associated water functions, and exemplifying common development trends (Brazil, India, and Burkina Faso). Landscape-scale resilience is explored using five indicators to assess the relative state of landscapes, ranging from a degraded state with low capacity to generate provisioning services that benefit human well-being and societies, to maximised production in agro-ecosystems, with high levels of direct economic benefits for human well-being and society.

Most of the resilience is embedded in knowledge and management capacity

The three cases show that substantial transformation has taken place both towards undesired degraded states from initially pristine conditions and, more recently, towards more sustainable practices and productive states. However, all three systems have been affected, in particular by the loss of biodiversity indicators and tentatively also by increased GHG emissions, compared to the initial state. Biophysical resilience, interpreted as the capacity to generate biomass and access water, seems to have been maintained and even increased, although there are indications that water availability downstream may have been reduced in some cases. Thus, based on these typical examples of landscape development, there may be cause for concern regarding certain aspects of water functions in rapidly developing agro-ecological landscapes, although the initial issue of biomass production capacity is possibly a less (biophysical) immediate issue. In all three cases, water access, availability and management have been fundamental to the shift to a more productive state. In sum, most of the resilience of the agro-ecological systems is human determined and embedded in the knowledge and management capacity of these systems, whether internally or externally provided.

Scope to increase production and water functions in landscapes

There is still potential to better utilise water functions and consequently ecosystem services in most agricultural systems. Since many landscapes still exhibit a gap between potential and actual NPP, there is still room to enhance biomass through better management of various ecosystem services. The limitations are management related rather than biophysical. Good farming and good landscape management are knowledge intensive and local knowledge is rarely sufficient to deal with impacts across spatial and temporal scales.

Agricultural intensification is necessary and will continue to alter water resource functions at the landscape scale and possibly beyond. It needs to be achieved while remaining focused on landscape multifunctionality, and aiming for the best possible landscape configuration in relation to water partitioning while avoiding critical thresholds. Achieving sustainability and resilience in productive and highly human-influenced landscapes must be context specific and valued internally as well as externally in order to maintain desired trajectories.

References

Aeschbach-Hertig, W. and Gleeson, T. (2012). Regional strategies for the accelerating global problem of groundwater depletion. *Nature Geoscience*, **5**, 853–861.

Alkama, R., Marchand, L., Ribes, A. and Decharme, B. (2013). Detection of global runoff changes: results from observations and CMIP5 experiments. *Hydrology and Earth System Sciences Discussions*, **10**, 2117–2140.

Andréassian, V. (2004). Waters and forests: from historical controversy to scientific debate. *Journal of Hydrology*, **291**, 1–27.

Andreini, M., Schuetz, T., Senzanje, A. *et al.* (2009). Small multi-purpose reservoir ensemble planning. CPWF Project Report Series: PN46. CGIAR Challenge Program on Water and Food. Available at: http://hdl.handle.net/10568/3773.

Barreto, L., Van Eupen, M., Kok, K. *et al.* (2012). The impact of soybean expansion on mammal and bird, in the Balsas region, north Brasilian Cerrado. *Journal for Nature Conservation*, **20**, 374–383.

Barron, J. (2012). Soil as a water resource: some thoughts on managing soils for productive landscapes meeting development challenges. Agro Environ 2012. Wageningen University, Wageningen, the Netherlands.

Barron, J., Cambridge, H., Rebelo, L., Ouattara, K. and Pare, S. (forthcoming). Impact on water partitioning of landuse change in a semiarid watershed under agricultural development in West Africa.

Barron, J. and Keys, P. (2011). Watershed management through a resilience lens. In *Integrated Watershed Management in Rainfed Agriculture*, ed. Wani, S. P., Rockström, J. and Sahrawat, K. L. Boca Raton: CRC Press, pp. 391–420.

Bartley, R., Speirs, W. J., Ellis, T. W. and Waters, D. K. (2012). A review of sediment and nutrient concentration data from Australia for use in catchment water quality models. *Marine Pollution Bulletin*, **65**, 101–116.

Batlle-Bayer, L., Batjes, N. H. and Bindraban, P. S. (2010). Changes in organic carbon stocks upon land use conversion in the Brazilian Cerrado: a review. *Agriculture, Ecosystems & Environment*, **137**, 47–58.

Bell, R. W., Schofield, N. J., Loh, I. C. and Bari, M. A. (1990). Groundwater response to reforestation in the Darling Range of Western Australia. *Journal of Hydrology*, **119**, 179–200.

Berdegué, J. A. and Fuentealba, R. (2011). Latin America: the state of smallholders in agriculture. Paper presented at the IFAD Conference on New Directions for Smallholder Agriculture, 24–25 January 2011. International Fund for Agricultural Development, Rome.

Blöschl, G. (1996). *Scale and Scaling in Hydrology*. Vienna: Technische Universität Wien.

Bolliger, A., Magid, J., Amado, J. C. T. *et al.* (2006). Taking stock of the Brazilian 'zero-till revolution': a review of landmark research and farmers' practice. In *Advances in Agronomy*, ed. Donald, L. S. Waltham, MA: Academic Press, pp. 47–110.

Bortolon, E. S. O., Mielniczuk, J., Tornquist, C. G., Lopes, F. and Bergamaschi, H. (2011). Validation of the Century model to estimate the impact of agriculture on soil organic carbon in Southern Brazil. *Geoderma*, **167–168**, 156–166.

Brauman, K. A., Daily, G. C., Duarte, T. K. e. and Mooney, H. A. (2007). The nature and value of ecosystem services: an overview highlighting hydrologic services. *Annual Review of Environmental Resources*, **32**, 67–98.

Brown, A. E., Zhang, L., McMahon, T. A., Western, A. W. and Vertessy, R. A. (2005). A review of paired catchment studies for determining changes in water yield resulting from alterations in vegetation. *Journal of Hydrology*, **310**, 28–61.

Callow, J. and Smettem, K. (2009). The effect of farm dams and constructed banks on hydrologic connectivity and runoff estimation in agricultural landscapes. *Environmental Modelling & Software*, **24**, 959–968.

Cambridge, H. and Barron, J. (forthcoming). *Application of SWAT for impact assessment of agricultural water management interventions in the Nariarle watershed, Burkina Faso.* SEI Technical Report: Stockholm Environment Institute.

Castro, N. M. D. R., Auzet, A.-V., Chevallier, P. and Leprun, J.-C. (1999). Land use change effects on runoff and erosion from plot to catchment scale on the basaltic plateau of Southern Brazil. *Hydrological Processes*, **13**, 1621–1628.

Chiew, F. H., Vaze, J., Viney, N. *et al.* (2008). Rainfall-runoff modelling across the Murray–Darling Basin. A report to the Australian Government from the CSIRO Murray–Darling Basin Sustainable Yields Project.

Chilton, J. and Seiler, K. (2006). Groundwater occurrence and hydrogeological environments. In *Protecting Groundwater for Health*, ed. Schmoll, O., Howard, G., Chilton, J. and Choru, I. London: IWA Publishing, pp. 21–47.

Coe, M., Latrubesse, E., Ferreira, M. and Amsler, M. (2011). The effects of deforestation and climate variability on the streamflow of the Araguaia River, Brazil. *Biogeochemistry*, **105**, 119–131.

Coram, J., Dyson, P., Houlder, P. and Evans, W. (2000). *Australian groundwater flow systems contributing to dryland salinity.* Bureau of Rural Sciences for National Land and Water Resources Audit. Available at: http://www.anra.gov.au/topics/salinity/pubs/national/salinity_gfs_report/report/pdf/report.pdf.

Costa, M. H., Botta, A. and Cardille, J. A. (2003). Effects of large-scale changes in land cover on the discharge of the Tocantins River, Southeastern Amazonia. *Journal of Hydrology*, **283**, 206–217.

de Bruin, A. and Barron, J. (2012). AWM interventions and monitoring and evaluation 2: developing indicators and thresholds based on stakeholder consultations at watershed level. Project report: Stockholm Environment Institute, Stockholm.

de Bruin, A., Mikhail, M., Noel, S. and Barron, J. (2010). AWM interventions and monitoring and evaluation: potential approaches at the watershed level. Stockholm Environment Institute, Stockholm.

de Groot, R. S., Alkemade, R., Braat, L., Hein, L. and Willemen, L. (2010). Challenges in integrating the concept of ecosystem services and values in landscape planning, management and decision-making. *Ecological Complexity*, 7, 260–272.

Dearing, J. A., Braimoh, A. K., Reenberg, A., Turner, B. L. and van der Leeuw, S. (2010). Complex land systems: the need for long time perspectives to assess their future. *Ecology and Society*, **15**, 21.

Douxchamps, S., Ayantunde, A. and Barron, J. (2012). Evolution of agricultural water management in rainfed crop-livestock systems of the Volta basin. CGIAR Challenge Program for Water and Food Working Papers: 4. International Livestock Research Institute, Nairobi. Available at: http://hdl.handle.net/10568/21721.

Ellis, E. C. (2011). Anthropogenic transformation of the terrestrial biosphere. *Philosophical Transactions of the Royal Society A: Mathematical, Physical and Engineering Sciences*, **369**, 1010–1035.

European Commission Joint Research Centre (2003). *Global Land Cover 2000 database.* European Commission Joint Research Centre, Brussels. Available at: http://bioval.jrc.ec.europa.eu/products/glc2000/glc2000.php (accessed 7 April 2006).

FAO, IUCN and UNEP (2013). Ecolex: the gateway to environmental law. FAO, IUCN and UNEP. Available at: http://www.ecolex.org/.

Faurès, J. M., Bernardi, M. and Gommes, R. (2010). There is no such thing as an average: how farmers manage uncertainty related to climate and other factors. *International Journal of Water Resources Development*, **26**, 523–542.

Favreau, G., Cappelaere, B., Massuel, S. *et al.* (2009). Land clearing, climate variability, and water resources increase in semiarid southwest Niger: a review. *Water Resources Research*, **45**.

Food and Agriculture Organization (2011). *The State of the World's Land and Water Resources for Food and Agriculture (SOLAW): Managing Systems at Risk.* Rome and London: Food and Agriculture Organization and Earthscan.

Food and Agriculture Organization (2013). AquaStat online database. Available at: http://www.fao.org/nr/water/aquastat/main/index.stm (accessed multiple dates).

Forrester, J. and Taylor, R. (2012). A transdisciplinary approach to modelling complex social-ecological problems in coastal ecosystems. Complexity Science at the Social Science Interface Conference, Royal Society Centre, UK, Royal Society Centre.

Garg, K. K., Karlberg, L., Barron, J., Wani, S. P. and Rockström, J. (2012a). Assessing impacts of agricultural water interventions in the Kothapally watershed, Southern India. *Hydrological Processes*, **26**, 387–404.

Garg, K. K., Wani, S. P., Barron, J., Karlberg, L. and Rockström, J. (2012b). Up-scaling potential impacts on water flows from agricultural water interventions: opportunities and trade-offs in the Osman Sagar catchment, Musi sub-basin, India. *Hydrological*

Processes, available at: http://onlinelibrary.wiley.com/doi/10.1002/hyp.9516/abstract;jsessionid=FB22A4233FCD7039DF0181A8815FBDA9.d02t01.

Gorgens, A. and Van Wilgen, B. (2004). Invasive alien plants and water resources in South Africa: current understanding, predictive ability and research challenges. *South African Journal of Science*, **100**, 27–33.

Government of Rajasthan (2013). Report on less/no inflow in Ramgarh Dam (District Jaipur). Technical Committee Constituted by Government of Rajasthan, Jaipur, India.

Haberl, H., Erb, K. H., Krausmann, F. *et al.* (2007). Quantifying and mapping the human appropriation of net primary production in Earth's terrestrial ecosystems. *Proceedings of the National Academy of Sciences of the United States of America*, **104**, 12942–12945.

Herrmann, S. M. and Tappan, G. G. (2013). Vegetation impoverishment despite greening: a case study from central Senegal. *Journal of Arid Environments*, **90**, 55–66.

Hillier, J., Brentrup, F., Wattenbach, M. *et al.* (2012). Which cropland greenhouse gas mitigation options give the greatest benefits in different world regions? Climate and soil-specific predictions from integrated empirical models. *Global Change Biology*, **18**, 1880–1894.

Ilstedt, U., Malmer, A., Verbeeten, E. and Murdiyarso, D. (2007). The effect of afforestation on water infiltration in the tropics: a systematic review and meta-analysis. *Forest Ecology and Management*, **251**, 45–51.

Joshi, P., Jha, A., Wani, S., Sreedevi, T. and Shaheen, F. (2008). Impact of watershed program and conditions for success: a meta-analysis approach. Global Theme on Agroecosystems Report: 46. International Crops Research Institute for the Semi-Arid Tropics. Available at: http://impact.cgiar.org/impact-watershed-program-and-conditions-success-meta-analysis-approach.

Kerr, J. (2002). Watershed development, environmental services, and poverty alleviation in India. *World Development*, **30**, 1387–1400.

Keys, P., Barron, J. and Lannerstad, M. (2012). *Releasing the Pressure: Water Resource Efficiencies and Gains for Ecosystem Services.* Nairobi and Stockholm: United Nations Environment Programme and Stockholm Environment Institute.

Kiersch, B. (2002). Land-water linkages in rural watersheds. FAO Land and water bulletin: 9. Food and Agriculture Organization, Rome. Available at: http://www.fao.org/docrep/004/y3618e/y3618e00.htm.

Klink, C. A. and Machado, R. B. (2005). Conservation of the Brazilian Cerrado. *Conservation Biology*, **19**, 707–713.

Lamarque, P., Quétier, F. and Lavorel, S. (2011). The diversity of the ecosystem services concept and its implications for their assessment and management. *Comptes Rendus Biologies*, **334**, 441–449.

Landers, J. N. (2001). Zero tillage development in tropical Brazil: the story of a successful NGO activity. Food and Agriculture Organization, Rome. Available at: http://www.fao.org/docrep/004/Y2638E/Y2638E00.HTM.

Landers, J. N. (2007). Tropical crop-livestock systems in conservation agriculture: the Brazilian experience. Food and Agriculture Organization, Rome. Available at: ftp://ftp.fao.org/docrep/fao/010/a1083e/a1083e.pdf.

Lehner, B., Reidy Liermann, C., Revenga, C. *et al.* (2011). Global reservoir and dam database, version 1 (GRanDv1), revision 01. NASA Socioeconomic Data and Applications Center, New York. Available at: http://sedac.ciesin.columbia.edu/data/collection/grand-v1/sets/browse.

Liebe, J. (2002). Estimation of water storage capacity and evaporation losses of small reservoirs in the Upper East Region of Ghana. Master of Science thesis, University of Bonn, Germany.

Locatelli, B. and Vignola, R. (2009). Managing watershed services of tropical forests and plantations: can meta-analyses help? *Forest Ecology and Management*, **258**, 1864–1870.

Mahe, G., Paturel, J.-E., Servat, E., Conway, D. and Dezetter, A. (2005). The impact of land use change on soil water holding capacity and river flow modelling in the Nakambe River, Burkina Faso. *Journal of Hydrology*, **300**, 33–43.

Manson, S. M. (2008). Does scale exist? An epistemological scale continuum for complex human–environment systems. *Geoforum*, **39**, 776–788.

McDonnell, J., Sivapalan, M., Vaché, K. *et al.* (2007). Moving beyond heterogeneity and process complexity: a new vision for watershed hydrology. *Water Resources Research*, **43**.

Millennium Ecosystem Assessment. (2005). *Ecosystems and Human Well-being: Synthesis.* Washington DC: Island Press.

Minella, J. P. G., Merten, G. H., Walling, D. E. and Reichert, J. M. (2009). Changing sediment yield as an indicator of improved soil management practices in Southern Brazil. *CATENA*, **79**, 228–236.

Molden, D. (ed.) (2007). *Water for Food, Water for Life: A Comprehensive Assessment of Water Management in Agriculture.* London: Earthscan.

Montanari, A. and Uhlenbrook, S. (2004). Catchment modelling: towards an improved representation of the hydrological processes in real-world model applications. *Journal of Hydrology*, **291**, 159.

Nilsson, C., Reidy, C. A., Dynesius, M. and Revenga, C. (2005). Fragmentation and flow regulation of the world's large river systems. *Science*, **308**, 405–408.

Olson, D. M., Dinerstein, E., Wikramanayake, E. D. *et al.* (2004). Terrestrial ecoregions of the world. A new map of life on Earth: a new global map of terrestrial ecoregions provides an innovative tool for conserving biodiversity. World Wildlife Fund for Nature. Available at: http://worldwildlife.org/publications/terrestrial-ecoregions-of-the-world (accessed 1 March 2006).

Ouattara, K., Pare, S., Savadogo Kaboure, S. *et al.* (2012). Baseline assessment of current livelihood strategies in Nariarlé watershed, Burkina Faso. SEI Project report. Stockholm Environment Institute, Stockholm.

Oudin, L., Andréassian, V., Lerat, J. and Michel, C. (2008). Has land cover a significant impact on mean annual streamflow? An international assessment using 1508 catchments. *Journal of Hydrology*, **357**, 303–316.

Ouedraogo, I., Savadogo, P., Tigabu, M. *et al.* (2011). Trajectory analysis of forest cover change in the tropical dry forest of Burkina Faso, West Africa. *Landscape Research*, **36**, 303–320.

Paré, S. (2008). Land use dynamics, tree diversity and local perception of dry forest decline in southern Burkina Faso, West Africa. Doctoral thesis, Swedish University of Agricultural Sciences, Uppsala, Sweden.

Pereira, P., Martha, G., Santana, C. and Alves, E. (2012). The development of Brazilian agriculture: future technological challenges and opportunities. *Agriculture & Food Security*, **1**, 4.

Piao, S., Friedlingstein, P., Ciais, P. *et al.* (2007). Changes in climate and land use have a larger direct impact than rising CO_2 on global river runoff trends. *Proceedings of the National Academy of Sciences*, **104**, 15242–15247.

Rocha, E. O., Calijuri, M. L., Santiago, A. F., de Assis, L. C. and Alves, L. G. S. (2012). The contribution of conservation practices in reducing runoff, soil loss, and transport of nutrients at the watershed level. *Water Resources Management*, **26**, 3831–3852.

Royal Netherlands Meteorological Institute (2013). KNMI Climate explorer, GHCN precipitation, station 65503. Royal Netherlands Meteorological Institute. Available at: http://climexp.knmi.nl/getprcpall.cgi?id=someone@somewhere&WMO=65503&STATION=OUAGADOUGOU_AERO&extraargs=.

Sanou, K. (2008). Communalisation et gestion des ressources en eau à l'échelle du bassin du Nayarle. Thesis, Université de Ouagadougou, Burkina Faso.

Scanlon, B. R., Jolly, I., Sophocleous, M. and Zhang, L. (2007). Global impacts of conversions from natural to agricultural ecosystems on water resources: quantity versus quality. *Water Resources Research*, **43**.

Schreider, S. Y., Jakeman, A. J., Letcher, R. A. *et al.* (2002). Detecting changes in streamflow response to changes in non-climatic catchment conditions: farm dam development in the Murray–Darling basin, Australia. *Journal of Hydrology*, **262**, 84–98.

Schulze, R. (2000). Transcending scales of space and time in impact studies of climate and climate change on agrohydrological responses. *Agriculture, Ecosystems & Environment*, **82**, 185–212.

Seitzinger, S., Mayorga, E., Bouwman, A. *et al.* (2010). Global river nutrient export: a scenario analysis of past and future trends. *Global Biogeochemical Cycles*, **24**, GB0A08.

Sivapalan, M. (2005). Pattern, process and function: elements of a unified theory of hydrology at the catchment scale. In *Encyclopedia of Hydrological Sciences*, ed. Anderson, M. G. London: John Wiley, pp. 193–219.

Strauch, M., Lima, J. E. F. W., Volk, M., Lorz, C. and Makeschin, F. (2013). The impact of best management practices on simulated streamflow and sediment load in a Central Brazilian catchment. *Journal of Environmental Management*, available at: http://dx.doi.org/10.1016/j.jenvman.2013.01.014.

Strzepek, K., Boehlert, B., Strzepek, K. and Boehlert, B. (2010). Competition for water for the food system. *Philosophical Transactions of the Royal Society B: Biological Sciences*, **365**, 2927–2940.

Syvitski, J. P. M., Peckham, S. D., Hilberman, R. and Mulder, T. (2003). Predicting the terrestrial flux of sediment to the global ocean: a planetary perspective. *Sedimentary Geology*, **162**, 5–24.

Tetzlaff, D., McDonnell, J., Uhlenbrook, S. *et al.* (2008). Conceptualizing catchment processes: simply too complex? *Hydrological Processes*, **22**, 1727–1730.

Tornquist, C. G., Gassman, P. W., Mielniczuk, J., Giasson, E. and Campbell, T. (2009). Spatially explicit simulations of soil C dynamics in Southern Brazil: integrating Century and GIS with i_Century. *Geoderma*, **150**, 404–414.

United Nations Global Environment Monitoring System: Water. (2008). *Water Quality for Ecosystem and Human Health*. Nairobi: United Nations Environment Programme.

van Noordwijk, M., Poulsen, J. G. and Ericksen, P. J. (2004). Quantifying off-site effects of land use change: filters, flows and fallacies. *Agriculture, Ecosystems & Environment*, **104**, 19–34.

Venot, J.-P. and Cecchi, P. (2011). Valeurs d'usage ou performances techniques: comment apprécier le rôle des petits barrages en Afrique subsaharienne? *Cahiers Agricultures*, **20**, 112–117.

Wagener, T., Sivapalan, M., Troch, P. and Woods, R. (2007). Catchment classification and hydrologic similarity. *Geography Compass*, **1**, 901–931.

Walling, D. E. (2006). Human impact on land–ocean sediment transfer by the world's rivers. *Geomorphology*, **79**, 192–216.

Wani, S. P., Rockström, J. and Sahrawat, K. L. (eds) (2011). *Integrated Watershed Management in Rainfed Agriculture*. Boca Raton, FL: CRC Press.

Wantzen, K. M., Siqueira, A., Cunha, C. N. d., de Sá, P. and de Fátima, M. (2006). Stream-valley systems of the Brazilian Cerrado: impact assessment and conservation scheme. *Aquatic Conservation: Marine and Freshwater Ecosystems*, **16**, 713–732.

Zinn, Y. L., Lal, R. and Resck, D. V. S. (2005). Changes in soil organic carbon stocks under agriculture in Brazil. *Soil and Tillage Research*, **84**, 28–40.

Governance and pathways

Nguyen Thi Lan carries his farming tools along a rice field in Vietnam. The management of the Mekong River and its tributaries is crucial for the social–ecological system in the Mekong Delta, where 80% of the population is engaged in rice cultivation.

Governance for navigating the novel freshwater dynamics of the Anthropocene

The focus of this chapter is the transformation of water governance required for living with the new water dynamics in the Anthropocene, and the expansion needed from the conventional blue water perspective. It discusses the integrated governance and management required for resilience building, including desirable landscape patterns of ecosystem services, and how to overcome past governance failures from a sustainability perspective. It stresses that resilience building comprises precautionary management for stewardship of precipitation, the need for cross-sectoral integration, and the overcoming of inertia of rigid or dysfunctional bureaucracies, making good use of emerging windows of opportunity.

8.1 The Anthropocene era requires novel governance and management

People and societies are embedded parts of the biosphere and depend on its functioning and life-support systems. However, in the Anthropocene era, humanity is also shaping the biosphere globally (Steffen *et al.*, 2007; Steffen *et al.*, 2011) in what are referred to as social–ecological systems (Berkes and Folke, 1998). When the functioning of these intertwined systems is altered, there is a high risk that adaptations to local conditions, such as the optimisation of agriculture to historical precipitation regimes, will no longer be sufficient.

The previous chapters have demonstrated that water plays a central role as the bloodstream of the biosphere, connecting people and places, affecting livelihoods and influencing social–ecological resilience. Clearly, we are operating in a new terrain in which truly integrated, complex systems interact at multiple scales from the local to the global, and we are confronted with turbulent times (Folke *et al.*, 2011; Walker and Meyers, 2004).

Planetary resilience is paramount to the world's ability to cope with the multiple dynamics and changes that are taking place at both the global and the local levels (Folke *et al.*, 2010; Rockström *et al.*, 2009b). There are calls for planetary stewardship to ensure global sustainability (Power and Chapin III, 2009; Reid *et al.*, 2010; Westley *et al.*, 2011). In this context, the Anthropocene era presents genuinely novel challenges for the governance and management of freshwater. We need to become active stewards of the biosphere, which we are part of and dependent on.

8.1.1 An essential distinction

A distinction needs to be made between management and governance, which are often mistakenly used synonymously. Resource management refers to the activities of analysing and monitoring, and developing and implementing measures to keep the state of a resource within desirable bounds (Pahl-Wostl, 2009). Management is about bringing together existing knowledge from diverse sources into new perspectives for practice (Westley, 1995). In contrast, governance sets the rules under which management operates. By governance systems we mean the patterns of interaction by actors with conflicting objectives and the instruments chosen to steer social and environmental processes in a particular policy area (Galaz *et al.*, 2008). Institutions, the norms and rules, are a central component (North, 1990), as are interactions between actors and the multi-level institutional setting, creating complex relationships between people and ecosystem dynamics (Pahl-Wostl, 2006). Such complex relations are captured in the concept of **adaptive governance for resilience and transformation**, which addresses these dynamic interactions and the capacity to manage freshwater and sustain ecosystem services in the face of uncertainty and change (Dietz *et al.*, 2003; Folke *et al.*, 2005; Pahl-Wostl, 2009).

8.1.2 Expansion from blue water governance

Water management has often focused on blue water for economic and social development, largely ignoring the links to and the implications for ecosystems. IWRM extends the focus to interacting sectors – from basins down to micro catchments (GWP, 2000). Theoretically, IWRM encapsulates 'all flows', but implementation has focused on blue water. The significance of interregional water sharing for food production, including the embedded role of freshwater in commodity flows and the trade in virtual water, is gaining increased attention (see Chapters 4 and 5).

More recently, green water has come into focus, especially in relation to rainfed agriculture (Falkenmark and Rockström, 2006; Rockström *et al.*, 2009a) and lately also to the trade-offs in bundles of ecosystem services in landscapes and catchments (Gordon *et al.*, 2008). In the era of global change, the focus of this latter concept is expanding into making productive and resilient use of precipitation. How can rain be put to productive use before it is lost to the catchment? This involves balancing between terrestrial ecosystems of various kinds and runoff for rivers and aquifers in upstream areas, and between irrigation, wetlands, cities and aquatic ecosystems downstream (Falkenmark and Folke, 2010). How can productivity and resilience be combined? What type of governance structures can support integrated landscape management, including the aspects of rainfall and ecosystem services?

8.1.3 Living with a new water dynamic

In this book, the water governance challenge is brought into unexplored territory and emphasis is put on the need to manage the water cycle in a holistic manner (see Figure 1.14 in Chapter 1). This raises the critical governance challenge of devising management mechanisms to enable fair and sustainable allocation of blue and green freshwater resources between different uses in dynamic landscapes, whether for sustaining livelihoods and poverty alleviation in poor regions or for stewardship of diverse bundles of ecosystem services in the developed world. In addition, such allocations must be put in the context of novel water teleconnections (see Chapter 4) and of both envisaged and surprising changes in rainfall patterns.

The challenge of 'sustaining rainfall' expands current water management in recognition of the fact that blue and green water flows and the generation of ecosystem services in catchments for livelihoods, well-being and economic development draw on continental-scale precipitationsheds (Keys *et al.*, 2012b) (see Chapter 7). Some of the rainfall may be generated through moisture recycling within the catchment and some may originate from recycling from other catchments or be derived from ocean evaporation (see Chapter 4). All are significantly influenced and altered by a diversity of human actions, incentives and decisions, ranging from local land-use change to the human imprint on the climate system. The Anthropocene era and global change are likely to play out in ways not previously experienced. The scale and speed of change are unprecedented and call for collective action and new governance arrangements to manage the dynamic role of freshwater in the biosphere, including a global perspective on water management (Pahl-Wostl *et al.*, 2008).

Furthermore, managing rainfall implies building the social–ecological capacity to live with new water dynamics, including changes in the variability of rainfall patterns exposing regions to changes in the frequency, magnitude and duration of droughts, storms, floods and other shocks and surprises (e.g. McSweeney and Coomes, 2011). How to deal with the increasing variability of droughts in dryland areas, where people already operate at the lower limit of meagre rainfall, is addressed in Chapter 6.

The new dynamics draw attention to water-related non-linear changes, tipping points and thresholds at the regional scale (Chapter 3), exposing vulnerabilities and challenges but also opportunities for social–ecological change. In a globally interconnected world where everyone lives in everyone else's backyard, drivers of change such as increasing human numbers, urbanisation, patterns of migration, diet change, emerging markets and the rapid spread of information technologies may combine with shocks, such as ecological crises, rapid shifts in food or fuel prices and volatile financial markets, to trigger water-related tipping points. These new interactions present a range of

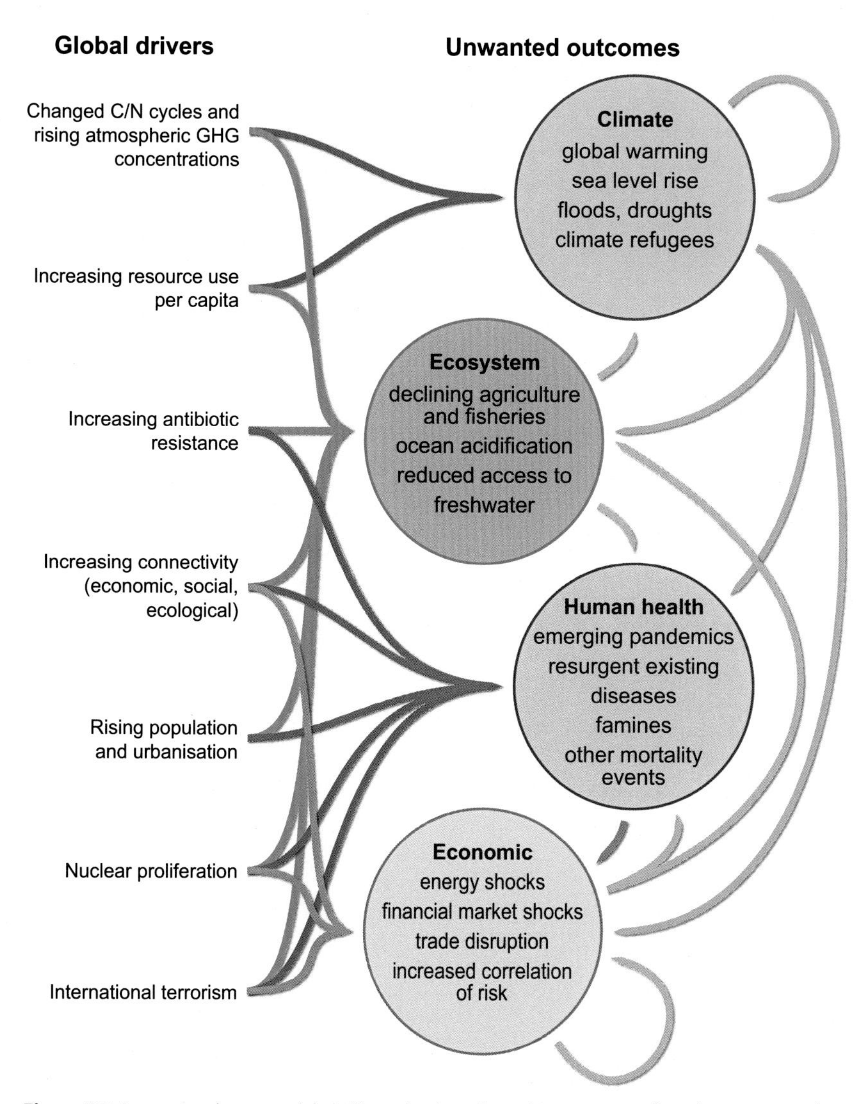

Figure 8.1 Interactions between global drivers, shocks and surprising outcomes from the economic, climate, health and ecosystem domains (adapted from Folke *et al.*, 2011).

institutional and political leadership challenges, which up to now have not been sufficiently elaborated by either crisis management researchers or institutional scholars (Galaz *et al.*, 2012). Figure 8.1 provides examples of interactions between global drivers, shocks and surprising outcomes from the economic, climate, health and ecosystem domains. Domains such as these are often treated separately, but are becoming increasingly interdependent, shaping and being shaped by water dynamics from the biosphere to the local levels.

8.1.4 Overview of approaches

Some societies show resilience in the face of shock and surprise – they adjust, reorganise and develop without significant impairment of their functioning. Others suffer, have difficulty coping with change and may shift into or deepen a pathway of unsustainable development (Tompkins and Hurlston, 2012; Walker *et al.*, 2009a). They may become caught in social–ecological traps, in which water and its use and misuse can play a significant role (Chapter 3).

Societies may also break down the resilience of the current system, or transform and build social–ecological resilience in a new pathway to development. Resilience can be seen in the ways in which communities respond to dynamics, change and progress along their development path (Chapin *et al.*, 2010; Folke *et al.*, 2010).

This book illustrates the essential role of water as the bloodstream of the biosphere, of which humans are an integrated part. Understanding the broader perspective of water in the dynamics of the biosphere for livelihood, societal development and social–ecological resilience becomes a new and important dimension in the local and regional management and governance of water resources. Water resilience, or the role of water in building resilience, is fundamental to a sustainable future for humanity.

8.2 IWRM: the dominant paradigm

This section provides an overview of current and traditional approaches to water management and governance and their implementation, and suggests ways forward for dealing with the novel challenges emerging in the Anthropocene era for the stewardship of freshwater for human well-being, resilience building and global sustainability.

Historically, water governance and management have been shaped by a technical tradition and a prediction and control paradigm, dealing with clearly defined problems and providing technical, end-of-pipe solutions. The aim has been to control single key variables such as river flow, and to reduce their variability in order to make resource dynamics more predictable. Despite demonstrable successes in several regions in dealing with problems such as hygiene and eutrophication, the drawbacks and long-term negative implications of such approaches have become increasingly evident. Providing technical fixes instead of dealing with problems at source is becoming increasingly expensive. Aquatic ecosystems are increasingly being degraded. Dealing with problems in isolation has often brought short-term success while generating new problems for the long term (Holling and Meffe, 1996). As a response to such perceived shortcomings, more integrated approaches started to emerge that aimed to balance economic, social and environmental sustainability in water management.

8.2.1 IWRM triggered by the Dublin principles

IWRM gained momentum with the adoption of the Dublin principles at the World Summit in Rio de Janeiro in 1992 (GWP, 2000). Over time, these principles developed into a new water management paradigm comprised of three elements: (1) an integrated approach across sectors and different uses and users; (2) balancing the three pillars of sustainability – economic, social and environmental concerns; and (3) participatory approaches and the involvement of women. IWRM clearly recognises the importance of soft strategies to deal with governance and management systems in addition to the hard strategies that emphasise infrastructure and technical fixes. More emphasis is also given to demand management rather than just developing the water supply. To some extent, this implicitly addresses blue–green water linkages and perspectives, even though the terms are seldom used by the IWRM community.

8.2.2 Implementation barriers

Although IWRM was strongly promoted as a path-breaking paradigm, progress with operationalisation has been slow and incremental, and has not led to major transformations in the governance and management of freshwater (GWP, 2000). A recent report commissioned by UN-Water draws positive conclusions (UN-Water, 2012). It finds, based on the results of a global survey, that the majority of countries have adopted IWRM principles into their laws and policies. However, implementation on the ground to translate principles into management practice and finally into an improved state of water resources and the sustainable use of water remains slow. Several barriers to implementation impede progress.

The prevalence of a predict and control paradigm

Traditional thinking can still be found, in particular regarding developing new supply rather than managing demand using soft strategies. For example, despite its promotion of the water-saving society, the Chinese government embarked on major infrastructure investment for south–north transfers from the Yangtze to the Yellow River without first assessing the potential for increasing the social value of existing water use in the Yellow River basin (Xia and

Pahl-Wostl, 2012). Similar plans in Spain related to water transfer from the Ebro were prevented by a national movement and strong public protest (Albiac *et al.*, 2006).

The transformation from traditions of reactive flood defence to more adaptive management regimes has proved difficult in most river basins. Established regimes seem highly resistant to change. Components of the current water governance and management system reinforce each other, create path dependency and induce inertia to adopting new perspectives on and directions for water management (Pahl-Wostl, 2007). These 'lock-ins' are further examples of the rigidity traps described in Chapter 3. As is the case in any trapped system, transformation to new social–ecological pathways and systems requires profound shifts in institutions, technologies and personnel, as well as the ecological, economic and social processes they influence in setting the basin's trajectory (see Box 8.2).

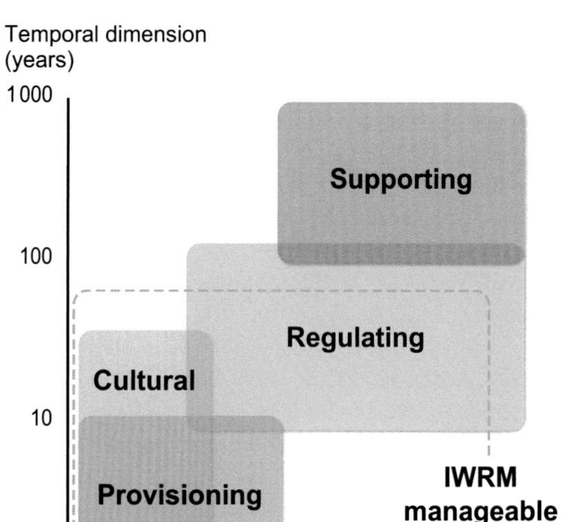

Figure 8.2 Spatial and temporal dimensions of ecosystem services in relation to the manageable area under IWRM (Keys *et al.*, 2012a). Provisioning services concern the role of water in food and timber production, regulating services include moisture feedback for local rainfall and the role of water in climate regulation, cultural services are features such as the role of water in sustaining cultural values and landscapes, and supporting services include sustaining rainfall for soil fertility.

Institutional fragmentation

Institutional barriers to sectoral integration are still a major impediment to the implementation of IWRM (UN-Water, 2012). Administrative borders seldom coincide with catchment or basin boundaries, and spatial planning and water management are largely disconnected. In addition, agricultural policy and priorities often dominate water management and environmental objectives, often overshadowing other ecological or societal aspects. However, broader sectoral integration is not necessarily facilitated by the basin principle. Although it increases the fit with hydrological boundaries, it can also weaken the interplay with other sectors that are still governed according to the old administrative boundaries (Moss, 2006).

Managing only part of the water cycle

The application of IWRM still tends to focus largely on blue water issues, not recognising the contribution of green water. Green water flows are essential to the generation of bundles of terrestrial ecosystem services – from food to flood amelioration to the role of freshwater as the bloodstream of the biosphere in relation to rainfall patterns and water teleconnections (see Chapter 4).

In practice, the application of IWRM often only includes parts of the water cycle and does not take into account the full complexity of the social–ecological systems to be managed. There have been proposals to overlap IWRM and ecosystem services as classified in the Millennium Ecosystem Assessment (2005), see Figure 8.2. Rockström *et al.* (1999) provide a global estimate of the role of freshwater in the generation of terrestrial ecosystem services.

IWRM is still an abstract concept

IWRM remains an elusive concept (Jeffrey and Gearey, 2006). This is particularly obvious when comparing countries on different rungs of the development ladder. Financial possibilities, public policy and public demand for the implementation of IWRM principles can coincide, e.g. in Western Europe (Box 8.1). By contrast, it is idealistic to expect developing countries, with burning problems such as acute food security crises, poverty alleviation emergencies and a generally large demand to secure livelihoods, will prioritise the development of ambitious IWRM schemes (AMCOW, 2012; UN-Water, 2012). This is particularly unlikely where the link with future potential benefits to tangible problems on the ground is not immediately apparent (see e.g. Pahl-Wostl *et al.*, 2012). However, there are pockets of positive progress, such as the Limpopo basin secretariat and the recent Zambezi treaty.

Although IWRM has proved helpful and may serve as a facilitator for local and national action, it has not yet led to profound changes in the stewardship of freshwater as part of the biosphere. Critics perceive this as the failure of a flawed concept (Biswas, 2004). However, the problems should not be mainly attributed to the concept but rather to its conservative interpretation in water policy and practitioner communities. Furthermore, paradigm shifts take decades (Pahl-Wostl *et al.*, 2011; in review). The adoption of IWRM as a leading paradigm has not overcome technocratic thinking and the prevailing mental models, which emphasise linear thinking, optimisation, rational decisions based on perfect information, predictability and the ability to control. Major institutional barriers to change reside in fragmented governance frameworks and established practice, routines and beliefs, which are reflected in the infrastructure. However, a new playing field for a more dynamic and adaptive water management is now emerging.

8.2.3 Land–water disconnect

Although spatial planning and IWRM are still largely disconnected, there are some signs of integration, as in the EU's Water Framework Directive (WFD) (Box 8.1), or in South Africa's, albeit increasingly criticised (e.g. Schreiner and Hassan, 2010), 'Working for Water Programme'. Efforts in Australia to curb its escalating soil salinisation problem are another example. Nonetheless, increased pressure for resources and ecosystem services on a human-dominated planet seldom connect freshwater with land-use change. Global drivers often overwhelm local efforts, which is reflected in weak institutional responses, 'land grabbing' and local and regional social–ecological vulnerability to fluctuating world market prices. The current trend for further simplification of landscapes through, for example, extensive soybean and palm oil production, with major shifts in governance towards large international players, will require global governance efforts to be redirected so that markets and incentives are reconnected to freshwater and the biosphere.

Box 8.1 Innovation in European Water Policy: the Water Framework Directive and the Floods Directive

Claudia Pahl-Wostl, Stockholm Resilience Centre and Institute of Environmental Systems Research, University of Osnabrück

The EU WFD (Bundesministerium für Umwelt, 2010) entered into force in 2000 and began a new era for EU water policy. The directive promotes an integrated approach to achieving 'good status' for all European waters (surface water and groundwater) by prescribing water management at the river basin scale. The new policy framework ended a phase of increasing fragmentation of water policy in terms of both objectives and means. The WFD is also the first major EU directive to formally prescribe the involvement of stakeholders and the wider public. The official information from the European Commission states the clear need for strong participation by all involved parties for the successful implementation of the WFD. The WFD promotes sectoral integration and encourages transboundary cooperation on international river basins. Management plans are to be revised every 15 years, supporting an adaptive approach to developing and implementing measures.

More recently, the EU Flood Risk Management Directive (FRMD) on the assessment and management of flood risks entered into force in November 2007. This directive was a response to the severe floods in many parts of Europe between 1998 and 2004, including the flood catastrophes in the Danube and Elbe basins in the summer of 2002. The FRMD requires EU member states to assess whether water courses and coastlines are at risk from flooding, to map flood risk, and to take adequate and coordinated measures to reduce potential impacts. The FRMD also reinforces the right of the public to have access to this information and to participate in the planning process. The implementation process requires harmonisation with the WFD.

The EU WFD and FRMD are promising steps towards an improvement in the ecological status of European waters, which a recent WFD classification demonstrated to be urgently required. The analysis revealed major ecological deficits. For example, while 88% of the surface water bodies in Germany had achieved good chemical status, only 10% of the water had good ecological status, 34% was classified as poor and 23% was awarded bad ecological status (BMU, 2010). Quality improvements can be achieved using technical measures, whereas improving the ecological status requires a shift towards the more holistic landscape management argued for in this chapter.

8.3 Governance frontiers

Many of the water challenges on the agenda today are likely to persist – from providing water for basic livelihoods, ecosystems, poverty alleviation, sanitation and human health, to supporting infrastructure and coping with erratic rainfall patterns. The new freshwater challenges that are the focus of this section are connected to the Anthropocene era and global change, including climate change.

8.3.1 New freshwater challenges

Debate about how better to capture real-world water dynamics has been triggered by the implications of climate change and the expected increase in uncertainties. Phrases such as 'stationarity is dead' and 'climate change changes the water rules' have appeared in the literature (Kabat *et al.*, 2003; Milly *et al.*, 2008). Most treaties and international agreements seem to lack the tools for dealing with contemporary challenges such as flood control and water quality, as well as adequate mechanisms for addressing changing social, economic or climate conditions. New approaches that could be incorporated into existing treaties to allow for flexibility in the face of climate change include: adjustable allocation strategies and water quality standards; response strategies to extreme events; amendment and review procedures; and institutions for joint management (Cooley and Gleick, 2011).

This book addresses: (1) water-mediated trade-offs between diverse human activities and securing ecosystem services in human-dominated landscapes; (2) water teleconnections; and (3) changes in the frequency, magnitude and duration of disturbances such as storms, floods and drought. It also expands the focus to making productive and resilient use of precipitation, and developing back-up systems and insurances to deal with complexity and uncertainty. In a globalised world, taking on the seemingly overwhelming task of dealing with fluctuating and unpredictable rainfall through the stewardship of larger precipitationsheds will challenge governance capacity. It is to be hoped that collaboration and new forms of governance will emerge that cross national borders and link remote landscapes. While great effort has been devoted to studying local to regional-scale vulnerabilities, hardly any attention has been paid to the new vulnerabilities arising from large-scale changes and connectivity in the GWS (Alcamo *et al.*, 2008; Sadoff and Grey, 2005). Needless to say, the gap between governance of the management of freshwater and attempts to implement IWRM is huge.

8.3.2 A new conceptual framework

Falkenmark and Folke (2010) developed a conceptual framework for a way to think about the stewardship of precipitation in an attempt to make the governance and management challenge presented by the wider role of freshwater in society more comprehensible. What are the means for catching rain and rainfall pulses before they are lost to the landscape? How can upstream consumptive water use by terrestrial ecosystems of various kinds, as well as runoff generation for rivers and aquifers, be balanced with downstream water allocations between wetlands, cities and aquatic ecosystems to achieve harmony across river basins? Their framework makes a number of principal proposals.

Recharge area activities

Activities in the recharge area, although scale dependent, relate to two principal changes in water flows. Changes to vegetation increase infiltration and rainwater harvesting and alter water partitioning in the upper part of a basin, that is, increased total evapotranspiration reduces the blue water flows for downstream uses. Alterations to groundwater recharge are mainly a rerouting of blue water flows from surface to sub-surface flows, or vice versa.

- Manage *vegetation* to secure crop production (currently 50–60% of global crop production is rainfed), timber production and greenhouse-gas sequestration using forestry or other perennial vegetation, productive grasslands for livestock, local moisture feedback cycles and numerous other terrestrial ecosystem services.
- Manage *infiltration* to secure green water availability, improve the structure of and organic material in soils and safeguard the numerous terrestrial ecosystem services dependent on water holding capacity; and secure drought protection and good crop yields, with trade-offs between bundles of ecosystem services.
- Harvest, *store* and manage *rainwater (in soil, aquifers, tanks, dams and reservoirs)* to secure local water availability for irrigation and supplementary irrigation during dry spells, and to meet household water needs. Storage of runoff makes water

available for critical phases of crop production and other ecosystem services.

- Manage *groundwater recharge* to secure protection against flash floods and base flow generation, and thereby accomplish both a reduction in undesired flooding and an increase in sub-surface flows, thus enhancing blue water availability over time.

Discharge area activities

Water management options, or potential management problems, in discharge areas are directly constrained or enhanced by the land and water uses in recharge areas.

- Manage *flood flow* by storing surplus water in reservoirs to secure flood protection, minimise inundations and flood damage, and maximise downstream water security for human needs.
- Manage *groundwater seepage* in water fed wetlands to secure wetland ecosystems and their role as stepping stones for migrating birds (Lundberg and Moberg, 2003) of significance in ecosystem dynamics and development and in regulating ecosystem services.
- Manage *reservoir outflow* to secure flood pulses for inundation-dependent flood plains used for pastures, for aquatic ecosystems and to support pasture-oriented food production and related ecosystem services, and for flushing of pollutants and silt.
- Manage *trade-offs* between human withdrawals for cities and industry and aquatic ecosystems to secure adequate environmental flow.

A hydro-unit based approach

In the context of recharge and discharge areas, it is possible to develop governance frameworks of 'hydro-unit'-based land management under different land uses and with different partitioning characteristics. Such an approach emphasises critical trade-offs in water use in catchments/landscapes and how to make better use of the bloodstream of the biosphere by creating water partitioning synergies between different uses, thereby enhancing overall catchment productivity (see also Chapter 7). The landscape could tentatively be divided into such hydro-units, which would require explicit attention to upstream/downstream freshwater interactions and to blue/green water trade-offs, with the main focus on green water management upstream and blue water availability downstream (Falkenmark and Folke, 2010).

A focus on precipitationsheds

A focus on precipitation and precipitationsheds combines the blue and green water branches for direct water use and for ecosystems services in the context of sustaining human well-being and societal development. The aim is to take the steps required to make a truly integrated approach more tangible. The challenge of sustaining and managing rainfall takes water stewardship a step further and will require adaptive and resilient social–ecological systems framed to live with change and with water fluctuations – systems that can collaborate and connect management across multiple scales and for diverse precipitationsheds.

The knowledge and understanding needed for such activities are in many cases lacking. A new structure for monitoring regional and global water dynamics will be needed to be able to prepare for and deal with freshwater–climate interactions and fluctuations in catchments and drainage basins in order to secure ecosystem services, including food production. The capacity of social–ecological systems to be stewards of ecosystem services and societal development needs to be strengthened in the face of these dynamics.

Recent bottom-up efforts

In that sense, it is encouraging to witness the many bottom-up efforts, social networks, bridging organisations, umbrella institutions and multi-level governance systems emerging worldwide to deal with diverse blue-green freshwater challenges, ecosystems services, climate change and fluctuating environments. Land and water actors are joining together for improved stewardship. In the agricultural, wetland river landscape of Kristianstad Vattenrike, southern Sweden, for example, recently designated as a UNESCO biosphere reserve, the challenges of floods, water quality, food production, tourism and conservation values are combined in flexible and diverse management supported by an adaptive governance system (Hahn, 2011; Hahn *et al.*, 2006). India's catchment development efforts use water-soil issues as a significant entry point for livelihood improvement, and the empowerment of local people has been key to their success (Joshi *et al.*, 2008; Wani *et al.*, 2011). Multi-level and often polycentric water governance systems, such as the well-known Bali temple-based rice irrigation system (Lansing, 1991), the Spanish huertas (Ostrom, 1990) and the irrigation systems in Nepal (Lam, 1994), have been in existence for some time.

However, in the Anthropocene era, bottom-up efforts based on historical local conditions will in many places have to be complemented with new knowledge and improved understanding of how systems impact and are impacted, beyond the area of historical interest. This will be essential to avoid potentially negative externalities and provide insights into the new approaches, technologies and opportunities that will be necessary in local landscape management, including of water resources. In addition, there is still no explicit global leadership on water resources and transformative change in the long run. Global water governance efforts need to recognise, connect and combine with international agenda setting, to provide overall support that frames the creativity of emergency into pathways promoting sustainability (Young *et al.*, 2008).

8.3.3 Emerging governance challenges

It is obvious that such a broad perspective requires new forms of adaptive, multi-level governance structures. It is in this context that adaptive water management to deal with changing situations, and uncertain and unexpected change involving multi-level, polycentric systems, comes into play (Ostrom, 2010; Pahl-Wostl, 2009). Such systems of adaptive governance are better suited to the stewardship of freshwater in dynamic landscapes. They rely not on panaceas or simple technical or institutional recipes for change, but on systemic and context-sensitive change of the entire governance and management system (Pahl-Wostl, 2009). Numerous recommendations, often reliant on simplistic universal remedies, have been put forward for water governance reform without testing their appropriateness in different contexts. A recent comprehensive analysis of complex water governance and management systems worldwide showed that polycentric governance regimes, characterised by a distribution of power but also effective coordination structures, performed better in achieving water-related goals (Pahl-Wostl *et al.*, 2012). However, this analysis confirmed earlier findings that economic and institutional development has led to meeting needs of the human population at the expense of the environment (Vörösmarty *et al.*, 2010). Effective and polycentric governance structures have attenuated but not reversed this trend. Such governance failures from a sustainability perspective can be overcome by adopting a dynamic landscape approach and a managing the water cycle perspective. This goal is ambitious but achievable.

Desirable landscape patterns

A desirable state, as a management goal, should be determined by defining the desirable spatial and temporal pattern of ecosystem services to be sustained at the scale of a river basin. This implies negotiating and reducing potential trade-offs between upstream and downstream regions, e.g. upstream consumptive uses reducing downstream flows, and the distribution of scarce water resources for different uses. Calling for the management of precipitationsheds might be interpreted as introducing a new scale for water management. However, the spatial boundaries of precipitationsheds are dynamic and not as well defined and identifiable as the spatial boundaries of river basins. Rather than introducing a new institution, the governance of basins needs to take account of these fundamental cross-scale linkages and feedbacks. Human dependence on moisture feedbacks, within basins and across basins, the virtual water flows in the food trade and the role of freshwater for not only provisioning but also regulating and supporting ecosystem services in the landscapes clearly illustrate how water, as the bloodstream of the biosphere, flows across both societal and biophysical domains, and underlines the need for a broad social–ecological governance approach.

Management for sustaining rainfall has one important caveat, which argues in favour of a precautionary management style. It cannot operate in a reactive mode at spatial scales where crossing thresholds is irreversible. Clearing the Amazon basin cannot be reversed. Global leadership is required to guarantee the functioning of major elements of the hydrological cycle. Furthermore, more dynamic bilateral agreements must be implemented under a global framework directive. Basin governance needs to build resilience and adaptive capacity to shocks and surprises, but can do so only within a safe operating space of variability. This needs to be guaranteed by such global and supra-basin agreements.

Resilience building

Building resilience means adopting an inward-looking perspective. What this means for the regional and local levels, where the majority of the management strategies identified above need to be implemented, is a key question. The argument put forward in this volume is that the county and province levels are the appropriate scales to develop visions of a landscape and the kind of ecosystem services to be sustained to achieve a desirable ecohydrological situation. However, as is argued above, actions at the local and regional scales must take into account the effects and impacts across larger scales. In the Anthropocene era, these effects are not just occasional or irregular events, but increasingly pervasive. The global continually influences the local and the regional.

Given the need for cross-sectoral integration and to be able to monitor and adapt to new insights, network governance seems to be a promising approach for integrative governance settings able to steer such a process. Landscape committees or associations that support platforms for knowledge generation and innovation might be promising coordination structures at the local level. They would coordinate planning and revise plans in case of undesirable developments. Such structures could overcome the inertia of the rigid bureaucracies that characterise many industrialised countries, where increasing cross-sectoral coordination has already led or might still lead to overregulation. Many developing countries face the opposite problem – of dysfunctional bureaucracies, weak governance structures, low levels of transparency and a high prevalence of patronage networks (see Box 8.4). Here, such structures could support building governance capacity from the bottom.

Governance transformation

A major remaining research challenge is to develop guidance on how to implement adaptive and integrated governance schemes in diverse contexts. Such schemes must recognise and respect the role of freshwater in the biosphere and meet the good governance principles of legitimacy, representativeness, effectiveness, efficiency, transparency, accountability and equity. This will be particularly important during phases of transformation (Chapin *et al.*, 2010; Olsson *et al.*, 2004; Olsson *et al.*, 2008; Olsson *et al.*, 2006). As a result, a focus on the governance of transformation towards more adaptive and integrated approaches is urgently needed.

Although scholars recognise the need for fundamental change, and call for sustainable transitions to deal with the challenges (Clark and Dickson, 2003; Lambin *et al.*, 2001), there are few empirically based insights into the social–ecological dynamics that make such shifts possible (Folke *et al.*, 2005). Crises, for example, require short-term responses and can, but need not necessarily, be a driver for long-term change.

Thus, there is still a lack of understanding of how to steer away from undesired ecological and social regimes, and towards new and improved social–ecological trajectories that sustain and enhance ecosystem services and human well-being in the longer term (Westley *et al.*, 2011). Transitions to more integrated approaches that address linked social–ecological systems require systemic shifts that involve changes in perceptions and mental maps, network configurations, steering mechanisms and interaction patterns among actors, as well as institutional and organisational structures at multiple levels (Geels and Schot, 2007; Olsson *et al.*, 2004; Pahl-Wostl, 2009). Arguments have been made that climate change may be a strong driver of more sustainable landscape development paths (e.g. Smith and Barchiesi, 2009). However, the pathways chosen will strongly depend on processes of social learning to overcome path dependence and structural constraints, and provide an enabling environment for agents of change (Pahl-Wostl, 2006, 2007, 2009). Such shifts have been described in both terrestrial and marine systems and the triggers may range from real or perceived crises from climate-related changes to political shocks (Gelcich *et al.*, 2010; Olsson *et al.*, 2006). In such situations, prediction and optimisation are of little use, and will have to be replaced by risk spreading and insurance strategies to maintain options and sustain social–ecological systems in the face of the surprise, unpredictability and complexity that characterise the Anthropocene era (Polasky *et al.*, 2011).

8.4 Signs of a paradigm shift and managing change

In recent decades, innovative legislation has been implemented in many countries, adopting an integrated approach by including improving the state of the environment as a management goal, involving the public, balancing different interests and leaving space for learning and adaptive approaches. Examples include the EU WFD (see Box 8.1) and the National Water Act in South Africa, which has been the subject of criticism (see Box 8.4). However, implementation has not yet kept pace with the ambitious goals of such new legal frameworks.

8.4.1 Flood management as an example

An illustrative example of the nature of a deep paradigm shift, and the drivers and barriers promoting and hindering change, is provided by the development of flood management and governance in recent decades (Pahl-Wostl *et al.*, in press). Historically, flood management was dominated by the objective of reducing flood hazards and increasing the safety of people and infrastructure on floodplains. Taming rivers and conquering swampy, disease-prone floodplains for human habitation and agricultural expansion were celebrated as major successes of engineering and technology.

However, over time the drawbacks also became increasingly visible. Despite a reduction in the number of flood incidents, flood damage increased because more assets were located in exposed areas. This further increased the demand for flood protection (Figure 8.3). In order to reduce flood hazards, rivers are regulated by dams and dykes are built. This leads to a reduction in floodplain biodiversity and a reduction in the benefits provided by floodplain services, such as water retention, resulting in increased peak flows. At the same time, yet more assets are built in the flood

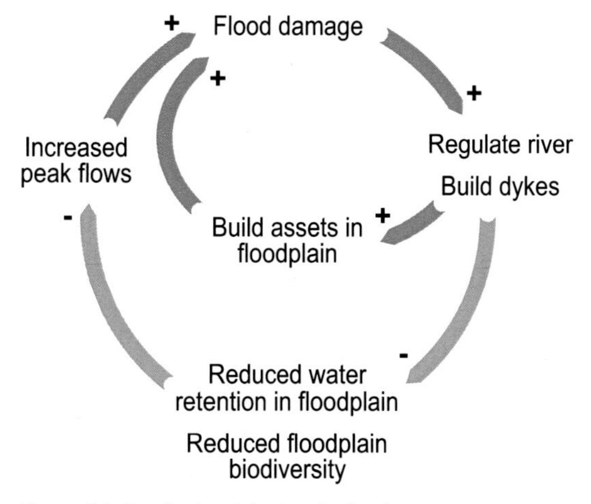

Figure 8.3 Feedbacks reinforcing the flood management paradigm.

plain. This leads to a higher likelihood of severe flood damage in case of flooding events, which imposes further pressure on flood protection and leads to the building of higher dykes in a classic example of a rigidity trap (see Chapter 3). The whole system is also vulnerable to climate change. Sensitivity to an increase in extreme flood events is high and adaptation options are reduced due to path dependence. Such path

Box 8.2 Competing governance paradigms in water management: the Tisza River

Jan Sendzimir, University of Natural Resources and Applied Life Science, Vienna (BOKU)

Water governance regime change has become an issue in Hungary following repeated failures of conventional management policies to handle a series of floods on the Tisza River starting in 1997. A 'shadow network' of activists and academics has emerged to point out how current river management appears trapped in a hopeless downward spiral of coping reactions that never build enough momentum to improve the situation. Increasing public participation catalysed by the shadow network pushed the water policy debate towards more experimentation with alternatives, but implementation appears to have stalled. A key ministry official, J. Váradi, invited the shadow network to participate in the national policy debate and shepherd alternative ideas into a more experimental river policy (VTT2).

The pivotal role of leadership became apparent when Váradi's departure from the parliamentary committee created a power vacuum. Neither the Ministry of the Environment and Water nor the shadow network could take over the leadership role, and the 'old guard' river management fraternity again gained power and promoted conventional river defence strategies. Such a rapid reversal of a major policy innovation shows the dominance of the formal policy process over incremental improvements in established practices. Without enabling legislation that supports translating informal agreements into binding commitments, devoted and sustained leadership and social networks, reframing or even causing transformative change in water management and governance may only be temporary and reversible, despite the inroads made by bottom-up participation. The resilience of the current conventional management regime appears to prevent it from experimenting and adapting. To navigate towards a transformation may thus involve managing social–ecological resilience, first by lowering it in moderately risky experiments to allow reconfiguration and then strengthening a more adaptive configuration of the new components of the social–ecological water regime (see Figure 8.b2.1).

Box 8.2 (*cont.*)

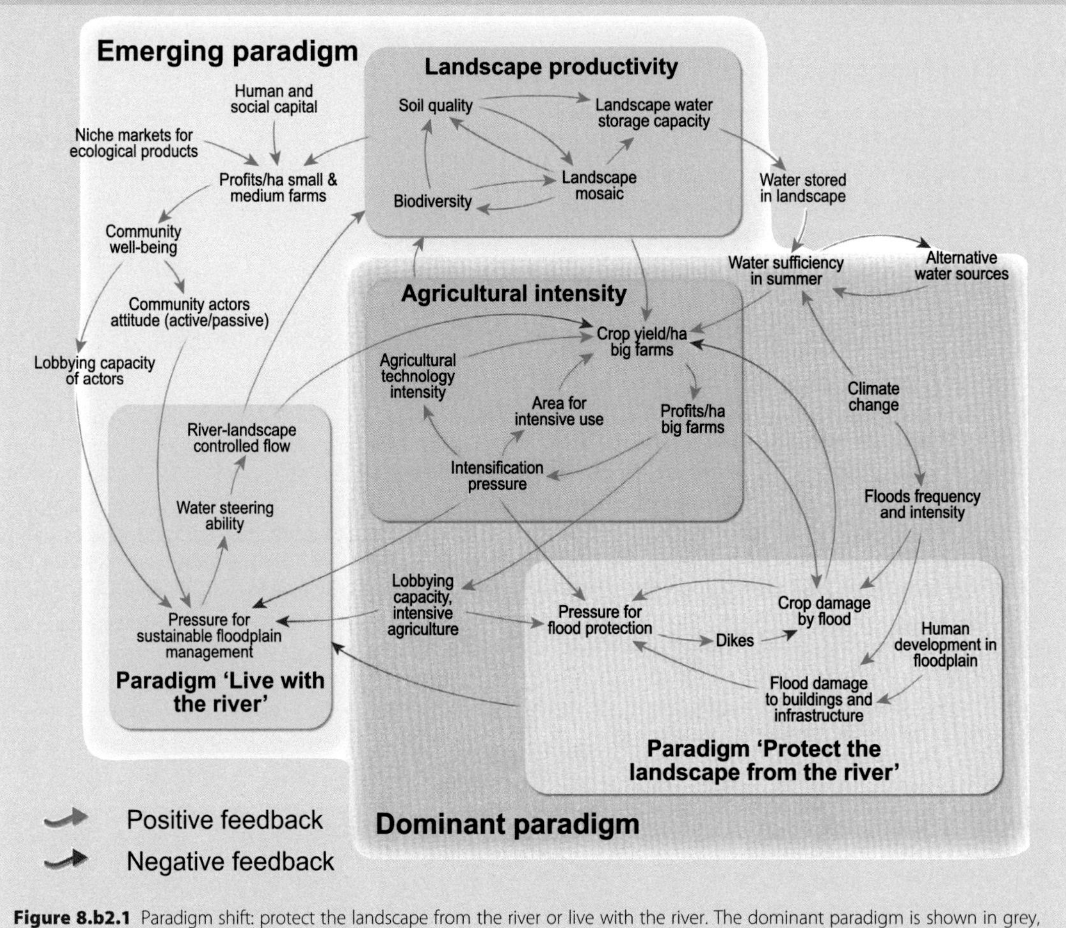

Figure 8.b2.1 Paradigm shift: protect the landscape from the river or live with the river. The dominant paradigm is shown in grey, the emerging paradigm in light blue. Eventually, there may be a paradigm shift between the two (Sendzimir *et al.*, 2008).

dependence is manifested not only in large-scale investment in long-lived infrastructure, but also in the whole governance structure that has co-evolved with a certain flood management paradigm (Box 8.2). Only recently has the loss of important ecosystem services, such as natural buffering capacity, corridor function and species conservation, been realised and become the focus of restoration attempts. Technical solutions have been proposed for reinstating these services by building facilities for flood retention, storage and bypass, supported by the general increase in environmental awareness in industrialised countries. This overall shift to reducing risk and exposure rather than increasing flood protection is a major guiding principle of the EU FRMD on the assessment and management of flood risks, which entered into force in November 2007 (see Box 8.1).

8.4.2 Room for the river

One of the most pronounced changes can be observed in the Netherlands, where the government has asked for a radical rethink of water management in general, and flood management in particular. The resulting policy stream, initiated through the *Ruimte voor de Rivier* (Room for the River) policy, has strongly influenced other areas of government policy. Greater emphasis is now given to the integration of water management and spatial planning with the regulating services provided by landscapes, and natural flooding regimes are highly valued. This requires a revision of land-use practices and reflects a gradual movement towards integrated landscape planning in which water is recognised as a natural and structural element. Impacts related to global change are predominantly linked to the water system through increased exposure to floods and droughts. However, in landscaping and land-use planning, water is still most often seen as a secondary concern and the delivery of water-related services is taken for granted.

Considering water as a key structuring element or guiding principle for landscape management and land-use planning requires technology, integrated systems thinking and the art of thinking in terms of mutual interests, between the different authorities, experts, interest groups and the public. In the Netherlands, water policymakers and managers have started to stress the importance of water as a structuring element in land-use planning. Moreover, the societal debate over plans to build in deep-lying polders or other hydrologically unfavourable spots, and new ideas about floating cities, indicate a considerable level of social engagement by both public and private parties around the issue of sustainable landscapes and water management. However, while new ideas have been adopted in policy terms, implementation takes time, as it is necessary to overcome considerable social resistance.

Flood management, as it is often practised, is a prime example of how natural capital has been eroded instead of making use of natural infrastructure to build adaptive capacity and resilience to deal with unexpected developments and surprises. The value of natural capital and associated ecosystem services remains unrecognised. This implies that the increasing trade-offs between human and environmental needs are not inevitable, but depend on how and by whom the benefits derived from nature are conceptualised. However, the obstacles encountered also highlight the barriers to change.

Crises, real or perceived, have proved important for triggering insights and opening windows of opportunity for change. However, the immediate response to a crisis does not support reflection and learning unless innovative ideas have already matured in parallel with established practices (see Figure 8.b2.1 in Box 8.2). As is described in Box 8.3, the shift in the management of the Everglades ecosystem from a vast subtropical wetland to a highly altered, multiple-use social–ecological system, was not a linear process but

Box 8.3 Crises and transformations in the Everglades

Lance Gunderson, Emory University

Large water-control projects have transformed the Everglades ecosystem in Florida from a vast subtropical wetland into a highly altered multiple-use social–ecological system. This transformation was not a linear process but turbulent and punctuated by change events that were perceived as crises. Each crisis precipitated actions (or a series of actions) that resulted in shifts to new governance strategies and management practices. The crises were created either by external environmental fluctuations (e.g. floods, droughts) or by human activities making the area less resilient to such fluctuations. In the twentieth century, these resulted in at least four major eras of water management (see Figure 8.b3.1).

The first two eras were a result of flooding linked to high rainfall. The third era was related to drought, and the fourth resulted from attempts to rectify latent or previously unattended ecosystem problems. For example, hurricanes struck the region in 1926 and 1928, claiming more than 2500 lives. The Army Corps of Engineers joined forces with the state to control flooding around the lake and made land accessible for farming. Agricultural produce price support combined with the new flood control programme to give farmers the perception of security they needed to double sugarcane production, thereby reinforcing the path. The major drought in 1970–71 made the prevailing way of doing business untenable and the Everglades were so dry that 300 000 hectares burned. Later, the 1981 drought, a 'one-in-200-year event', caused a serious drop in water levels, followed by an extreme El Niño Southern Oscillation event that produced unusually heavy rains. The different strategies and paradigms had different capacities to deal with the events.

The lesson of the Everglades calls for adaptive approaches to learning from experience, and the social–ecological resilience and adaptive capacity to shift pathways and avoid lock-in and traps in water management and governance. These lessons become critical in a time of climate change.

Box 8.3 (*cont.*)

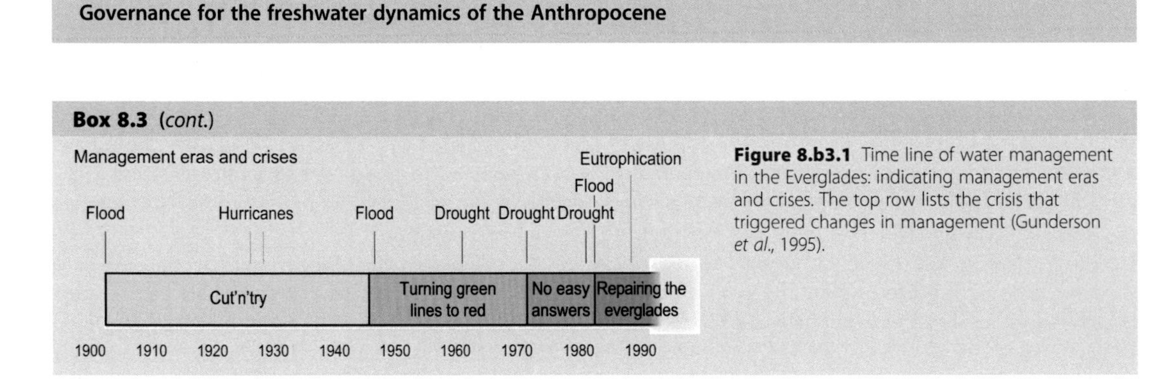

Figure 8.b3.1 Time line of water management in the Everglades: indicating management eras and crises. The top row lists the crisis that triggered changes in management (Gunderson *et al.*, 1995).

turbulent and punctuated by change. These events were perceived as different crises and triggered the transformation.

8.4.3 Transformation into new stability landscapes

There is a need to move away from regulating environmental flows to meet certain goals, such as flood protection or to produce natural resources such as dairy produce or fruit, and towards meeting the challenge of multi-level collaborative societal responses to a broader set of feedbacks and thresholds in social–ecological systems.

Land and water management are crucial to the Goulburn–Broken catchment in the Murray–Darling Basin in Australia, for example, where widespread dryland cropping, grazing, and irrigated dairy and fruit production produce a quarter of the State of Victoria's export earnings (Walker *et al.*, 2009b, see also Box 3.1). At first glance, economically lucrative activities seem to be thriving, supported by irrigation. If the analysis is broadened to take a social–ecological approach to account for the capacity of the landscape to sustain the ecosystem services and values in the region, however, the picture looks quite different. Widespread clearing of native vegetation and high levels of water use for irrigation have resulted in rising water tables, creating severe salinisation problems. The problems are so severe that the region is facing crossing serious social–ecological thresholds with possible knock-on effects. Crossing such thresholds could result in irreversible change in the region (Walker *et al.*, 2009b).

Hence, strategies for adaptability that are socially desirable can lead to vulnerable social–ecological systems and persistent undesirable states such as poverty traps or rigidity traps (Scheffer *et al.*, 2009), as is described in Chapter 3. Will the adaptability of people and the governance of the Goulburn–Broken catchment be sufficient to deal with the ongoing environmental change and interacting thresholds? Can this social–ecological system avoid being pushed into a poverty trap, or does the system need to be transformed into a new stability landscape, forcing people to change deeply held values and their identity (Walker *et al.*, 2009b)?

In the Makanya agro-ecosystem, Tanzania, people are faced with a poverty trap reflected in reduced soil fertility, more frequent dry spells, reduced availability of moisture and increasing crop failures. In such a situation, institutions and practices that stimulate small-scale water system innovations could help break undesirable feedbacks and shift the present trajectory to a new stability landscape of fewer crop failures and improved field water balances (Enfors, 2013) (see Chapter 6). Declining agricultural productivity in several Latin American countries linked to land degradation prompted farmers to begin experimenting with unconventional methods of land management, in particular low-till alternatives to ploughing that enhanced soil organic matter and fertility. The experimental learning approach on a small scale, with processes for improvements and cross-scale learning, caused a transformation in the whole farming system. Currently, more than 25 million hectares of agricultural land is under a no-tillage regime in Brazil alone, and in Latin America as a whole the transition from conventional plough-based agriculture to no-till systems has reached a scale at which it is possible to talk of an agrarian revolution (Fowler and Rockström, 2001). Digging the hole deeper versus shifting to new pathways is further discussed in Box 8.2 in relation to the Tisza River and the governance paradigm of protecting people from the river versus living with river dynamics (Sendzimir *et al.*, 2008).

8.4.4 Windows of opportunity

Major shifts or transformations (Folke *et al.*, 2010) to new ways of managing landscapes have taken place in different parts of the world, combining stewardship of water, ecosystem services and social and economic development. It is often the case that unexpected internal or external change that creates a window of opportunity is a prerequisite for allowing social–ecological transformation to gain momentum. Three phases have been identified for such a trajectory: preparing for transformation, navigating the transition and building the resilience of the new direction (see Figure 8.4). A window of opportunity links the first and second phases. Crises, perceived or real, often trigger transformations (Folke *et al.*, 2009).

In southern Sweden there has been a major transformation from dispersed, uncoordinated governance to pathways of adaptive governance in the lower Helgeå river basin in the face of flooding, declining conservation values, water quality concerns and the conflicting needs of fisheries, agriculture and tourism in a semi-urban landscape. The transformation emerged from bottom-up initiatives by a few committed leaders, or 'change agents' or institutional entrepreneurs (Westley *et al.*, in press). Building a shared vision, trust and a connection between diverse networks, they created a mutual sense of purpose and meaning – not just rules, roles and resources (Westley, 2001). From this, a bridging organisation emerged within the formal institutional structures as well as a lot of trust and engagement among the actors and levels below. However, the initiative could not have taken off without support from the levels above. When change opened a window of opportunity, the initiative was able successfully to move across levels and gain support from the local, regional and national levels, supported by enabling legislation. The value of the region is now recognised internationally and it has recently become one of UNESCO's Biosphere Reserve Areas (Hahn, 2011; Hahn *et al.*, 2006; Olsson *et al.*, 2004).

Rapid and large-scale political change has acted as a driver for the transformation of the water sector in countries such as South Africa, but has not had much effect in others – such as Uzbekistan. Box 8.4 presents a positive example of how the features and conditions of the social–ecological system in South Africa were critical to enabling society to use the window of opportunity that emerged with the end of apartheid.

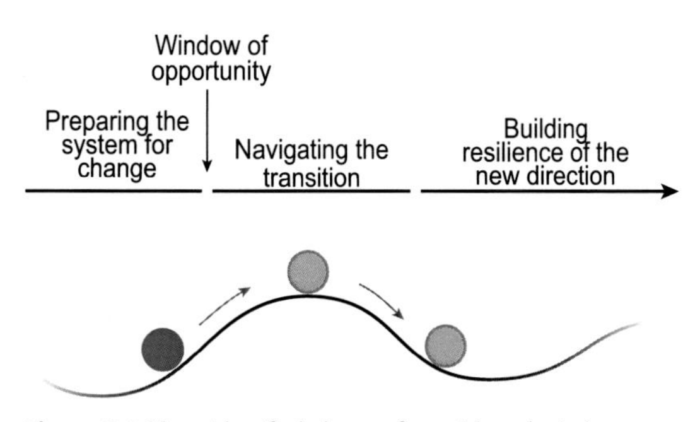

Figure 8.4 Three identified phases of a social–ecological transformation – preparing for transformation, navigating the transition, and building resilience of the new direction. A window of opportunity links the first and second phases (adapted from Folke *et al.*, 2009).

As a result, water governance was transformed towards better stewardship of natural resources, although the benefits and governance of the transformation are constantly being challenged. In contrast, the lack of some of the enabling features and the negative impact of others, such as rigid informal networks, acted as a barrier in Uzbekistan (Schlüter and Herrfahrdt-Pähle, 2011). As a consequence, the opportunities provided by the break-up of the Soviet Union and subsequent independence were missed.

Box 8.4 Socio-political shocks and transformation in the water sector: South Africa and Uzbekistan

Maja Schlüter, Stockholm Resilience Centre

Elke Herrfahrdt-Pähle, German Development Institute

Transforming the water sector to improve water governance is difficult because of the inertia of existing regimes and their general reluctance to change. A window of opportunity is usually needed to initiate a transformation (Chapin *et al.*, 2010; Olsson *et al.*, 2004). Large-scale socio-political shocks such as the end of apartheid or the break-up of the Soviet Union can open such a window. However, whether this opportunity to improve the stewardship of natural resources is taken depends on some crucial features of the resilience of the social–ecological system as well as its preparedness for change. These features include the existence and enhancement of leadership, good social relations, networking capabilities, available knowledge, an enabling institutional setting and social learning processes that experiment with new development pathways. Other features, such as rigid networks, might have to be broken up to facilitate change. If existing social–ecological feedbacks

Box 8.4 (*cont.*)

begin to generate fewer benefits because of ecological degradation or political or economic change, the structures may change, causing transformation. The cases of the water sector reforms in South Africa and Uzbekistan illustrate how interactions between these features have led to a transformation in South Africa but prevented one in Uzbekistan.

Enabled social learning

When apartheid ended in South Africa in 1994 there was a looming ecological crisis of water scarcity as well as increasing impacts of climate change on water resources. Apartheid had created a dual economy and deprived people of access to water, with huge effects on livelihoods resulting in widespread poverty and poor health status. Like elsewhere, a technical control paradigm dominated water management. During the apartheid era, local and traditional knowledge and innovation with respect to dealing with scarce water resources were lost. However, once the window of opportunity was open, the involvement of South African and foreign experts as well as local and foreign best practices helped to overcome barriers and enabled social learning to build new knowledge and prepare for a transformation. Leadership was provided by a visionary and motivational minister of water affairs and forestry, Mr Kader Asmal, who managed to build trust and connect actors. A group of water managers had already begun thinking about a new water law in the 1980s, laying the groundwork for an exploration of new pathways to transition with networks of water managers (de Coning, 2006). As a result, the window of opportunity was used to develop a National Water Act in 1998. It provided an enabling environment for the development of new institutions to improve water management, such as Catchment Management Agencies with representative boards. However, the consolidation of the institutional changes is currently being challenged by implementation problems caused by the openness of the new water law, and the rotation of staff in the respective agencies which prevents the development of leadership and the building of trust (Herrfahrdt-Pähle, 2012).

Prevented social learning

The external shock of the break-up of the Soviet Union in 1991 transformed the Soviet Republic of Uzbekistan into an independent country. The new country inherited severe legacies such as overused water resources and massive environmental degradation (Schlüter *et al.*, 2006). The country's economy has been dependent on revenues created by intensive cotton agriculture. Environmental degradation and the socio-economic transition after independence created economic losses and health impacts with negative effects on livelihoods, leading to poverty and out migration. In contrast to South Africa, the window of opportunity created by the political shock did not lead to a transformation. The new water code, which was rapidly developed in 1993, largely followed the blueprint of the former Soviet law. The major factors which prevented a transformation in Uzbekistan were reinforcing feedbacks that trapped the system in its current state, e.g. informal networks that worked to maintain the status quo for those who still benefited from it (Schlüter and Herrfahrdt-Pähle, 2011). A strong hierarchy and low levels of trust and ownership prevented the social learning that is necessary to activate social memory and create new knowledge to prepare for a transformation (Schlüter *et al.*, 2010). Seventy years of Soviet hegemony left the technical control paradigm dominant. Traditional and local knowledge were largely lost (Mollinga, 2010) and could not be mobilised. In addition, high levels of specialisation among actors in the agricultural sector caused a fragmentation of knowledge. The institutional setting was characterised by the dominance of agricultural over the water sector, effectively preventing change in the latter. Thus, cognitive and structural factors together with the lack of leadership because of a continuing presence of Soviet-trained officials in the ministry acted as barriers to institutional change in the water sector. Initiated half-hearted reforms were often little more than a renaming of old institutions or top-down imposition of largely ineffective (at least in the intended sense) new institutions, such as water user associations (Sehring, 2009).

8.4.5 A three-phase transformation process

Studies of transformations of social–ecological systems to ecosystem-based management have generated a conceptual model that characterises the transformation process as the emergence of adaptive governance of social–ecological systems (Chapin *et al.*, 2010; Folke *et al.*, 2009; Folke *et al.*, 2005) (Figure 8.4). The nature of such change can be captured by the concept of triple-loop learning (Pahl-Wostl, 2009). This approach posits that societal

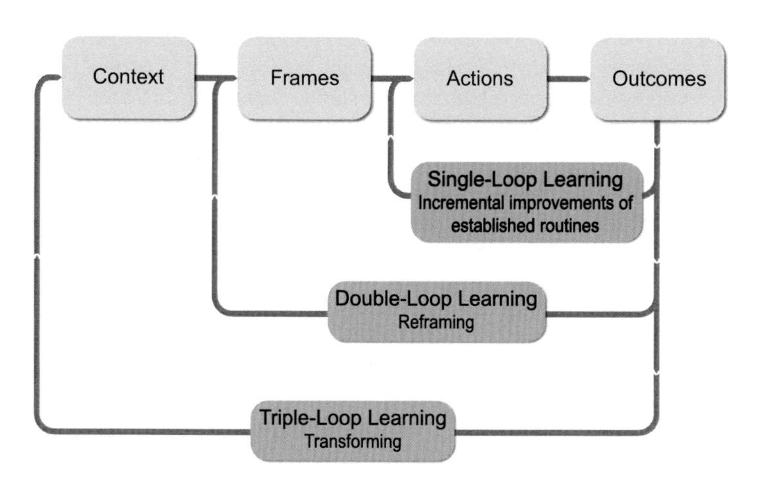

Figure 8.5 Concept of triple-loop learning (Pahl-Wostl, 2009).

transformations that improve water stewardship and resilience can be described as multi-level and multi-loop processes (Figure 8.5).

Single-loop learning refers to an incremental improvement of action strategies and daily routines that does not question underlying assumptions. Double-loop learning refers to a revisiting of assumptions (e.g. about cause–effect relationships) within a value-normative framework. In triple-loop learning, underlying values, beliefs and world views are reconsidered if the assumptions within a world view are no longer valid. Translated to the landscape in Figure 8.5, double-loop learning is required to prepare the system for change. Triple-loop learning is required to get the system over the crest towards a new state. Triple-loop learning allows for a re-examination of the underlying ideological and value systems that keep the system trapped in its current state. Societal learning is an exploratory search process where actors experiment with innovation and try to overcome or even remove the constraints and boundaries they encounter. Windows of opportunity support a wider societal recognition of innovations and reframing, and allow a background discourse to move in the foreground and be picked up by many actors. Such shifting discourses are a necessary but not a sufficient step for profound transformation.

A paradigm shift in water resources management has moved to the discourse but not yet to the transformation stage (cf. Pahl-Wostl *et al.*, 2011). To cross the threshold to transformative change requires a concerted effort on and across different levels. The governance of system transformations connecting still fragmented initiatives to transformative change is a critical challenge (see Figure 8.4). Such transformational change to a recognition of freshwater as the bloodstream of the biosphere and to improved stewardship of our role in the biosphere is critical. System-wide transformations are required when considering planetary boundaries, and as such broader governance for water resilience is an essential pathway to sustainability (Folke *et al.*, 2010; Westley *et al.*, 2011).

8.5 Governance of transformation: a multi-level challenge

In Western flood management there has been a shift in recent decades from the 'control of floods' paradigm to 'living with floods' (Pahl-Wostl *et al.*, 2011, in review) and dynamic coasts. The latter are an involuntary fact of life in many coastal and riverine areas in the developing world, and this is expected to be exacerbated by the challenges of climate change. Although some developed states are moving in the direction of structural reorganisation and a transformation of society, barriers and institutional inertia still have to be overcome. Too many managers still view water issues as sector issues, and the more integrated governance approaches are predominantly focused on blue water alone. Approaches to dealing with water scarcity and drought seem to be largely locked into the current frame of reference, with an emphasis on improving the efficiency of current water-use patterns. However, a reframing has taken place in the shift from the sole emphasis on water supply and resource development to managing water demand (Gleick, 2003).

Although the systems view of agriculture and water is expanding, little evidence can yet be detected for a more profound movement in governance towards an entirely new paradigm of the role of water in biosphere stewardship, in confronting planetary freshwater boundaries or governing precipitation-sheds, thereby sustaining rainfall and building social–ecological resilience in water stewardship of green and blue flows and their associated ecosystem services across landscapes and regions. Such a paradigm shift would encourage land-use practices that are guided by water availability rather than trying to maintain prevailing land-use practices that are unlikely to be sustained in the long term.

Furthermore, floods and droughts often continue to be treated as disconnected problems. Countries that deal more effectively with floods largely ignore droughts, and vice versa, even when there is a significant likelihood that the other extreme will occur (Huntjens *et al.*, 2011). A systemic approach that aims to build the resilience of the social–ecological system will overcome such a fragmented perspective, recognising that freshwater is a significant factor in the global life-support system and for human well-being and societal development. That is why the metaphor of water as the bloodstream of the planet is a central feature of this book, but a metaphor in itself is obviously not sufficient to achieve improved stewardship.

This book emphasises that water governance needs to make the interdependence of the social and the ecological explicit – not just in one place but across regions and globally – and to address it. It is a simple fact of the challenge of the Anthropocene era – a planet shaped by humanity. How can this be done in practice? A promising approach is multi-level governance that allows the emergence of local, bottom-up processes while framing them for global sustainability. The EU WFD has this potential, and it is reasonable to assume that many new umbrella institutions, cross-boundary and bridging organisations, and social–ecological networks will emerge from the community-of-practice across regions and globally. Two potential examples are transitions basins for dryland management and global networks of social–ecological basins.

It is also reasonable to assume that major transformations will surface at different scales – in line with the rapidly increasing understanding that human well-being and societal development urgently require a reconnection of people, institutions and policies with the biosphere – and account for the critical role of freshwater in this context. Such transformations will not always be smooth and there will be radically different points of departure for governing and managing change in different parts of the world.

More closed systems with extensive patronage networks require quite a different approach to developing countries with dysfunctional bureaucracies, but may be more conducive to building capacity from the bottom up and gaining support. Inertia in Western bureaucracies and mindsets needs to be overcome in order for such transformations to happen there. Nonetheless, a globalised world with new forms of connectivity in information flows, knowledge and exchange holds the potential for rapid redirections and new sustainable pathways for humanity. Multi-level, flexible governance systems that allow for learning and adaptive management of freshwater within and across dynamic landscapes will undoubtedly be needed to support such transformations.

Summary

New governance challenges

The terms management and governance are often mistakenly used synonymously. Governance sets the rules under which management operates. The lens of water management has long focused on blue water for economic and social development, largely ignoring links to and the implications for ecosystems. IWRM extended this focus to interacting sectors in catchments, but remains concerned with blue water. More recently, green water flows have come into focus, especially in relation to rainfed agriculture, and lately also to the trade-offs between bundles of ecosystem services. The water governance challenge is now entering new terrain, emphasising the need to manage the water cycle in a holistic manner. Sustaining rainfall implies building the social–ecological capacity to live with new water dynamics, such as changes in the variability of rainfall patterns. In spite of strong promotion, the adoption of the IWRM approach, based on the Dublin principles, has not yet led to major transformations in the governance or management of freshwater. Several barriers appear to have impeded progress, including institutional barriers and the fact that IWRM approaches do not recognise the contribution of green water. Spatial planning and IWRM remain largely disconnected

A new conceptual framework around the stewardship of precipitation

In an attempt to make the governance and management challenge of the wider role of freshwater in society more comprehensible, a new conceptual framework has been developed as a way to think about the stewardship of precipitation. What are the means for catching rain and rainfall pulses before they are lost to the landscape? Can a delicate balance be achieved between upstream and downstream needs? The possibility is envisaged of developing governance frameworks for hydro-unit-based land management under different land uses and with different rainwater partitioning characteristics. This would require explicit attention to upstream/downstream freshwater interactions and blue/green trade-offs.

Such a broader perspective would need new forms of multi-level governance structures that are adaptive and have multi-level, polycentric systems. The state towards which we need to manage should be determined by defining the desirable spatial and temporal patterns of the ecosystem services to be sustained at the river basin scale. Basin governance needs to build resilience and adaptive capacity to shocks and surprises, but it can do so only within a safe operating space of variability, which needs to be guaranteed by global and supra-basin agreements.

Signs of paradigm shifts

There are signs of paradigm shifts and of managing change. Innovative legislation has been implemented in many countries, adopting an integrated approach that includes the state of the environment as a management goal. One example is the EU WFD. Implementation, however, has not kept pace with the ambitious goals of such new legal frameworks. An illustrative example of the nature of a deep paradigm shift is provided by the changes observed in flood management, where the new concept of 'room for the river' requires the revision of land-use practices and a gradual move towards integrated landscape planning in which water is recognised as a natural structural element. Crises, real or perceived, have proved important for triggering insights, and opening windows of opportunity and windows for change.

Triple-loop learning

Studies of transformations of social–ecological systems into ecosystem-based management have generated a conceptual model based on the concept of triple-loop learning. Single-loop learning refers to an incremental improvement in action strategies without questioning underlying assumptions. Double-loop learning refers to a revisiting of assumptions within a value-normative framework. In triple-loop learning, values, beliefs and worldviews are reconsidered but a triple loop is required to get the system over the crest to a new state. The paradigm shift in water resources management has changed the discourse but not yet reached the transformation stage.

Encourage land-use practices guided by water availability

No evidence can yet be detected of a more profound movement in governance towards an entirely new paradigm, such as confronting planetary freshwater boundaries or governing precipitationsheds, sustaining rainfall and building social–ecological resilience in the stewardship of green and blue water flows and their associated ecosystem services. Such a paradigm would encourage land-use practices that are guided by water availability rather than trying to maintain prevailing land-use practices that cannot be sustained in the long run. Floods and droughts are generally still treated as disconnected problems.

As is emphasised throughout this book, governance of water as the bloodstream of the planet needs to make the interdependence of the social and the ecological explicit, not only in one place but across regions and globally – and to address this interdependence. Promising approaches include multi-level governance that allows for the emergence of local, bottom-up processes while framing them for global sustainability. Inertia in Western bureaucracies and mindsets needs to be overcome for this transformation to take place. A globalised world with new forms of connectivity in information flows, knowledge and exchange holds the potential for rapid redirections and new, sustainable pathways for humanity.

References

African Ministers' Council on Water (2012). *Status report on the application of integrated approaches to water resources management in Africa.* African Ministers' Council on Water, Abuja, Nigeria. Available at: http://www.amcow-online.org/index.php?option=com_content&view=article&id=262&Itemid=141&lang=en.

Albiac, J., Hanemann, M., Calatrava, J. and Uche, J. (2006). The rise and fall of the Ebro water transfer. *Natural Resources Journal*, **46**, 727.

Alcamo, J. M., Vörösmarty, C. J., Naiman, R. J. *et al.* (2008). A grand challenge for freshwater research: understanding the global water system. *Environmental Research Letters*, **3**, 010202.

Berkes, F. and Folke, C. (1998). *Linking Social and Ecological Systems: Management Practices and Social Mechanisms for Building Resilience.* Cambridge: Cambridge University Press.

Biswas, A. K. (2004). Integrated water resources management: a reassessment. *Water International*, **29**, 248–256.

Bundesministerium für Umwelt, NuR (2010). *Water Framework Directive: the way towards healthy waters.* Bundesministerium für Umwelt, Naturschutz und Reaktorsicherheit, Berlin. Available at: http://www.uba.de/uba-info-medien-e/4021.html.

Chapin, F. S., Carpenter, S. R., Kofinas, G. P. *et al.* (2010). Ecosystem stewardship: sustainability strategies for a rapidly changing planet. *Trends in Ecology and Evolution*, **25**, 241–249.

Clark, W. C. and Dickson, N. M. (2003). Sustainability science: the emerging research program. *Proceedings of the National Academy of Sciences*, **100**, 8059–8061.

Cooley, H. and Gleick, P. H. (2011). Climate-proofing transboundary water agreements. *Hydrological Sciences Journal*, **56**, 711–718.

de Coning, C. (2006). Overview of the water policy process in South Africa. *Water Policy*, **8**, 505–528.

Dietz, T., Ostrom, E. and Stern, P. C. (2003). The struggle to govern the commons. *Science*, **302**, 1907–1912.

Enfors, E. (2013). Social-ecological traps and transformations in dryland agro-ecosystems: using water system innovations to change the trajectory of development. *Global Environmental Change*, **23**, 51–60.

Falkenmark, M. and Folke, C. (2010). Ecohydrosolidarity: a new ethics for stewardship of value-adding rainfall. In *Water Ethics: Foundational Readings for Students and Professionals*, ed. Brown, P. G. & Schmidt, J. J. Washington DC: Island Press, pp. 247–264.

Falkenmark, M. and Rockström, J. (2006). The new blue and green water paradigm: breaking new ground for water resources planning and management. *Journal of Water Resources Planning and Management-Asce*, **132**, 129–132.

Folke, C., Carpenter, S. R., Walker, B. *et al.* (2010). Resilience thinking: integrating resilience, adaptability and transformability. *Ecology and Society*, **15**, 20.

Folke, C., Chapin III, F. S. and Olsson, P. (2009). Transformations in ecosystem stewardship. In *Principles of Ecosystem Stewardship: Resilience-based Natural Resource Management in a Changing World*, ed. Chapin, T., Kofinas, G. P. & Folke, C. New York: Springer, pp. 103–125.

Folke, C., Hahn, T., Olsson, P. and Norberg, J. (2005). Adaptive governance of social-ecological systems. *Annual Review of Environmental Resources*, **30**, 441–473.

Folke, C., Jansson, Å., Rockström, J. *et al.* (2011). Reconnecting to the biosphere. *AMBIO: A Journal of the Human Environment*, **40**, 719–738.

Fowler, R. and Rockström, J. (2001). Conservation tillage for sustainable agriculture: an agrarian revolution gathers momentum in Africa. *Soil and Tillage Research*, **61**, 93–108.

Galaz, V., Biermann, F., Crona, B. *et al.* (2012). 'Planetary boundaries' – exploring the challenges for global environmental governance. *Current Opinion in Environmental Sustainability*, **4**, 80–87.

Galaz, V., Olsson, P., Hahn, T., Folke, C. and Svedin, U. (2008). The problem of fit between ecosystems and governance systems: insights and emerging challenges. In *The Institutional Dimensions of Global Environmental Change: Principal Findings and Future Directions*, ed. Young, O., King, L. A. & Schroeder, H. Boston: MIT Press, pp. 147–186.

Geels, F. W. and Schot, J. (2007). Typology of sociotechnical transition pathways. *Research Policy*, **36**, 399–417.

Gelcich, S., Hughes, T. P., Olsson, P. *et al.* (2010). Navigating transformations in governance of Chilean marine coastal resources. *Proceedings of the National Academy of Sciences*, **107**, 16794–16799.

Gleick, P. H. (2003). Global freshwater resources: soft-path solutions for the 21st century. *Science*, **302**, 1524–1528.

Global Water Partnership (2000). Integrated water resources management. TAC background paper: 4. Global Water Partnership, Stockholm.

Gordon, L. J., Peterson, G. D. and Bennett, E. M. (2008). Agricultural modifications of hydrological flows create ecological surprises. *Trends in Ecology and Evolution*, **23**, 211–219.

Gunderson, L. H., Light, S. S. and Holling, C. (1995). Lessons from the Everglades. *BioScience*, **45**, 66–73.

Hahn, T. (2011). Self-organized governance networks for ecosystem management: who is accountable? *Ecology and Society*, **16**, 18.

Hahn, T., Olsson, P., Folke, C. and Johansson, K. (2006). Trust-building, knowledge generation and organizational innovations: the role of a bridging organization for adaptive comanagement of a wetland landscape around Kristianstad, Sweden. *Human Ecology*, **34**, 573–592.

Herrfahrdt-Pähle, E. (2012). Integrated and adaptive governance of water resources: the case of South Africa. *Regional Environmental Change*, **13**, 551–561.

Holling, C. S. and Meffe, G. K. (1996). Command and control and the pathology of natural resource management. *Conservation Biology*, **10**, 328–337.

Huntjens, P., Pahl-Wostl, C., Rihoux, B. *et al.* (2011). Adaptive water management and policy learning in a changing climate: a formal comparative analysis of eight water management regimes in Europe, Africa and Asia. *Environmental Policy and Governance*, **21**, 145–163.

Jeffrey, P. and Gearey, M. (2006). Integrated water resources management: lost on the road from ambition to realisation? *Water Science and Technology*, **53**, 1–8.

Joshi, P., Jha, A., Wani, S., Sreedevi, T. and Shaheen, F. (2008). Impact of watershed program and conditions for success: a meta-analysis approach. Global Theme on Agroecosystems Report: 46. International Crops Research Institute for the Semi-Arid Tropics, Patancheru, India. Available at: http://impact.cgiar.org/impact-watershed-program-and-conditions-success-meta-analysis-approach.

Kabat, P., van Schaik, H., Appleton, B. and Veraart, J. (2003). Climate changes the water rules: how water managers can cope with today's climate variability and tomorrow's climate change. Dialogue on Water and Climate, Delft, the Netherlands. Available at: http://www.unwater.org/downloads/changes.pdf.

Keys, P., Barron, J. and Lannerstad, M. (2012a). *Releasing the pressure: water resource efficiencies and gains for ecosystem services.* United Nations Environment Programme and Stockholm Environment Institute.

Keys, P. W., van der Ent, R. J., Gordon, L. J. *et al.* (2012b). Analyzing precipitationsheds to understand the vulnerability of rainfall dependent regions. *Biogeosciences*, **9**, 733–746.

Lam, W. F. (1994). Lessons from the experience of irrigation governance in Nepal: institutions, collective action, and public policy. In *Institutions, Incentives and Irrigation in Nepal*, ed. Benjamin, P., Lam, W. F., Ostrom, E. and Shivakoti, G. Washington DC: U.S. Agency for International Development, pp. 102–116.

Lambin, E. F., Turner, B. L., Geist, H. J. *et al.* (2001). The causes of land-use and land-cover change: moving beyond the myths. *Global Environmental Change*, **11**, 261–269.

Lansing, J. S. (1991). *Priests and Programmers: Technologies of Power in the Engineered Landscape of Bali.* Princeton, NJ: Princeton University Press.

Lundberg, J. and Moberg, F. (2003). Mobile link organisms and ecosystem functioning: implications for ecosystem resilience and management. *Ecosystems*, **6**, 87–98.

McSweeney, K. and Coomes, O. T. (2011). Climate-related disaster opens a window of opportunity for rural poor in northeastern Honduras. *Proceedings of the National Academy of Sciences*, **108**, 5203–5208.

Millennium Ecosystem Assessment. (2005). *Ecosystems and Human Well-being: Synthesis.* Washington DC: Island Press.

Milly, P., Julio, B., Malin, F. *et al.* (2008). Stationarity is dead. *Science, American Association for the Advancement of Science*, **319**, 573–574.

Mollinga, P. P. (2010). Hot water after the cold war: water policy dynamics in (semi-) authoritarian states. *Water Alternatives*, **3**, 512–520.

Moss, T. (2006). Solving problems of 'fit' at the expense of problems of 'interplay'? The spatial reorganisation of water management following the EU water framework directive. In *Integrated Water Resources Management: Global Theory, Emerging Practice, and Local Needs*, ed. Mollinga, P. P., Dixit, A. and Athukorala, K. Thousand Oaks: Sage Publications, pp. 64–108.

North, D. C. (1990). *Institutions, Institutional Change and Economic Performance.* Cambridge: Cambridge University Press.

Olsson, P., Folke, C. and Hahn, T. (2004). Social-ecological transformation for ecosystem management: the development of adaptive co-management of a wetland landscape in southern Sweden. *Ecology and Society*, **9**, 2.

Olsson, P., Folke, C. and Hughes, T. P. (2008). Navigating the transition to ecosystem-based management of the Great Barrier Reef, Australia. *Proceedings of the National Academy of Sciences*, **105**, 9489–9494.

Olsson, P., Gunderson, L. H., Carpenter, S. R. *et al.* (2006). Shooting the rapids: navigating transitions to adaptive governance of social-ecological systems. *Ecology and Society*, **11**, 18.

Ostrom, E. (1990). *Governing the Commons: The Evolution of Institutions for Collective Action.* Cambridge: Cambridge University Press.

Ostrom, E. (2010). Polycentric systems for coping with collective action and global environmental change. *Global Environmental Change*, **20**, 550–557.

Pahl-Wostl, C. (2006). The importance of social learning in restoring the multifunctionality of rivers and floodplains. *Ecology and Society*, **11**, 10.

Pahl-Wostl, C. (2007). Transitions towards adaptive management of water facing climate and global change. *Water Resources Management*, **21**, 49–62.

Pahl-Wostl, C. (2009). A conceptual framework for analysing adaptive capacity and multi-level learning processes in resource governance regimes. *Global Environmental Change*, **19**, 354–365.

Pahl-Wostl, C., Becker, G., Sendzimir, J. and Knieper, C. (in press). From flood protection to integrated flood management: a multi-level societal learning process towards sustainability.

Pahl-Wostl, C., Gupta, J. and Petry, D. (2008). Governance and the global water system: a theoretical exploration. *Global Governance: A Review of Multilateralism and International Organizations*, **14**, 419–435.

Pahl-Wostl, C., Jeffrey, P., Isendahl, N. and Brugnach, M. (2011). Maturing the new water management paradigm: progressing from aspiration to practice. *Water Resources Management*, **25**, 837–856.

Pahl-Wostl, C., Lebel, L., Knieper, C. and Nikitina, E. (2012). From applying panaceas to mastering complexity: toward adaptive water governance in river basins. *Environmental Science & Policy*, **23**, 24–34.

Polasky, S., Carpenter, S. R., Folke, C. and Keeler, B. (2011). Decision-making under great uncertainty: environmental management in an era of global change. *Trends in Ecology and Evolution*, **26**, 398–404.

Power, M. E. and Chapin III, F. S. (2009). Planetary stewardship. *Frontiers in Ecology and the Environment*, **7**, 399.

Reid, W. V., Chen, D., Goldfarb, L. *et al.* (2010). Earth system science for global sustainability: grand challenges. *Science*, **330**, 916–917.

Rockström, J., Falkenmark, M., Karlberg, L. *et al.* (2009a). Future water availability for global food production: the potential of green water for increasing resilience to global change. *Water Resources Research*, **45**.

Rockström, J., Gordon, L., Folke, C., Falkenmark, M. and Engwall, M. (1999). Linkages among water vapor flows, food production, and terrestrial ecosystem services. *Conservation Ecology*, **3**, 5.

Rockström, J., Steffen, W., Noone, K. *et al.* (2009b). A safe operating space for humanity. *Nature*, **461**, 472–475.

Sadoff, C. W. and Grey, D. (2005). Cooperation on international rivers: a continuum for securing and sharing benefits. *Water International*, **30**, 420–427.

Scheffer, M., Bascompte, J., Brock, W. A. *et al.* (2009). Early-warning signals for critical transitions. *Nature*, **461**, 53–59.

Schlüter, M. and Herrfahrdt-Pähle, E. (2011). Exploring resilience and transformability of a river basin in the face of socioeconomic and ecological crisis: an example from the Amudarya River basin, Central Asia. *Ecology and Society*, **16**, 32.

Schlüter, M., Hirsch, D. and Pahl-Wostl, C. (2010). Coping with change: responses of the Uzbek water management regime to socio-economic transition and global change. *Environmental Science & Policy*, **13**, 620–636.

Schlüter, M., Rüger, N., Savitsky, A. G. *et al.* (2006). TUGAI: an integrated simulation tool for ecological assessment of alternative water management strategies in a degraded river delta. *Environmental Management*, **38**, 638–653.

Schreiner, B. and Hassan, R. (eds) (2010). *Transforming Water Management in South Africa: Designing and Implementing a New Policy Framework*. Heidelberg, Germany: Springer.

Sehring, J. (2009). Path dependencies and institutional bricolage in post-Soviet water governance. *Water Alternatives*, **2**, 61–81.

Sendzimir, J., Magnuszewski, P., Flachner, Z. *et al.* (2008). Assessing the resilience of a river management regime: informal learning in a shadow network in the Tisza River Basin. *Ecology and Society*, **13**, 11.

Smith, D. M. and Barchiesi, S. (2009). Environment as infrastructure: resilience to climate change impacts on water through investments in nature. Perspective Document for the 5th World Water Forum. World Water Council, Istanbul.

Steffen, W., Crutzen, P. J. and McNeill, J. R. (2007). The Anthropocene: are humans now overwhelming the great forces of nature? *Ambio*, **36**, 614–621.

Steffen, W., Persson, A., Deutsch, L. *et al.* (2011). The Anthropocene: from global change to planetary stewardship. *Ambio*, **40**, 739–761.

Tompkins, E. L. and Hurlston, L.-A. (2012). Public-private partnership in the provision of environmental governance: a case of disaster management. In *Adapting Institutions: Governance, Complexity and Social-ecological Resilience*, ed. Boyd, E. F. Cambridge: Cambridge University Press, pp. 171–190.

UN-Water (2012). Status report on the application of integrated approaches to water resources management. United Nations Environment Programme, Nairobi. Available at: http://www.unwater.org/rio2012/report/press-release-19june.html.

Vörösmarty, C. J., McIntyre, P., Gessner, M. O. *et al.* (2010). Global threats to human water security and river biodiversity. *Nature*, **467**, 555–561.

Walker, B., Barrett, S., Polasky, S. *et al.* (2009a). Looming global-scale failures and missing institutions. *Science*, **325**, 1345–1346.

Walker, B. and Meyers, J. A. (2004). Thresholds in ecological and social-ecological systems: a developing database. *Ecology and Society*, **9**, 3.

Walker, B. H., Abel, N., Anderies, J. M. and Ryan, P. (2009b). Resilience, adaptability, and transformability in the Goulburn–Broken catchment, Australia. *Ecology and Society*, **14**.

Wani, S. P., Rockström, J. and Sahrawat, K. L. (eds) (2011). *Integrated Watershed Management in Rainfed Agriculture*. Boca Raton, FL: CRC Press.

Westley, F. (1995). Governing design: the management of social systems and ecosystems management. In *Barriers and Bridges to the Renewal of Ecosystems and Institutions*, ed. Gunderson, L., Holling, C. S. and Light, S. New York: Columbia University Press, pp. 391–427.

Westley, F. (2001). The devil in the dynamics: adaptive management on the front lines. In *Panarchy: Understanding Transformations in Human and Natural Systems*, ed. Gunderson, L. H. and Holling, C. S. Washington DC: Island Press, p. 508.

Westley, F., Olsson, P., Folke, C. *et al.* (2011). Tipping toward sustainability: emerging pathways of transformation. *Ambio*, **40**, 762–780.

Westley, F., Tjörnbo, O., Schultz, L. *et al.* (in press). A theory of transformative agency in linked social-ecological systems. *Ecology and Society*.

Xia, C. and Pahl-Wostl, C. (2012). The process of innovation during transition to a water saving society in China. *Water Policy*, **14**, 447–469.

Young, O. R., King, L. A. and Schroeder, H. (2008). *Institutions and Environmental Change: Principal Findings, Applications, and Research Frontiers*. Cambridge, MA: MIT Press.

Forest view, in the Danum Valley rainforest of Sabah, Malaysia. This lowland rainforest area is home to a wide variety of birds and mammals, including orangutans, Sumatran rhino and the rare Borneo pygmy elephant.

Pathways to the future

This chapter summarises the role of water in achieving global sustainability and human prosperity on an increasingly crowded planet - characterised by rising interdependence, turbulence, social–ecological interactions and uncertainty. Of particular concern is the increased likelihood of human-induced water-related tipping points in local and regional environments and even in the Earth System as a whole. These often surprising events are reflections of the loss of social–ecological resilience for dealing with change, and could have severe implications for societies and human well-being. The chapter addresses the shifts in governance and management cultures required of strategies for water resilience, and stresses that freshwater is the key to resilience in social–ecological systems. Maintaining redundancy in landscapes through a high degree of biodiversity and a rich mosaic of different land-use types is a key strategy for building resilience and sustaining rainfall and the 'wetness' of landscapes.

9.1 Bloodstream management

9.1.1 Key messages and building blocks

This chapter addresses these challenges in integrating the different streams of analysis contained in this volume - from the social–ecological pressures on the planet to the IWRM practices available to communities across the world to improve land and water management. It attempts to weave together a 'waterway forward' for integrated land and water resource management in the new Anthropocene era. It does so within the framework of the two key messages of the book. First, that human development and wealth originates from social–ecological systems, which all require freshwater for their ability to generate ecosystem functions and services, and to provide social–ecological resilience. From this follows that misuse of freshwater resources challenges not only human well-being, but also social–ecological resilience and the capacity of societies to deal with change, persist and continue to develop.

These messages further develop our understanding of the role of water in human prosperity. They constitute the basic analytical building blocks of a widened paradigm of integrated land and water resource management that addresses three critical dimensions:

1. The need for an integrated systems perspective, with an emphasis on green and blue water resources for land and water resource management.
2. The role of water in ecosystem services and social–ecological resilience, which provides landscapes with the ability to generate water-related wealth for human communities through the supply of freshwater, food production, and the provision of diverse landscapes and a stable environment.
3. The cross-scale water interlinkages from the catchment to the biosphere in which water plays a dual function - on the one hand affected by the impacts of global changes such as shifts in rainfall patterns due to climate change, while on the other hand acting as a driving force which directs the outcome of future change, such as the drying out of rainforests triggering a loss of habitats.

We have defined this push for IWRM as a strategy of *enhancing water resilience for human prosperity and global sustainability*. Water resilience implies stewardship of freshwater to persist and adapt within a desired state of dynamic landscapes (in simple terms 'stay where you want to be' when the world around you is changing), to strategies and capacities to transform water management, in situations of crises, in order to enable transitions toward new trajectories

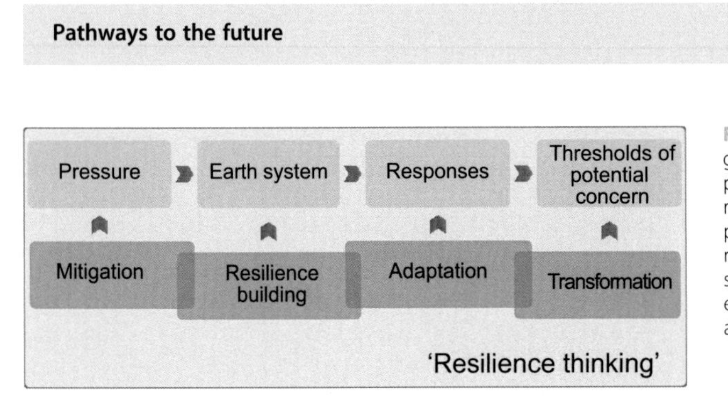

Figure 9.1 The interconnected challenges facing global water resource management, from human pressures to impacts on the Earth System, influence responses from societies and potential regime shifts of particular concern in social–ecological systems. The range of actions on water resilience for global sustainability along this change continuum encompasses mitigation, local resilience building, adaptation and transformation.

of development for human well-being and resilience. At its heart, this requires a widening of our water resource focus from managing water supply, or what is basically runoff, to *managing and sustaining rainfall*. Thus far, the source of water, i.e. precipitation, has largely been seen as a given static factor. In the Anthropocene era, we must get used to a new water reality in that the way we influence the climate system and the way we manage the biosphere (e.g. forest and land management) – in which water plays a key role – in one time frame has implications for the availability of rainfall in the next time frame. In a schematic way, governing and managing water for resilience encompasses a range of actions from mitigation to resilience building, adaptation and transformations (see Figure 9.1).

The sections below summarise insights from Chapters 1–8, followed by a reflection of the implications of a wider water resilience approach to the governance and management of water resources within the context of human development and global sustainability.

9.2 A new global water situation in the Anthropocene era

As is argued throughout this book, humanity's water predicament has shifted to a significantly new situation. Until recently, the emphasis on IWRM has focused on the management and distribution of a predefined, albeit variable, water resource. The input of water – as rainfall, groundwater or surface runoff – could conveniently be assumed to be an external state variable determined by natural forces beyond human influence.

The acceleration of global environmental change has effectively forced us to shift this perspective. Over the past 5–10 years, overwhelming scientific evidence of human pressures on Planet Earth conclusively shows that the modern world is hitting, or starting to hit, the ceiling of many critical Earth System processes. The aggregate impacts of 7 billion people, multiplied by the environmental implications of modern human enterprise, are generating unexpected dynamics at the scales above the local scale of activity. The management of large basins such as Lake Chad and planning for agriculture in regions such as Southern Africa are increasingly influenced not only by how well IWRM is implemented locally, but also by how regional climate systems are affected by regional and global environmental changes, and how these in turn interact with local to regional land-use change (e.g. affecting moisture feedback from forests, a key source of rainfall), and how these interactions in turn affect monsoon systems and thus rainfall patterns for farmers, urban dwellers and businesses on the ground.

This book describes this escalating human pressure on Planet Earth as a 'quadruple squeeze' involving four parallel processes: (1) demographic pressures from a population that will approach 9 billion by 2050 as well as increasing affluence, linked to growing demand for food, water, energy and other ecosystem services; (2) the anthropogenic climate crisis, which is already on its way to warming the planet beyond 2°C (as compared to pre-industrial temperatures); (3) the human-induced environmental crisis, which over the past 50 years has undermined ecosystem services faster than at any point in human history; and (4) ecosystem and biosphere stress and risks of abrupt – often irreversible – changes, due to loss of resilience, which reduce our operating space on Earth (cf. Chapter 1). Scientific evidence indicates that this quadruple squeeze has become so prominent that we have entered a new geological epoch, the Anthropocene, in which humanity constitutes a geological force for change at the planetary scale. This is a new insight; the first scientific synthesis of the evidence for an Anthropocene era appeared in 2004 (Steffen *et al.*, 2004). It is also an insight that has profound implications for water and human development.

The water cycle has been altered by each of the global squeeze processes: through shifts in rainfall patterns due to climate change, changes in rainfall partitioning in landscapes due to ecosystem change and overextraction of natural resources, and by regime shifts, which cause abrupt shifts in biosphere conduits of water, be it through collapse of wetlands shifting flow of blue water, or abrupt shifts in green water flows if forest systems flip over to savannah systems. At the same time, water is a determinant of the degree of stability or instability in each squeeze process, supplying water to society for food production, sanitation and other uses, enabling carbon sinks and regulating the climate, generating habitats and producing biomass, and regulating resilience in landscapes. Water is thus both a victim (often a first victim) of global environmental change, and a driver of change and a regulator of Earth resilience; in short, water truly operates as the bloodstream of the biosphere.

9.2.1 Three reasons for heightened concern about water

A set of overarching reasons for deeper concern about water in the Anthropocene era provides the basis for advancing a new framework for water governance and management. The first reason is the pace and scale of the impact. The notion of the Anthropocene era is based on the recent exponential rise in human pressures on Earth System processes, which began in the mid-1950s but over the past 10 years has generated regional to global impacts such as the accelerated melting of the world's ice covered water towers – like the Himalayan glaciers – and growing evidence of extreme water-related weather events. These exponential pressures span essentially all the ecosystem and environmental processes that matter for human well-being – from land-use change, to greenhouse-gas emissions to the dominance of and reliance on nutrient cycles. A significant insight from Chapters 2 and 4 is that water interacts with all these processes, while itself being put under unprecedented pressure – blue water withdrawals, for instance, have been rising exponentially from approximately 1500 km^3/year in the 1950s to 4000 km^3/year in 2010 (Oki and Kanae, 2006; Shiklomanov and Rodda, 2004), with current consumptive use amounting to some 2600 km^3/year. At the same time green water use from cropland amounts to some 7000 km^3/year and 15 000 km^3/year from permanent grazing land, together amounting to about one-third of total terrestrial green water flows (Oki and Kanae, 2006; Rost *et al.*, 2009), of which the rest, moreover, is committed to generate key ecosystem services in the biosphere (Rockström *et al.*, 1999). The first key insight, therefore, is that all human pressures are interconnected with water as the bloodstream of the biosphere, and a related concern about the capacity of the biosphere to sustain societal development. The continued increase in human demands on natural resources and ecological space is manifested in growing pressures on finite freshwater resources (Chapter 3). In the next 40 years, the population of the world is projected to rise from 7 billion to approximately 9 billion people (UN DESA, 2011), increasing the need for water to produce food by a staggering 2300–4100 km^3/year (based on UN medium population projections and the range depending on low or high water productivity assumptions) (Rockström *et al.*, 2009a). Even the most optimistic outlook (of 2300 km^3/year) amounts to approximately the current annual consumptive water use in irrigation globally (2600 km^3/year).

The second key reason for concern follows on the first. Water not only interacts with the exponential rise in pressures on climate, land, chemicals, air and so on, but is also to a very large extent the 'first victim' of change. As Charlotte de Fraiture, at the time a scientist at the International Water Management Institute (IWMI), pertinently pointed out at World Water Week in Stockholm in 2005: 'if climate mitigation is about gases, then climate adaptation is about water'. There is much truth in this, and not only for climate change. Global environmental changes – be they changes in global mean temperature or linked to deforestation and land degradation – all first and foremost have a direct impact on freshwater flows, with immediate impacts on human well-being. This impact places a large burden on human livelihoods, generally through constraints on producing food and sustaining health. Climate change has many implications related to changes in temperature, such as the displacement of species and the proliferation of diseases. Perhaps the often most immediate effect is on regional climate systems and weather patterns, triggering droughts, floods and dry spells, causing shifts in rainfall patterns and inducing shifts in sea levels (IPCC, 2012). Water is therefore, at the local scale, a victim of the Anthropocene era.

The third reason for increased water concern in the Anthropocene era is the risk of crossing tipping points when transgressing planetary boundaries (as analysed in Chapters 2 and 4). Water together with land provides ecosystem services and supports social–ecological resilience in landscapes and river basins, i.e. the capacity to not only deliver food and other ecosystem services for human societies, but also sustain development in a desired state in the face of change. Increasingly, scientific evidence shows that large biomes – such as rainforests, savannahs, grasslands and temperate forests – are at risk of abruptly shifting into less productive states if pushed too far by changes in climate, deforestation or loss of biodiversity.

Water plays a dual role in such processes. Water is a control variable for resilience, in the sense that it contributes to building resilience by generating biomass, sustaining biodiversity and organic matter, and regulating temperatures. It is also a state variable affected by sudden changes. For example, deforestation in a rainforest could, due to loss of resilience in combination with a trigger in the form of a series of extreme droughts due to climate change, abruptly shift into a savannah state, manifested as reduced moisture flows in a drier, more open forest ecosystem state with high air turbulence. This new state is a drier state, subject to a new 'water logic', where moisture feedback and humidity levels are lower and where water availability falls. Reversing the process, by shifting from a savannah to a forest state, may be very difficult indeed, if not impossible.

Therefore, in the Anthropocene era, three new factors must be taken into account in local and regional water resource planning and management:

- There are exponential pressures from human enterprise on multiple environmental processes, ranging from climate change to land and forest degradation and biodiversity loss, all of which influence water resources at the local to regional scale.
- As well as being a driver of environmental change, water is also the first victim of such change and thus strongly associated with extreme shocks (such as sudden rises in food prices) with large-scale social impacts, which requires particular attention in terms of investment in adaptation and resilience building.
- Social and environmental changes are occurring more and more rapidly on a regional to global scale and there is a growing risk of crossing major tipping points in the biosphere, with possibly abrupt and irreversible impacts on water resources at the local to regional scales.

9.2.2 The new global water dynamics

These are critical factors in the new context in which water resources will be governed and managed in the coming decades. A critical aspect of these global dynamics is that the social magnitude of change, and its increased pressures on available freshwater resources, is on a par with continuing global environmental change. In no sector is the rapidly changing water situation as clear as it is for agriculture.

Agricultural development has benefited from a harmonious water cycle for the past 10 000 years. We had barely entered the warm, interglacial Holocene period – an extremely stable geological epoch with a global mean temperature that oscillated by ±1°C, when we invented agriculture. This was the key to the transformation from semi-nomadic hunter-gatherer societies to densely populated sedentary civilisations – from the Mesopotamian irrigation society, to the Maya, Inca, Egyptian, Greek and Roman empires, and today's modern economies. Agriculture is the backbone of modern civilisations and stable access to freshwater is one of the key preconditions for agricultural development (Falkenmark, 1989). The industrial revolution has propelled much of the world into a new, industrialised state of human well-being, especially since the 1950s. It was built on fossil fuels releasing immense amounts of cheap energy for industrial and other processes, combined with breakthroughs in agriculture, and the development of commercial fertilisers based on the Haber–Bosch process of transforming inert N_2 from the atmosphere to reactive nitrogen for crops. This was the time of the 'great acceleration of the human enterprise' (Steffen *et al.*, 2007), where the exponential increase in the human pressures on the Earth System began (Figure 1.2, Chapter 1) and the human population rapidly increased from 3 billion to 7 billion today.

During this period of acceleration, global withdrawals of blue water, predominantly for irrigated agriculture, grew by a factor of six – almost twice as fast as population growth (Chapter 1). This extraordinary development, resulting in an almost triple increase in world food production, was made possible by the predictability and stability of global and regional hydrological cycles, which in turn remained in equilibrium thanks to the resilience of the Earth System. This was determined by the configuration of the biosphere – that is, a stable distribution of forests, grasslands, wetlands, peatlands, as well as the biodiversity hosted in these biomes – and in the oceans, atmosphere and cryosphere – and the stability of the Earth's glaciers and large ice sheets.

The rapid modernisation of water resource development across the world (over 5000 km^3 of water is stored behind dams each year) was made possible by the stable water conditions in the Holocene period. Not surprisingly, this predictable state of freshwater on the planet influenced the evolution of our approach to governing and managing water resources – the evolution of the modern approach to IWRM – which is based on the assumption that freshwater availability as rainfall supply or runoff distribution is a predetermined state variable that we cannot and do not influence. Only in the past 10–20 years has this changed. We are now leaving the Holocene period, having transgressed the estimated planetary boundary for climate in the early 1990s, and have potentially pushed the Earth System beyond its stable range of environmental conditions over the past 10 000 years. We are entering the Anthropocene epoch, an era of changing water resource conditions due to rapid global environmental change.

Moreover, this transition is occurring at a critical social juncture. We have pushed water resource development to extremes over the past 100 years. Today, several of the world's large rivers no longer reach the ocean due to basin closure (Molden *et al.*, 2007). Despite tremendous engineering feats, over 3 billion people still suffer various degrees of water scarcity (Kummu *et al.*, 2010). There is already a significant degree of intensified freshwater extraction and use in many places in the world, causing major social problems due to the unsustainable overuse of water, and reflected, for example, in falling groundwater tables in rural India and the drying of rivers in Australia and the United States.

We have been successful in favourable Holocene water conditions. In the Anthropocene era, we will need to address growing social needs and urgent sustainability criteria in less favourable water conditions and increasingly turbulent times. This is the core reason why water resource management must now move from the realm of efficiency and optimisation to also include strategies for resilience building for a fluctuating and less predictable water future where surprise rather than reliability is the norm, and for integration to address scale interactions as well as interactions between water, the biosphere and human activities.

9.2.3 The risk of crossing thresholds of potential concern

Water resource constraints due to growing human pressures on limited freshwater resources have been, and will rightly continue to be, a core concern among water resource professionals and for the water resource agenda at large. In the Anthropocene era, the 'traditional' focus on the regional and global 'water crunch' will be complemented by the risk of crossing thresholds in the Earth System, which may cause abrupt, often irreversible, changes with significant, potentially deleterious, implications for human societies. As is shown in Chapters 2 and 4 in particular, there is growing scientific evidence of the risks of crossing thresholds, triggering abrupt regime shifts in the Earth System, due to climate change (Lenton *et al.*, 2008; National Academy of Sciences of United States of America, 2009) and to global loss of biodiversity (Barnosky *et al.*, 2012), to an overload of nutrients triggering tipping points in lakes (Carpenter *et al.*, 2011), to water-induced regime shifts in agricultural systems (Gordon *et al.*, 2008). Despite the growing evidence base of the risk of abrupt, non-linear changes in social–ecological systems due to social and environmental interactions and reinforcing feedbacks, there is limited knowledge of the risk to water-induced thresholds.

To assess the risk of water-induced thresholds requires an analysis of water partitioning points in the global hydrological cycle, in order to discern possible long-term risks of water-related abrupt shifts. Many types of land use and water use involve slow changes in both the partitioning and merging points in the water cycle, where water fluxes either divert or meet (Figure 9.2). *Land-use* activities may involve alterations to rainwater infiltration in recharge areas at both the upper and the lower partitioning points. They may also involve alterations to the vapour-flux system in areas where vapour flow from oceans meets green water flow from land. When it comes to *water use*, slow changes may occur when consumptive water use, i.e. a vapour-flux component, is shifted from river flow or groundwater flow by water outtake – especially for irrigation (Figure 9.2). Several of the environmental changes

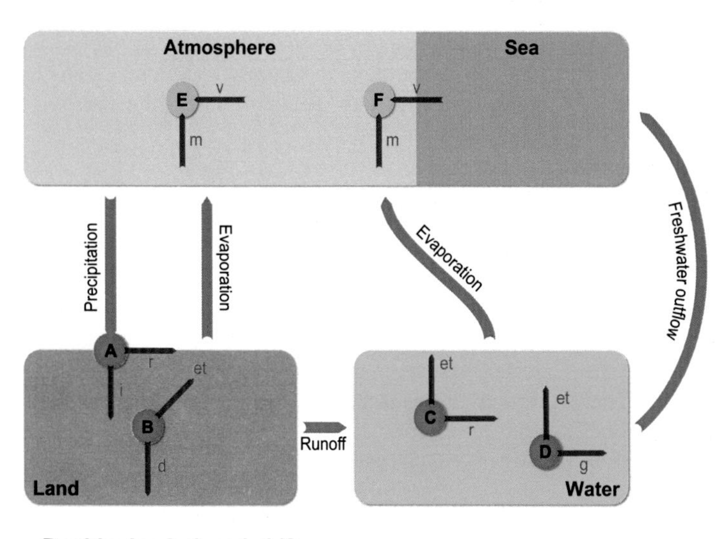

Partitioning induced shifts

Partitioning points

A	**Desertification**	Between surface runoff (r) and infiltration (i)
B	**Salinisation**	Between consumptive use (et) and deep percolation (d)
C	**River depletion**	Between consumptive use (et) and surface flow (r)
D	**Groundwater depletion**	Between consumptive use (et) and groundwater flow (g)

Merging points

E	**Savannisation**	Between vapour flow (v) and moisture feedback (m)
F	**Monsoon shift**	Between vapour inflow (v) and moisture feedback (m)

Figure 9.2 Six locations in the water cycle where slow changes may be ongoing in water partitioning and merging processes, and related ecohydrological changes in vulnerable regions. Partitioning induced shifts on land: (A) *land degradation/desertification* is linked to the upper partitioning point at the soil surface, reducing infiltration into the soil, (B) *salinisation* is linked to the lower partitioning point, where deforestation disturbs the partitioning between water uptake by the roots and percolation/groundwater recharge. Partitioning induced shifts in freshwater: (C) *river depletion* caused by an added diversion from the river flow by consumptive use of river water, mainly for irrigation, and (D) *groundwater depletion* linked to exploitative groundwater use, mainly for irrigation, that reduces groundwater outflow and river baseflow. Merging point for water flows inducing shifts due to atmospheric change: (E) *savannisation* is caused by deforestation, decreasing moisture feedback to the atmosphere, and therefore reduced feeding of the atmospheric vapour flow, and reflected in reduced regeneration of precipitation. Merging point for water flows inducing shifts due to ocean-land interactions: (F) *monsoon shift*, thought to be linked to the merging of marine vapour inflow and an accelerated terrestrial moisture feedback from large-scale irrigation systems.

resulting from these hydrological changes are well established, others are phenomena that have started to cause concern more recently.

Land use:

- A *land degradation/desertification* is linked to the upper partitioning point at the soil surface, reducing infiltration into the soil;
- B *salinisation* is linked to the lower partitioning point, where deforestation disturbs the partitioning between water uptake by roots and percolation/groundwater recharge;
- E *savannisation* is caused by deforestation, decreasing moisture feedback into the atmosphere, and therefore reduced feeding of the atmospheric vapour flow, reflected in reduced regeneration of precipitation;
- F *monsoon shift* is thought to be linked to the merging of marine vapour inflow and an accelerated terrestrial moisture feedback from large-scale irrigation systems.

Water use:

- D *groundwater depletion* is linked to overexploitation of groundwater for irrigation, involving a water diversion larger than the groundwater recharge;
- C *river depletion* is caused by an added diversion from the river flow by consumptive use of river water, mainly for irrigation.

(letters A–F refer to the merging/partitioning points in Figure 9.2)

All these ecohydrological changes manifest themselves as either *land productivity decline* or *water productivity decline* in hydro-climatically vulnerable areas to be associated with thresholds of potential concern. Both types interact and are of central importance to humanity's future due to their ability to undermine human life-support systems.

9.2.4 Facing up to the existence of tipping points and the risk of large-scale regime shifts

Understanding the risk of water-induced thresholds is at the heart of adding a resilience lens to water resource governance and management. It relates to

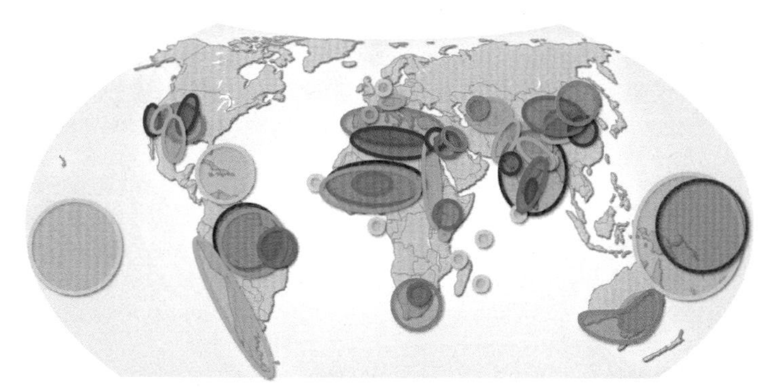

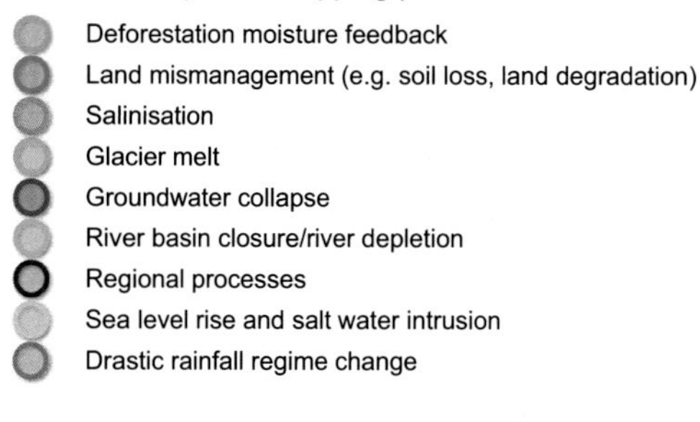

Figure 9.3 Aggregated global map of water-related TPCs, further elaborated in Chapter 3 (Figure 3.3).

the ability to analyse and act on the recognition of the existence of such thresholds, and the risks and opportunities related to crossing them – what we refer to as water-related thresholds of potential concern (Gordon *et al.*, 2008). This risk has profound implications for IWRM. In addition to the need to understand how social and environmental changes at higher scales affect the local management scale, e.g. how climate change affects catchment planning, it also adds several key priorities to the aim of management, from a general focus on productivity, efficiency and the optimisation of water resource allocation and use, to a stronger focus on assessing and dealing with risk and uncertainty, and investing in social and ecological redundancy to generate margins of security and insurance for uncertainty and surprise. There is also a focus on the spatial configuration of landscapes, such as heterogeneity, in order to have enough resilience to either avoid crossing unpleasant thresholds or shift and transform social–ecological systems and local and regional levels from undesirable conditions into more desirable states (Carpenter *et al.*, 2012; Folke *et al.*, 2010; Walker *et al.*, 2006).

Chapter 3 is a first attempt to map water-related thresholds of potential concern (TPCs) (Walker and Salt, 2006) for water resource governance and management in different regions of the world (see Figure 9.3 for an overview). These TPCs include the risk of large-scale regime shifts, or significant and rapid changes in how large systems such as the monsoon or savannahs operate, due to interactions among different drivers of change such as climate change, biodiversity loss and land-use change – generating shifts in water partitioning and water merging points (Figure 9.2). Figure 9.3 shows that several of the water-related TPCs originate from processes of regional to global change that are not directly related to water, such as the risk of shifts in the Indian monsoon due to climate change induced by emissions of greenhouse gases, and the Asian brown cloud induced by air pollution.

In addition to changes in the scale and pace of human pressures on all the components of the Earth System that affect water, a further critical change of relevance to water resource governance and management is growing interconnectedness across scales and systems. As is highlighted in Chapter 3, shifts in certain 'pivotal' biomes on Earth can generate domino effects on water resources in other parts of the world. One example of this is the potential loss of rainforests in Latin America and Central Africa, which could trigger warming in Southern Europe and Eastern China, as a result of water- and climate-related teleconnections in regional climate systems, thereby triggering shifts in rainfall patterns (Snyder *et al.*, 2004). Similarly, in regions such as China, Latin America and the Sahel, where rainfall is highly dependent on moisture feedback from terrestrial ecosystems, the management of forests in one region can shift rainfall in other regions (see Box 4.4). This suggests that nations and river basins are insufficient units for water governance and management, and that 'precipitationsheds' – the region providing the source for rainfall over a certain community, state or basin – may be a more appropriate unit of relevance (Keys *et al.*, 2012). Similarly, water management at any scale – be it the catchment or river basin – needs to actively connect with the global scale, recognising the growing evidence that rainfall in one region of the world is interconnected with the resilience of large biomes in other parts of the planet, such as a stable ice sheet in the Arctic, stable temperate forests in Eurasia and a stable Amazon rainforest.

9.2.5 A new ethics responding to water connectivity across scales

Furthermore, there is in the Anthropocene era a need to consider a new 'ethics of spatial water connectivity'. It is difficult to envisage how a world which is already under such profound water pressure due to overuse, with half the world's current population subject to water scarcity and rapidly rising water demands to feed 1 billion hungry and 2 billion new inhabitants by 2050, and where water conditions are rapidly worsening due to global environment change, will be able to provide water for food, health and well-being for all its people – especially if the 'unit of management' remains the nation state or the individual river basin.

There is an urgent need to acknowledge the fundamental role of 'virtual water' flows. They are already today very important, amounting to some 700 km^3/year, or approximately 25% of global consumptive blue water use for food production (Oki and Kanae, 2004). This percentage is set to increase in the future. Regions blessed with ample water resources and subject to less serious impacts of global environmental change will play an increasingly important role in providing food for regions with large deficits in their ability to provide water for food. Our estimates show that by 2050 approximately 5 billion people will be living in countries where – even with a 25% closure of the current yield gap – there will not be enough freshwater for current croplands to produce an adequate diet for all citizens of 2200 kcal/cap/day, including 5% meat-based foods. Of these, 3.5 billion might be able to afford to cover their food requirements using 'virtual water', but as many as 1.5 billion are likely to face a situation in which both water and money are constraining factors (Rockström *et al.*, 2012; see also Chapter 5).

In sum, the world faces a triple interlinked 'water problem': (1) rapidly changing water conditions in an increasingly turbulent and connected world where abrupt shifts in water availability cannot be excluded; (2) water-induced TPCs may become a reality; and (3) continuing human pressure on water demand in a world where local water resources are unlikely to be sufficient to support human well-being. Meeting this new Anthropocene water challenge will require collaborative efforts across scales, both below and above the traditional water governance scale of the river basin, and a strategic focus on complementing water productivity and efficiency with water resilience.

9.3 The need to reconnect with the biosphere

As is argued throughout this book, and is supported by several recent syntheses (Blue Planet Laureates, 2012; Folke and Rockström, 2011; Stockholm Memorandum, 2011), human pressures on the Earth System are socially and ecologically linked, generating accelerated changes in societies (Walker *et al.*, 2009).

9.3.1 Interacting social and environmental turbulence

One rapid social driver of change is that the majority of the world's population has started to move out of poverty, leading to the rapid rise of an affluent middle class with aspirations for material growth and increased incomes in a rapidly urbanising world. This is both a water and an equity challenge on a momentous scale. At the same time, the information technology, nanotechnology and molecular revolutions are accelerating with complex potentials.

The speed of connectivity and the feedbacks of globalisation are creating new and complex dynamics across levels and in domains such as finance, food, energy and disasters – with often surprising results (Folke *et al.*, 2011).

In addition, international institutions are becoming increasingly complex and fragmented through the evolution of a suite of public, private and hybrid forms of transnational collaboration (Andonova and Mitchell, 2010). This presents new governance challenges in relation to planetary stewardship and global sustainability (Biermann *et al.*, 2012).

One such example of growing and interacting social and environmental turbulence with strong links to water, demonstrating the world's new complexity, is the abrupt political upheavals across North Africa in 2011, collectively known as the Arab Spring. On one level, the revolutions that took place in Tunisia, Egypt and Libya, and the civil uprisings in Bahrain, Syria and Yemen, were social responses to decades of repression under infamous dictatorial regimes. These revolutions were, at least in part, enabled by the mass connectivity provided by widespread access to the internet and mobile telephones. On another level, these uprisings were influenced by a rapid spike in food prices across the region, which jumped by more than 100% in 2008, and then again in 2010. This rise in food prices may have been caused by the complex interactions of volatile oil prices, speculation on the food market and constraints on the world's grain markets, caused by weather-related events, linked to anthropogenic climate change, generating water scarcity. Critical among these weather-related events was the 2010 ban on grain exports in Russia, following crop losses due to an unprecedented heat wave and a rash of wildfires (Rimas and Fraser, 2010). Biofuel policies in the United States and Europe may also have contributed by increasing demand for crops that could otherwise have been used for food (Lagi *et al.*, 2012).

In addition to such dynamics, we are witnessing a more volatile global hydrological cycle, generating a higher frequency and amplitude of water shocks which occur in a world that is increasingly interconnected, with socio-political factors that interact with resource constraints and ecological changes to propagate disruptions across the world in unexpected ways (Carpenter *et al.*, 2012; Walker *et al.*, 2009). When this happens, social change can take place at a large scale, as interacting social and ecological processes play out faster. This can sometimes generate surprises, when processes begin with a slow onset – creeping over decades – but end up with an abrupt impact. In such a world, it is urgent that we understand how changes in precipitation, due to large-scale regional to global environmental change, can interact with local social change to cause surprising and rapid shifts in living conditions for communities.

9.3.2 A misleading mental disconnect between economic growth and biosphere functions

This relatively recent evidence of the interconnectedness of water across scales, between local partitioning of rainfall and global drivers influencing the availability of rainfall, and between water and human dimensions of change, conflicts with current worldviews and development paradigms, which seldom connect to real-world global dynamics. Perspectives and worldviews have been developed that mentally disconnect human progress and economic growth from fundamental interactions with the biosphere (Arrow *et al.*, 1995; O'Brien, 2009). This disconnect causes societies to lose sight of the fact that fossil-fuel-supported human expansion has led to an erosion of the long-term capacity of natural capital to sustain societal development (Folke *et al.*, 2002; Millennium Ecosystem Assessment, 2005). The focus on economic and social development means that the life-supporting environment has become external to society, so that people and nature are treated as two separate entities. The life-supporting environment has been defined as 'that part of the Earth that provides the physiological necessities of life, namely food and other energy, mineral nutrients, air and water', and the life-support system as 'the functional term for the environment, organisms, processes, and resources interacting to provide these physical necessities' (Odum, 1989). We completely fail to account for changes in the capacity of natural capital to sustain human well-being in measurements of progress such as GDP or the Human Development Index (HDI). We use models of natural resource extraction based on one resource at a time, and tend to treat the environment as a separate sector in policymaking and decision-making (Folke *et al.*, 2002).

Fortunately, however, things are changing. The policy agenda promoting a transition to a 'green economy', linked to efforts to put an economic value on ecosystem services and biodiversity through The Economics of Ecosystem services and Biodiversity (TEEB), led by the former senior banker at Deutsche Bank, Pavan Sukdev, helps to break this disconnect by internalising natural capital in the economy. So far there has been limited progress, as demonstrated by the difficulties in getting world governments to endorse the principles of a green economy at the United Nations Rio+20 Earth Summit in June 2012. Nonetheless, there is a growing recognition that the biosphere constitutes the basis for economic growth, and that the real value of ecosystem services and natural resources needs to be represented in the balance sheet of businesses and nations (UNU-IHDP and UNEP, 2012). One recent manifestation of the growing insight of the need for a new growth paradigm within the life-support systems on Earth, is the proposed unified framework for sustainable development proposed by science (Griggs *et al.*, 2013) and by the high-level multi-stakeholder group represented by the United Nations leadership council of the Sustainable Development Solutions Network (SDSN). Set up by UN Secretary-General Ban Ki-moon, and led by Jeffrey Sachs, professor in economics and development and head of the Earth Institute at Columbia University, it has the task of providing guidance to the process of transforming the UN MDGs (which come to an end in 2015) to SDGs. The SDSN proposes a new framework for human prosperity of growth and development within planetary boundaries, in order to ensure resilient provision of water, food and energy security for the world in the twenty-first century (SDSN, 2013).

The same argument applies to water. As we have shown, there is clear and positive movement towards a wider systems-based approach to water resources. Freshwater has largely been regarded as a natural resource and as an economic good to be extracted from rivers and groundwater (blue water) for households, industry and irrigation. IWRM extended the focus to interacting sectors in catchments, but was still mainly concerned with the blue water branch of the water cycle. More recently, water vapour or green water flows – which constitute 65% of the continental water cycle and are the core focus of this book – have gained increased attention, especially in relation to rainfed agriculture (Rockström, 1999; Rockström *et al.*, 2009a). So has the role of freshwater in trade-offs between bundles of ecosystem services and tipping points in dynamic landscapes (Gordon *et al.*, 2008; Raudsepp-Hearne *et al.*, 2010). In this latter context, the focus is broadening to combine thinking on productivity with resilience in managing and sustaining rain – another key focus of this book – which draws on a wide array of recent advances in water resource management. New approaches are also emerging, such as adaptive water governance (Pahl-Wostl, 2007; Pahl-Wostl *et al.*, 2011).

This ongoing shift in perspective reconnects water governance to life-supporting ecosystems, emphasising the role of water as the bloodstream of the biosphere with people as an embedded part (Falkenmark and Folke, 2003, 2010). Similar trends are seen in shifts towards ecosystem-based adaptive governance of dynamic landscapes and seascapes that incorporate forestry, agriculture and fisheries (e.g. Chapin *et al.*, 2009; Olsson *et al.*, 2008). The urgent challenge of the stewardship of social–ecological systems, their resilience and of ecosystem services from the local to the global levels is becoming clearer (e.g. Chapin *et al.*, 2010; Rockström *et al.*, 2009b; Westley *et al.*, 2011). The critical role of water, with human activities as part of the biosphere perspective, however, is still largely unexplored.

Obviously, societies are not only interconnected globally through political, economic and technical systems, but also through the Earth's biophysical life-support systems. Globalising human–environment interactions are characterised by increasing connectivity, speed, mobility, and scale (Young *et al.*, 2006). For example, shrimps farmed in ponds in Thailand for export to Asian and European markets are fed with fishmeal sourced from marine ecosystems all over the globe, including stocks from the North Sea. Globalisation has made it possible for the shrimp farming industry to make use of marine ecosystem capacity worldwide for its supply of fishmeal (Deutsch *et al.*, 2007). Numerous such interactions play out in all corners of the world.

Increases in connectivity, speed and scale can enhance the capacity of societies to adapt to changing circumstances. If globalisation operates as if it is disconnected from the biosphere, however, it may undermine the capacity of the life-supporting ecosystems to sustain such adaptations and transformations (Folke *et al.*, 2011). Reflexivity can help us move out of narrowly defined development pathways and traps, and reconnect to the biosphere (Voss *et al.*, 2006; Young *et al.*, 2006). This would imply a shift away from managing natural resources one by one and the environment as an externality, towards the stewardship of interdependent social–ecological systems as a prerequisite for human well-being (Chapin *et al.*, 2010; Ostrom, 2009). Hence, in a globalised society, there are no ecosystems without people and no people that do not depend on ecosystem work (Costanza, 1991; Ellis and Ramankutty, 2008). They are intertwined and ecosystem services are therefore generated by social–ecological systems (Berkes *et al.*, 2003).

As water is a prerequisite for all ecosystem services, it follows that water is at the core of social–ecological systems. The growing volatility of the supply of water due to global and regional environmental changes, and pressures on water due to growing social demands and unsustainable use make it clear that productivity- and efficiency-dominated approaches to water resource management must drop the assumption that the water source - rain - is static. They must be combined with resilience-based approaches to securing and enhancing human well-being, where water is used not only as efficiently as possible, but also in ways that safeguard the persistency, adaptability and transformability of desired social states for human societies.

9.3.3 An economy deeply dependent on a functioning biosphere

The conclusion therefore is that profound change is needed beyond the traditional water sector in order to successfully govern and manage water for resilience and human well-being in a rapidly changing and globalised world, and in ways that integrate social and environmental dimensions. We need to reconnect our societies and economies to the biosphere and become stewards of water in this context.

This shifts the agenda of human development away from considering ecosystems and the biosphere as a stable support function of the economy - as natural capital separate from human capital and financial/technical capital. We need a shift in paradigm from the economy as the perceived centre of the universe to the economy as a significant sub-system of the biosphere that is embedded in and deeply dependent on its functioning. The economy, its social institutions, its norms and the rules guiding human behaviour, need to act within a safe operating space of local, regional and planetary boundaries in order to have the resilience required to support human prosperity in a rapidly changing world (Figure 9.4).

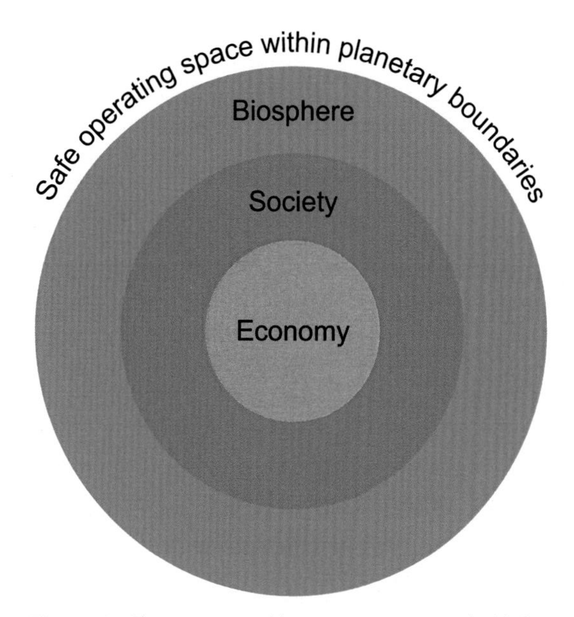

Figure 9.4 The economy and human society as an embedded part of the biosphere. Modified from Daly (1977) and Folke (1991).

In conclusion, social–ecological systems are dynamic and connected, from the local to the global, in complex webs of interactions subject to gradual and abrupt changes. Dynamic and complex social–ecological systems require strategies that build resilience instead of attempting to control for optimal production and short-term gain in environments that are assumed to be relatively stable (Folke *et al.*, 2002; Armitage *et al.*, 2008). The shift from treating people and nature as separate entities to seeing them as interdependent social–ecological systems provides exciting opportunities for societal development in collaboration with the biosphere, as part of a global sustainability agenda for humanity (Griggs *et al.*, 2013). The role of water in such a collaboration and agenda needs to become explicit.

9.4 Towards 'water stewardship for resilience'

Stewardship of freshwater for social–ecological resilience is a new approach to water management and governance. It implies a recognition of the broader role of freshwater in human affairs, extending the water lens from an input into industrial development, urbanisation and irrigation all the way through green water flows in diverse biomes to the bloodstream of the biosphere as a whole. This book shows how societies and economies are embedded parts of the biosphere. It has also provided ample illustrations of the central role of freshwater in the operation of the biosphere. Many would probably still approach such a broader view of water from an efficiency perspective, however, focusing on the allocation of freshwater in a system that is assumed to be fairly stable and subject to only marginal change. Management experts would probably seek to control and optimise the use of blue and green water flows for direct human use and for food production and other ecosystem services in social–ecological landscapes.

9.4.1 Learning to live with change

In contrast to such an approach, combining resilience thinking and freshwater stewardship is now, in this new epoch, about living with the dynamics of continual change, accepting that we are facing increasingly turbulent times as well as genuine uncertainty and surprise. We therefore need to progress insurance thinking, by building management and governance systems that are flexible and prepared for change. Resilience is about learning to live with change, and about the capacity to deal with change while continuing to develop. It is also about water stewardship that allows for alternative pathways – whether to adapt to change on the current pathway or to change on to other pathways, turning a crisis into an opportunity.

In the resilience literature, the efficiency pathway is often referred to as specific resilience, which can lead to social–ecological traps that are difficult to get out of (Cifdaloz *et al.*, 2010). In contrast, water stewardship for resilience should engage with strategies that enhance general resilience. General resilience refers to the capacity of social–ecological systems to adapt or transform in response to unfamiliar, unexpected and extreme shocks. Conditions that enable general resilience are diversity, modularity, openness, the possession of reserves, feedbacks, nestedness, monitoring, leadership and trust (Carpenter *et al.*, 2012).

Biggs *et al.* (2012) set out seven generic policy-relevant principles for enhancing general resilience in the face of disturbance and ongoing change in social–ecological systems. These principles are: maintain diversity and redundancy, manage connectivity, manage slow variables and feedbacks, foster an understanding of social–ecological systems as complex adaptive systems, encourage learning and experimentation, broaden participation, and promote polycentric governance systems. We believe that such conditions and principles should become part of water stewardship for resilience building in turbulent times.

A major reason for this is that in the new Anthropocene era, human prosperity will rely on our ability to stay within the safe operating space of planetary boundaries. This will require planetary stewardship of all the environmental processes that contribute to Earth resilience. Water, in its role as the bloodstream of the biosphere, is a determinant of the state of most planetary boundaries, most notably land use, biodiversity, nutrient flows, the climate system, and air and chemical pollution as a carrier of substances. This means that we require an integrated water stewardship for resilience across scales, from the local to the global. Moreover, all the evidence suggests that we need rapidly to move away from current dangerous trajectories. The most recent science suggests that we have already transgressed the planetary boundaries on climate change, rate of biodiversity loss and interference with the global nitrogen and phosphorus cycles, and that we are rapidly approaching the boundaries for freshwater and land-use change (Carpenter and Bennett, 2011; Rockström *et al.*, 2009b). This means that the world requires a rapid transition to global sustainability, which can be achieved only through an unprecedented and worldwide concerted effort to alter the current exponentially negative trends for global environmental change (Rio Declaration, 1992; Stockholm Memorandum, 2011). Thus, incremental change within a business-as-usual paradigm is not an option. To this we need to add that changes to the Earth System have gone so far that we are likely to face a future with even more turbulence, shocks and stresses, of which most, as is highlighted repeatedly in this book, will be water related.

9.4.2 A general approach: a shift towards adaptive water governance and management

An overarching strategy for water stewardship in the Anthropocene era will be to move from a water regime focused on productivity, optimisation and efficiency, to a regime that builds resilience. This will require investment in water strategies that promote social and ecological redundancy, that is, many different functional and response options in the face of shocks and change. This may sound basic and potentially even simple, but it entails a number of fundamental shifts in governance and management strategies. For example, managing for water resilience will require: (1) a focus on managing water and the ability to deal with trade-offs in order to sustain a wide diversity of water utilities in landscapes; (2) the current social focus on water to supply food, and on blue water supply to societies and industries; (3) a focus on blue water for environmental water flows; (4) a focus on green water flows for a wide range of ecological functions which require investment in biodiversity and spatial diversity in the configuration of land-use types; and (5) a focus on cross-scale management to sustain rainfall.

Institutional flexibility will be an increasingly important feature of water governance, providing the capacity to rapidly change plans as the water reality shifts, learn from the past and other experiences, and adapt to changing conditions. An adaptive water management and adaptive water governance approach (see Chapter 8) provides room for experimentation and learning, but needs to be connected across multiple levels of governance (Biermann *et al.*, 2012; Young, 2011). Interestingly, there are signs of the emergence of new forms of multi-level governance, with new actors starting to interact in diverse and polycentric governance systems (Galaz *et al.*, 2012; Ostrom, 2010). Key features are the ability to live with change, to manage water in ways that provide a multitude of options in the face of extreme hydrological events and to develop water systems that are as resilient as possible, not only in terms of persistency, but also in terms of being adaptable and, in the face of crises, flexible enough to allow for a change to a new trajectory.

An approach to water resilience that nurtures persistency must focus on sustaining rainfall and the ability to maintain the multitude of roles that water plays, as well as the multitude of functionalities that water enables the biosphere to play. The current fossil energy-based industrialised era allowed us to develop modern urban and agricultural systems based on large-scale solutions that simplified landscapes. For example, modern agricultural systems based on irrigation and monocultures of highly productive cereals have resulted in the loss of functional groups in landscapes – from diversity in soil biology capable of sustaining soil fertility to a lack of functional land-use types that help to retain moisture and provide moisture feedback. Similarly, modern cities have largely erased ecosystem diversity and engineered ways of diverting water flows away from urban areas. This has created brittle social–ecological environments vulnerable to outbreaks of disease and water quality and quantity problems, and which lack flexibility in the face of external shocks – such as when food prices rocket on the world market, creating passive and reactive responses in cities, and causing unrest and sometimes even food riots. Before the industrial era of landscape manipulation, the human cultures that survived sudden large shocks were those that had developed different social and ecological back-up systems, e.g. by sustaining forests, multiple water supply systems and a diversity of life-support systems for food, energy and freshwater (Berkes and Folke, 1998).

The difference today, of course, is the scale of impact. In the globalised era of human change, we need to sustain diversity for persistency at all scales from maintaining water regulating biomes at the Earth scale, to managing all remaining biodiversity in our catchments in ways that provide diversity for persistency – with an underlying strategy of sustaining future rainfall. One way of operationalising this in water governance is to manage water in ways that build functional diversity. In the current economic paradigm, it is sometimes difficult to defend such a water strategy. Redundancy rarely, if ever, receives a monetary value in cost–benefit analyses. There is however ample evidence that nurturing diversity as a strategy for general resilience building in the spatial configuration of catchments and river basins can generate critical water services and provide buffers in case of shocks. A classic example is the decision in New York to provide freshwater to the city by managing the ecological functions of the Catskill catchment or, in other words, to maximise high quality water supply from the forested catchment rather than investing in a conventional, downstream water-treatment plant (Heal, 2000).

Another key component of water stewardship for resilience emerges from the recognition that the global turbulence within the nexus of water, bio-resources and economic development is rapidly increasing as a result of the pressures of abrupt global environmental change, rapidly growing demand for biomass-based food and energy, and global economic volatility. As a consequence, the future will be increasingly difficult to predict, placing greater importance on water resource governance and planning that integrates and invests in the ability to act rapidly on new opportunities, and on the capacity to respond to surprise. This, in turn, will require a shift in governance and management 'culture' away from the current predominant paradigm based on planning and the allocation of predictable, stable water resources in efficient and low-cost ways, to a regime that incorporates redundancy, multiple options, flexibility and strategic risk assessments. Such a regime would be an integral component of strategies that combine efficiency with resilience, and the delivery of services with risk mapping and insurance strategies. A key component of such a widened water regime is the ability to continually explore new pathways for adaptation, innovation and transformation, recognising that the only thing we can be certain of about the future in the area of water resources is that it will be highly uncertain.

Our proposed shorthand for this strategy for water resource governance and management is 'water resilience'. Governance and management for water resilience embraces integration and flexibility, the importance of redundancy, and adaptation and transformation in the context of highly connected and complex social–ecological interactions from the local to the global scales, as well as the ability to innovate and deal with shocks, stresses and surprise while securing a positive development pathway. Governance for water resilience requires a systems-based approach that links social and ecological dimensions in new ways. The stewardship of landscapes is seen as the core strategy for the efficient and resilient supply of water-related ecosystem services – not the other way around. The current focus, however, tends to be on securing water services while protecting the environment. In a water resilience strategy, the active stewardship of local catchments, river basins and the entire Earth System forms the core strategy for IWRM, and thus for the sustainable provision of water services and other ecosystem functions and services.

9.4.3 A specific approach: water resilience to meet humanity's food needs

The core focus of our attempt to advance a new integrated resilience approach to water governance and management has been on how to govern and manage water for food in ways that build resilience and contribute to global sustainability, while providing adequate food for a rapidly developing and growing world population. The reason for this focus is obvious to all who follow the trajectories of global change and world development. Food is not only the basis for human well-being and economic development. We now face an unprecedented global challenge in the need to produce up to 70% more food over the coming 40 years and an associated staggering 2000–4000 km^3/year increase in consumptive water use. This must be achieved in a world where land, water and ecosystem constraints are rapidly increasing due to the combined effects of a gradual deterioration in water availability and abrupt effects on water resources linked to global environmental change.

The water–food–development nexus is the eye of the new global water storm facing humanity. Our assessment is that this great challenge will require a triply green revolution, where food productivity is rapidly enhanced ('green' for production increase) in ways that build resilience and are sustainable ('green' for sustainability) and that largely rely on a productivity revolution in the use of 'green' water in rainfed agricultural systems, which today cover 80% of the world's agricultural land area and offer the largest untapped potential in terms of yield gaps and the underutilisation of freshwater resources (Foley *et al.*, 2011). This book also emphasises, in line with, for example, the conclusions from the *Comprehensive Assessment on Water Management in Agriculture* (Molden *et al.*, 2007), that the most promising water strategies for a sustainable and resilient closure of yield gaps in agriculture lie in the interface between green-water-dependent rainfed farming and blue-water-dependent irrigation farming. Many of the most promising innovations to increase productivity and build water resilience in agricultural systems lie in combining blue water strategies – storing runoff in water harvesting and small-scale irrigation systems – with farming systems that largely operate on green water in rainfed systems in order to bridge dry spells and short periods of drought (see Chapter 6).

An integrated 'water accounting' approach can illustrate the operational integration required to achieve a sustainable triply green revolution that meets the water needs of both today and tomorrow, that is, for both current and future generations, e.g. by managing water in one rainy season to secure rainfall for the next rainy season through landscape management that secures adequate moisture feedback.

Our schematic, developed throughout this book, builds on the following logical relationships:

$$\text{Human well-being (HW)} = \text{social–ecological water resilience } (W_{\text{resilience}})$$

$$W_{\text{resilience}} = \text{sustaining rainfall (P)} = \text{green water } (W_{\text{green}}) + \text{blue water } (W_{\text{blue}})$$

$$W_{\text{green}} + W_{\text{blue}} = \text{social water supply (domestic, industry)} (W_{\text{society}}) + \text{food } (W_{\text{food}}) + \text{other ecosystem water (terrestrial and aquatic)} (W_{\text{ecosystems}}) + \text{moisture feedback } (W_{\text{feedback}})$$

This means that water resilience ($W_{resilience}$) can be defined in terms of green and blue water being managed in ways that sustain future rainfall while providing water for societies, food and ecosystems and ensuring enough moisture feedback (Figure 9.5):

$$W_{resilience} = W_{society} + W_{food} + W_{ecosystems} + W_{feedback}$$

This framework provides a new 'boundary' for water governance. It indicates that in order to achieve water resilience, we need to ensure that water is provided for all functions in landscapes – from communities to catchments, river basins, nations and regions. As $W_{resilience}$ is defined by the volume of sustained rainfall, it is by definition a finite amount, albeit one that dynamically changes over time as a result of natural variability and anthropogenic change. This means that trade-offs will arise between the four main components of water resilience shown above. Our assessment indicates that the largest trade-off will be between the rapidly growing demand on water for food, and the other components of social–ecological water resilience.

Fortunately, there appears to be a set of win–win synergies in which improved water resilience and human well-being can be achieved without deep

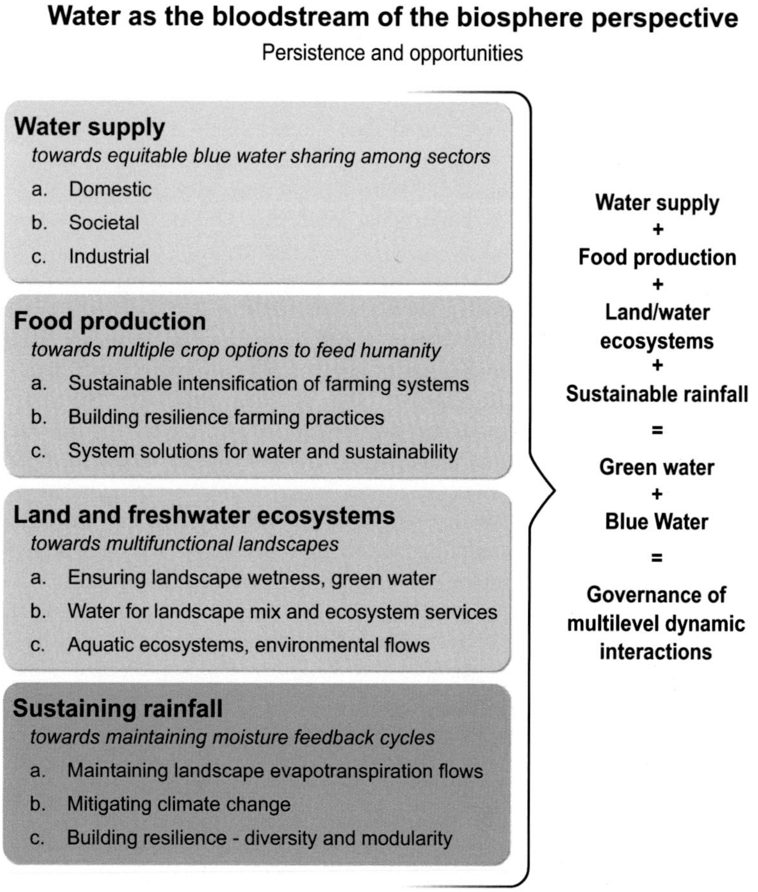

Water as the bloodstream of the biosphere perspective

Persistence and opportunities

Water supply
towards equitable blue water sharing among sectors
a. Domestic
b. Societal
c. Industrial

Food production
towards multiple crop options to feed humanity
a. Sustainable intensification of farming systems
b. Building resilience farming practices
c. System solutions for water and sustainability

Land and freshwater ecosystems
towards multifunctional landscapes
a. Ensuring landscape wetness, green water
b. Water for landscape mix and ecosystem services
c. Aquatic ecosystems, environmental flows

Sustaining rainfall
towards maintaining moisture feedback cycles
a. Maintaining landscape evapotranspiration flows
b. Mitigating climate change
c. Building resilience - diversity and modularity

Water supply + Food production + Land/water ecosystems + Sustainable rainfall = Green water + Blue Water = Governance of multilevel dynamic interactions

Figure 9.5 Conceptual framing of the water components to be incorporated into water stewardship for water resilience.

trade-offs between the components that define water resilience. The two most central synergies we want to highlight are vapour shifts and the robustness of moisture feedback. Vapour shift is the opportunity to produce 'more crop per drop', meeting growing water demand in agriculture by shifting current non-productive flows in terms of evaporation to productive flows in terms of plant transpiration. There is vast empirical evidence of vapour shift in agriculture, particularly when moving from low-productive and low-yielding agriculture in hot tropical regions with large evaporation losses, to high yielding systems (Molden, 2007). This is significant because the regions with the largest untapped yield potential, that is, the regions with the largest current yield gap, coincide with regions with large losses through evaporation and are the world's most hungry regions with the fastest population growth, and therefore the fastest growth in food demand. The second synergy is in the tentative indications that total green water flows from catchments and basins are relatively robust in terms of shifts in land-use type, that is, that similar amounts of green water flow are 'fed back' to the atmosphere to generate future rainfall. This means that trade-offs between different land-use types may have only partial impacts on future rainfall. At the same time, evidence shows that forests do indeed play a critical role in sustaining rainfall in regions where moisture feedback plays a dominant role in generating rainfall.

An overall conclusion, however, is that strategies for water resilience require water governance and management regimes that focus on ecosystem-based landscape management. This means management where all spatial land-use types are configured in ways that ensure favourable partitioning of rainfall into green and blue water flows, the generation of bundles of ecosystem services and safeguarding moisture feedback for future rainfall. The latter requires management strategies that connect scales from the local to the regional and the global. We have called this a strategy for sustaining 'wetness' in landscapes, which in turn requires active water stewardship of the spatial mosaic of land-use types across landscapes.

9.4.4 Adaptive capacity to cope with unpredictable weather extremes

The management of water has always, in all sectors, been a question of adapting to changing circumstances. The Earth System currently seems to be responding faster than anticipated to human impacts. The Arctic sea ice, for example, is melting faster than scientific assessments from just a few years ago estimated that it would. We can no longer exclude the possibility that the Arctic sea ice has crossed a tipping point where the positive albedo feedback – where light-coloured ice that reflects solar radiation melts and is replaced by a darker, more energy absorbent surface – together with warmer ocean currents and abrupt shifts in wind patterns, could be signs of a new dynamic of self-accelerated melting. There are also analyses of weather extremes, based on meteorological observations, that suggest that extreme weather occurrences, more or less non-existent in the 1950s – extremes of more than three standard deviations away from the average – are now becoming normal occurrences, affecting 10–15% of the land surface (Hansen *et al.*, 2012). This is extreme weather: 1 in 1000 year floods and droughts. Less than a decade ago, science in general indicated that such rapid increases in weather extremes, although likely, were not to be expected until beyond 2050. This indicates a 'stirring up' of the global water glass, with much more erratic rainfall and temperature extremes occurring earlier than predicted. Only 50 years ago, nowhere on Earth experienced such severe extremes. How will this new global predicament affect water governance and management? What kind of water stewardship do we need to build water resilience, integrate water with the social–ecological complexity facing humanity, while at the same time meeting rapidly growing demands for food and basic water needs?

Water is by definition a variable entity, characterised by natural variability in supply and demand. Nonetheless, the predominant water governance paradigm has largely focused on erasing variability and managing for a stable and predictable supply of water, equating water resource management with investments in water infrastructure for the storage and distribution of water. This has been a hugely successful strategy, as is outlined in Chapters 1 to 4, with an exponential growth of human-controlled water storage from less than 1000 km^3/year of storage capacity to 5000 km^3/year in only 100 years. Growing human pressures on finite water resources, growing water scarcity and a human-induced increase in abrupt shifts in water availability mean that the water landscape is rapidly changing. An already highly variable entity is becoming even more variable.

This means that adaptive water management (Pahl-Wostl *et al.*, 2007) is now a more important part of IWRM than ever before. Water has to be managed for persistency – sustaining rain – and for transformation – the ability to find new pathways for development in situations of crisis or regime shifts. Adaptive management of water has been defined as the capacity to manage water in ways that adapt to unavoidable changes while maintaining a desired social–ecological state. This can include the adaptive capacity of farmers to change soil and crop management practices (new inter-cropping strategies, changes in seed, dry planting strategies, etc.) in the face of rainfall variability while maintaining the same livelihood base. Building adaptive capacity for water resource management entails the ability to acquire knowledge of how to cope with new conditions while investing in social and institutional networks that allow a desired social state to persist despite shocks and stresses. There is ample evidence of multiple strategies for adaptive water management, all of which have the same universal red thread of developing flexibility, multiple options, networks and broad knowledge systems (Pahl-Wostl, 2007; see Chapter 8).

9.4.5 Water governance to deal with complexity, uncertainty, new risks and unexpected outcomes

A thread running throughout this book is the growing recognition of the ever increasing interconnectedness and turbulence that hit water first-hand – be it through abrupt shifts in rainfall, altered rainwater partitioning due to rapid land-use changes or rapid changes in water demand as a result of policy shifts (e.g. on biofuels) or demographic trends (e.g. urbanisation). The current water governance paradigm is ill-prepared to deal with this new reality, as it is predominantly centred on the notion of the stability and predictability of blue water supplies and an approach focused on optimisation and efficiency. Evidence is growing of the limitations on our ability to predict and control water systems, and the importance of the human dimension is becoming increasingly clear (Pahl-Wostl *et al.*, 2011). The current approach has served societies well under stable and predictable water conditions – in situations where the hydrograph of river flow over time can be relied on. Today's reality, however, under the pressure from abrupt global environmental change and human pressures on land and water resources, is very different.

Water governance must embrace complexity and uncertainty as integral parts of normality. This requires a change in governance from a focus on prediction and control, to strategies that are adaptive and flexible to changing conditions, and which incorporate learning (Pahl-Wostl, 2007). Our assessment is that this is nothing less than a paradigm shift in water governance, which will require a process of deep transformation.

A core challenge for IWRM is to better tackle the new challenges of interconnected social–ecological systems and global environmental change (Galaz *et al.*, 2008). Water governance to deal with new risks, uncertainty and the often unexpected outcomes of complex social–ecological interactions will involve the development of flexible institutions and policies that facilitate learning, adaptation and the ability to transform. As is noted in Chapter 8, there are indications of such a reframing of water governance and a paradigm shift in water policy. Progress with implementation remains slow (Pahl-Wostl *et al.*, 2011), and often fails to take account of resilience and complexity (Galaz, 2005; Olsson and Galaz, 2009). Key to fostering social learning, which is central to adaptive water governance, is strengthening cooperation structures and social networks in combination with advanced information management (Huntjens *et al.*, 2010).

It is also clear, as is discussed at length throughout this book, that water governance must now connect the scales from local communities, to catchments, river basins, regions and nations and the global scale. Despite all the evidence of cross-scale interdependence, where water availability at the local scale depends on water policies and the actions taken at higher scales, and vice versa, connecting scales in a coherent way remains a major challenge (Schlüter *et al.*, 2010).

9.4.6 Critical scale and integration gaps to fill in water governance and management

This book has addressed several critical gaps in water governance and management. Our conclusion is that filling the critical ones will entail moving from IWRM to integrated land and water resource management (ILWRM), in terms of:

- the need to integrate green and blue water into ILWRM;
- the need to sustain rainfall and wetness in landscapes as a strategic goal;

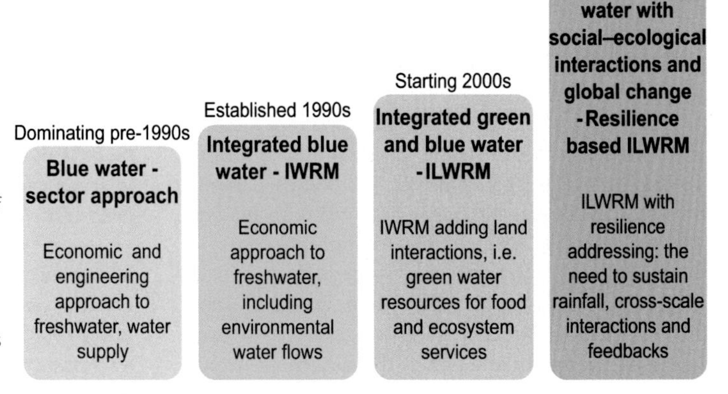

Figure 9.6 The evolution of water resource thinking within governance and policy over the past 30 years, starting with the predominantly 'blue' water-oriented engineering approaches of the pre-1990s, via the introduction of IWRM (and the economic value of water) in the early 1990s. This was followed by a more systems oriented approach to water and land in the early 2000s, with an emphasis on the 'blue' water branch of environmental water flows. In more recent years we see the emergence of a stronger integration of water, land and ecosystems, including both terrestrial and aquatic ecosystem services (green and blue water flows), as well as a stronger emphasis on cross-scale linkages and feedbacks, social–ecological interactions, resilience and sustainability. The blue water/sector approach was described in Stockholm Environment Institute (1997) and National Research Council (1991). Examples of important publications for the IWRM era are the 2009 World Water Development Report (WWAP, 2009) as well as Molden (2007). The first generation of ILWRM is elaborated by L'vovich (1979), Calder (1999) and Falkenmark and Rockström (2004).

- the need to connect scales, recognising that water resources locally depend on how catchments, basins, regions and the planet as a whole are 'managed';
- the need to recognise that human well-being derives from our stewardship of social–ecological systems, and that these in turn depend on water resources to deliver ecosystem services that form the basis for human well-being.

It must be remembered that:

- all these complex interactions in social–ecological systems determine the water resilience in our landscapes and societies;
- we live in an increasingly turbulent world in terms of all aspects of water resources for human well-being – from risks of extreme events to strains on the supply of water for food and other societal uses. Water risks have always been integral to water governance, but we now face an increasing frequency and rising amplitude of extreme water shocks due to human pressures and global change. We thus need to connect the Anthropocene era to water resource governance and management.

Over the past 20 years, since the birth of modern thinking on IWRM, major progress has been made with addressing several of the highlighted water gaps (Figure 9.6). In the 1990s, we started to move away from the dominant paradigm of the 1980s and early 1990s, of IWRM as solely a question of managing stable runoff water, and that the solutions to resource scarcity and unsustainable water use largely lay in putting the right price on water (as set out in the Dublin principles in the run-up to the UNCED 1992 conference in Rio). In the early 2000s, considerable efforts were made to integrate decades, if not centuries, of knowledge on the connections between soil and moisture, and between land and water in managing landscapes. At the same time, we started to see a rapid mainstreaming of the science on water for ecosystems, which however remains focused on environmental water flows to sustain aquatic habitats or blue water ecosystems. We are now starting to take the next evolutionary step towards an integrated paradigm of green and blue water resources in the context of social–ecological interactions and global change – what we in short define as resilience-based ILWRM.

9.4.7 Putting it all together: feeding humanity and sustaining rainfall and landscape wetness

Humanity faces an urgent need to recognise the unprecedented global momentum of two colliding forces that are fundamentally changing the global water arena. The first force is our entry into the Anthropocene era, where we as humanity now have the ability to shift the 'planetary control levers' that regulate the flows of water on Earth. This force has a dramatic effect on water resources across the world. No factor is so rapidly and profoundly affected by global environmental changes as water, resulting in more common and severe water shocks – be they droughts, floods, dry spells or extreme weather events.

The second colliding force is represented by the rapid growth of the global demand for water to feed a rapidly growing world economy. Population growth and increasing affluence will in the coming 40 years propel the world from a reality where most people are poor and live in rural areas to a world where most people are comfortable and live in cities. Water is both the bloodstream of the landscape and the fundamental driver of human well-being in terms of food, energy and health. We must prepare ourselves for growing global water turbulence as the two giants of water demand for 9 billion people and the water impacts from abrupt global environmental change collide.

This collision will not be smooth, with incremental implications on water flows and human societies. Instead, Earth resilience is at stake, i.e. the capacity of the Earth System to buffer human disturbances over long stretches of time, often decades, in order to avoid large incumbent risks of triggering major regime shifts, what we have called TPCs as defined by Biggs *et al.* (2003). The risks of crossing such water-related tipping points, and the implications for human societies, cannot be predicted. The likelihoods can be assessed, including the growing evidence of risks related to the abrupt change of rainforests into savannahs, when moisture feedback is lost due to deforestation or heating from climate change; shifts in the monsoon system, under the pressure from soot and climate change; shifts in freshwater lakes, due to eutrophication and loss of biodiversity; and collapses of agricultural systems, under the pressure from loss of landscape diversity.

It is new scientific evidence of these global dynamics that forms the basis for our justification for a new integrated paradigm for water resource management based on an understanding of resilience and social–ecological systems and the need to connect scales from the local to the planetary level. We propose a simple, integrated conceptual framework to guide this new water paradigm, which we presented in more detail in Chapter 1, and which is here synthesised in a summary framework (Figure 9.7). Our proposal is that we connect the strands of thinking: (1) green and blue water thinking, and land and water integration; (2) social–ecological systems and resilience thinking; and (3) global change and cross-scale interactions. A core strategy that emerges from this framework is the need to build water resilience, recognising that water is both a 'victim' and a driver of change. Stewardship of water can sustain ecosystem services and build robust (the persistency dimension of resilience) social–ecological systems that are able to resist change. Water management can also be carried out in ways that provide capacities to adapt to unavoidable changes, and allow communities to transform in the face of crises (the two additional dimensions of resilience).

Nowhere is this 'perfect water storm' of growing affluence within a growing world population, abrupt global changes and the risk of tipping points, as critical as in agriculture. Feeding humanity in this new water reality of growing water scarcity and increasing uncertainty over future water resources is a huge global challenge that, in our assessment, will require nothing less than a triply green revolution (see Chapter 6). We need sustainability and resilience to be the fundamental prerequisites for agricultural development in order to ensure food production from farming systems able to regenerate rainfall and withstand shocks and stresses. Sustainable intensification is necessary to meet growing social demands while ensuring long-term stability and productivity. Ultimately, in an ecologically constrained world, intensified sustainability is the avenue towards enabling persistent prosperity. Our conclusion is that we need to internalise water resilience as the basis for all efforts to increase agricultural productivity.

This in turn requires that all agricultural development must be based on an ecosystem approach that integrates ecological functions and the delivery of bundles of ecosystem services as an integral strategy for agricultural development. There is now increasing evidence that maintaining redundancy in landscapes – through a high degree of biodiversity and a rich mosaic of different land-use types – is a key strategy for building resilience and sustaining rainfall, i.e. sustaining forested areas as critical zones for moisture feedback. This explains our proposal, as a plausible hypothesis based on existing evidence and the existing knowledge base, that sustaining rainfall and maintaining wetness in landscapes in a world where we are able to shift rainfall patterns are the 'first principles' for water resource governance and management. This requires the adoption of strategies that build water resilience into social–ecological systems at local scales while connecting to the global scale.

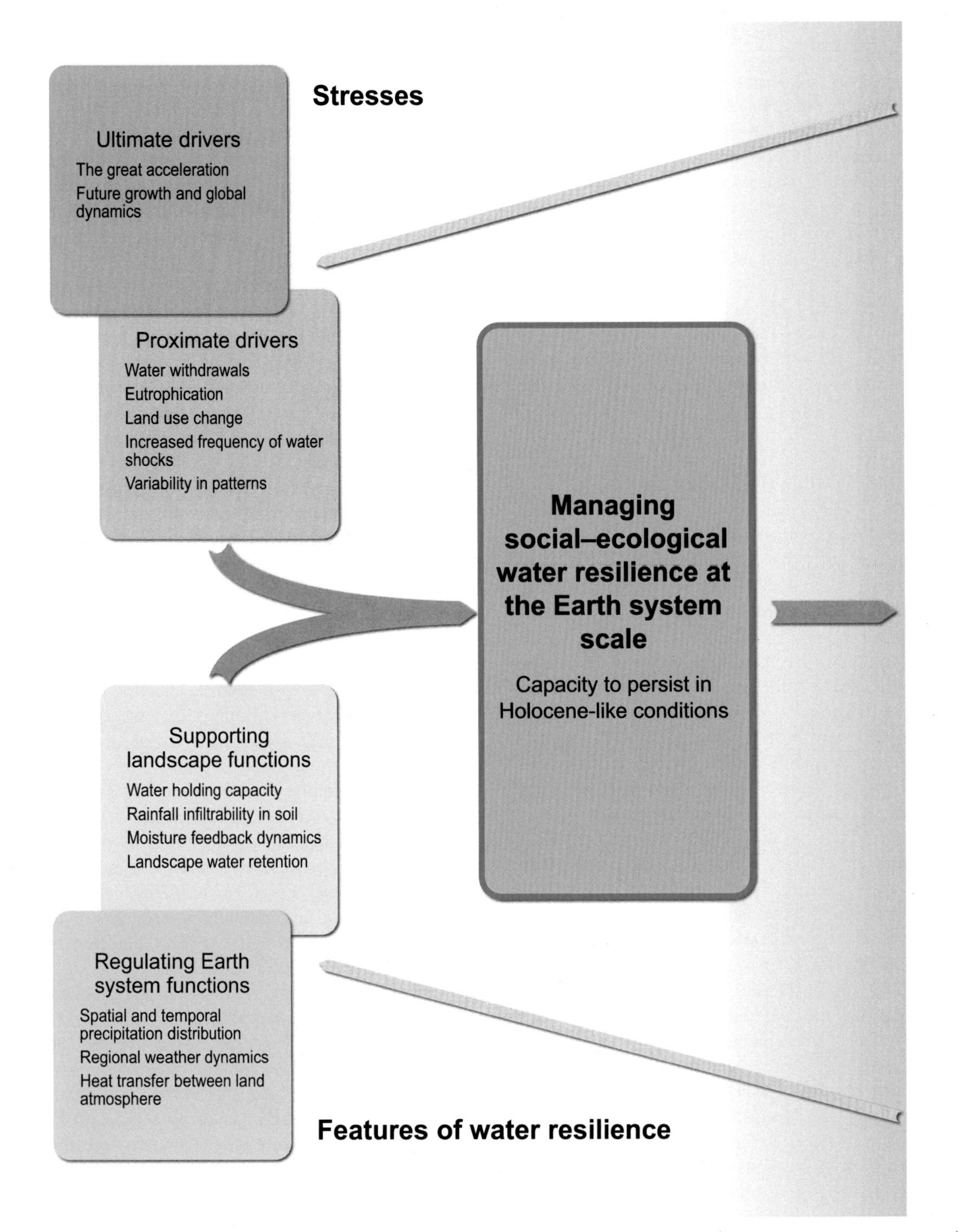

Figure 9.7 Synthesis of our proposed conceptual framework for a social–ecological systems approach to water resource governance and management for human prosperity, resilience and global sustainability in an era of abrupt changes.

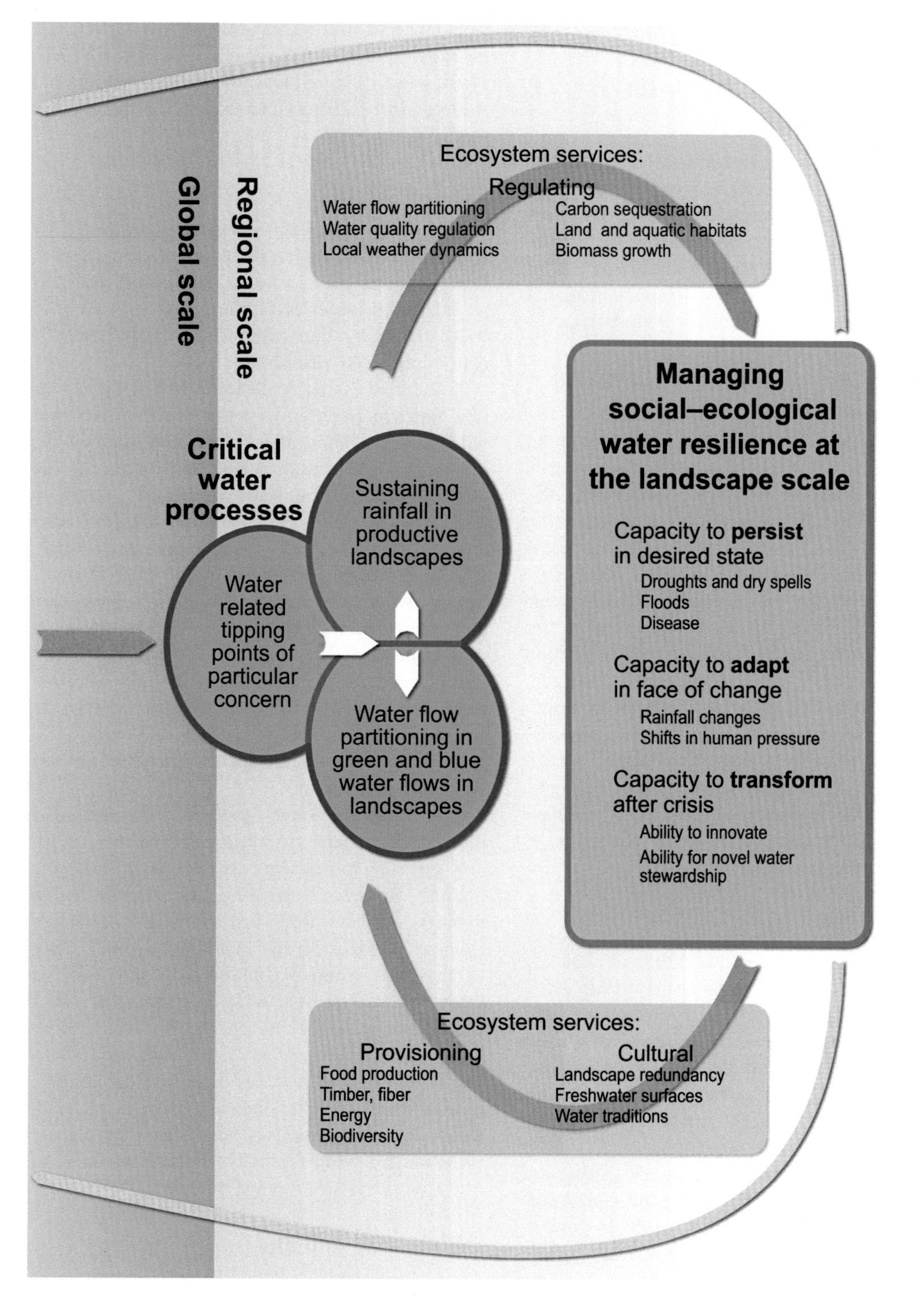

Figure 9.7 (cont.)

Summary

A new global water situation in the Anthropocene epoch

There are three overarching reasons for a new framework for water governance and management: (1) the exponential human pressures on all the ecosystem and environmental processes that matter for human well-being; (2) the fact that water is the 'first victim' of climate change; and (3) the risk of crossing tipping points when transgressing planetary boundaries.

The need to reconnect with the biosphere

The world now faces a doubly interlinked 'water drama': (1) rapidly changing water conditions in an increasingly turbulent and connected water world; and (2) growing human pressures on water demands. This involves risks of crossing thresholds, triggering abrupt regime shifts. While an affluent middle class is aiming for material growth and increased income, we have identified a trend towards a more volatile global hydrological cycle with a higher frequency and greater amplitude of water shocks. Changes in precipitation could interact with local social change to cause surprising and rapid shifts in living conditions for populations. The dominant worldview, however, mentally disconnects human progress and economic growth from fundamental interactions with the biosphere. A global sustainability agenda for humanity will have to involve a shift towards seeing people and nature as interdependent social–ecological systems.

How can this be done?

Staying within the safe operating space of planetary boundaries will require planetary stewardship of all the environmental processes that contribute to Earth resilience. An approach to water resilience must focus on sustaining rainfall and the ability to maintain water's many roles and functions in the biosphere. The global turbulence at the nexus of water, bioresources and economic development is rapidly increasing under the pressures of growing demand for biomass-based food and energy, and global economic volatility. This will require a shift in governance and 'management culture' from the current predominant paradigm, based on planning and the allocation of predictable water resources, to a regime that incorporates redundancy, multiple options, flexibility and strategic risk assessments. Governance for water resilience will require a systems approach, where stewardship of the landscape is seen as the core strategy for the efficient and resilient supply of ecosystem services.

Water resilience

One promising innovation to increase productivity is to combine blue water strategies with farming systems that operate on green water in rainfed systems. Using an integrated 'water accounting' approach, we illustrated the integration required to achieve a sustainable 'triply green' revolution.

We defined water resilience in terms of green and blue water managed in ways that sustain future rainfall while providing water for societies, food and ecosystems. This means trade-offs between the four main components of water resilience – water for society, food, ecosystems and moisture feedback – and benefitting from two win–win synergies – vapour shifts and secure robustness of moisture feedback. Strategies for water resilience require water governance and management regimes that focus on ecosystem-based landscape management that is configured to ensure favourable partitioning of rainfall into green and blue water flows. This strategy for sustaining the 'wetness' of landscapes requires active water stewardship of the spatial mosaic of land-use types across landscapes.

The current water governance paradigm is focused on managing a *stable and predictable supply of water* using infrastructure for its storage and distribution. Adaptive management, however, must manage water in ways that *adapt to unavoidable changes*. Its focus on the stability and predictability of blue water supplies, optimisation and efficiency, means that the current water governance paradigm is ill-prepared for dealing with the new reality. Water governance must embrace complexity and uncertainty and 'connect the scales' from local communities to catchments, river basins, regions/nations and the global scale. Strategies for adaptive water management must develop flexibility, multiple options, networks and broad knowledge systems.

Putting it all together

As the 'two giants' collide – water demand from 9 billion people and water impacts from a rapidly growing economy – Earth resilience will buffer human disturbance over decades until we trigger major regime shifts and reach tipping points of particular concern. The new water paradigm must connect: (1) green and blue water thinking, and land and water integration; (2) social–ecological systems and resilience thinking; and (3) global change and cross-scale interactions, and the core strategy must build water resilience by recognising that water is both a victim and a driver of change.

All agricultural development is based on an ecosystem approach that integrates ecological functions and the delivery of bundles of ecosystem services. The key strategy will be to build resilience and sustain rainfall, e.g. by sustaining forested areas as critical zones for moisture feedback and maintaining redundancy in landscapes through a high degree of biodiversity and a rich mosaic of different land-use types. In addition to being the bloodstream of the biosphere, water is the key to resilience in social–ecological systems.

References

Andonova, L. B. and Mitchell, R. B. (2010). The rescaling of global environmental politics. *Annual Review of Environment and Resources*, **35**, 255–282.

Armitage, D. R., Plummer, R., Berkes, F. *et al.* (1995). Adaptive co-management for social-ecological complexity. *Frontiers in Ecology and the Environment*, **7**, 95–102.

Arrow, K., Bolin, B., Costanza, R. *et al.* (1995). Economic growth, carrying capacity, and the environment. *Science*, **268**, 520–521.

Barnosky, A. D., Hadly, E. A., Bascompte, J. *et al.* (2012). Approaching a state shift in Earth's biosphere. *Nature*, **486**, 52–58.

Berkes, F., Colding, J. and Folke, C. (2003). *Navigating Social-ecological Systems: Building Resilience for Complexity and Change.* Cambridge: Cambridge University Press.

Berkes, F. and Folke, C. (1998). *Linking Social and Ecological Systems: Management Practices and Social Mechanisms for Building Resilience.* Cambridge: Cambridge University Press.

Biermann, F., Abbott, K., Andresen, S. *et al.* (2012). Navigating the Anthropocene: improving Earth System governance. *Science*, **335**, 1306–1307.

Biggs, H. C., Rogers, K. H., Du Toit, J., Rogers, K. and Biggs, H. (2003). An adaptive system to link science, monitoring and management in practice. In *The Kruger Experience: Ecology and Management of Savanna Heterogeneity*, ed. Du Toit, J. T., Biggs, H. C. and Rogers, K. H. Washington DC: Island Press, pp. 59–80.

Biggs, R., Schlüter, M., Biggs, D. *et al.* (2012). Toward principles for enhancing the resilience of ecosystem services. *Annual Review of Environment and Resources*, **37**, 421–448.

Blue Planet Laureates (2012). *Blue Planet Award Synthesis: Environment and Development Challenges: The Imperative to Act.* Nairobi: United Nations Environment Programme.

Calder, I. R. (1999). *The Blue Revolution: Integrated Land Use and Integrated Water Resources Management.* London: Earthscan.

Carpenter, S., Cole, J., Pace, M. *et al.* (2011). Early warnings of regime shifts: a whole-ecosystem experiment. *Science*, **332**, 1079–1082.

Carpenter, S. R., Arrow, K. J., Barrett, S. *et al.* (2012). General resilience to cope with extreme events. *Sustainability*, **4**, 3248–3259.

Carpenter, S. R. and Bennett, E. M. (2011). Reconsideration of the planetary boundary for phosphorus. *Environmental Research Letters*, **6**, 014009.

Chapin, F. S., Carpenter, S. R., Kofinas, G. P. *et al.* (2010). Ecosystem stewardship: sustainability strategies for a rapidly changing planet. *Trends in Ecology & Evolution*, **25**, 241–249.

Chapin, F. S., Kofinas, G. P., Folke, C. and Chapin, M. C. (2009). *Principles of Ecosystem Stewardship: Resilience-based Natural Resource Management in a Changing World.* Heidelberg: Springer.

Cifdaloz, O., Regmi, A., Anderies, J. M. and Rodriguez, A. A. (2010). Robustness, vulnerability, and adaptive capacity in small-scale social-ecological systems: the Pumpa Irrigation System in Nepal. *Ecology and Society*, **15**, 39.

Costanza, R. (1991). *Ecological Economics: The Science and Management of Sustainability.* New York: Columbia University Press.

Daly, H. E. (1977). *Steady-state Economics: The Economics of Biophysical Equilibrium and Moral Growth.* San Francisco, CA: W H Freeman.

Deutsch, L., Gräslund, S., Folke, C. *et al.* (2007). Feeding aquaculture growth through globalization: exploitation of marine ecosystems for fishmeal. *Global Environmental Change*, **17**, 238–249.

Ellis, E. C. and Ramankutty, N. (2008). Putting people in the map: anthropogenic biomes of the world. *Frontiers in Ecology and the Environment*, **6**, 439–447.

Falkenmark, M. (1989). Water scarcity and food production in Africa. In *Food and Natural Resources*, ed. Pimentel, D. and Hall, C. W. Waltham, MA: Academic Press, pp. 163–190.

Falkenmark, M. and Folke, C. (2003). Freshwater and welfare fragility: syndromes, vulnerabilities and challenges. Introduction. *Philosophical Transactions of the Royal Society of London Series B: Biological Sciences*, **358**, 1917–1920.

Falkenmark, M. and Folke, C. (2010). Ecohydrosolidarity: a new ethics for stewardship of value-adding rainfall. In *Water Ethics: Foundational Readings for Students and Professionals*, ed. Brown, P. G. and Schmidt, J. J. Washington DC: Island Press, pp. 247–264.

Falkenmark, M. and Rockström, J. (2004). *Balancing Water for Humans and Nature: The New Approach in Ecohydrology*. London: Earthscan.

Foley, J. A., Ramankutty, N., Brauman, K. A. *et al.* (2011). Solutions for a cultivated planet. *Nature*, **478**, 337–342.

Folke, C. (1991). Socio-economic dependence on the life-support environment. In *Linking the Natural Environment and the Economy: Essays from the Eco-eco Group*, ed. Folke, C. and Kåberger, T. Alphen aan den Rijn: Kluwer Academic Publishers, pp. 77–94.

Folke, C., Carpenter, S., Elmqvist, T. *et al.* (2002). Resilience and sustainable development: building adaptive capacity in a world of transformations. *AMBIO*, **31**, 437–440.

Folke, C., Carpenter, S. R., Walker, B. *et al.* (2010). Resilience thinking: integrating resilience, adaptability and transformability. *Ecology and Society*, **15**, 20.

Folke, C., Jansson, Å., Rockström, J. *et al.* (2011). Reconnecting to the biosphere. *AMBIO*, **40**, 719–738.

Folke, C. and Rockström, J. (2011). Third Nobel laureate symposium on global sustainability: transforming the World in an era of global change. *AMBIO*, **40**, 717–718.

Galaz, V. (2005). Social-ecological resilience and social conflict: institutions and strategic adaptation in Swedish water management. *AMBIO: A Journal of the Human Environment*, **34**, 567–572.

Galaz, V., Crona, B., Österblom, H., Olsson, P. and Folke, C. (2012). Polycentric systems and interacting planetary boundaries: emerging governance of climate change-ocean acidification-marine biodiversity. *Ecological Economics*, **81**, 21–32.

Galaz, V., Olsson, P., Hahn, T., Folke, C. and Svedin, U. (2008). The problem of fit between ecosystems and governance systems: insights and emerging challenges. In *The Institutional Dimensions of Global Environmental Change: Principal Findings and Future Directions*, ed. Young, O., King, L. A. and Schroeder, H. Boston, MA: MIT Press, pp. 147–186.

Gordon, L. J., Peterson, G. D. and Bennett, E. M. (2008). Agricultural modifications of hydrological flows create ecological surprises. *Trends in Ecology & Evolution*, **23**, 211–219.

Griggs, D., Stafford-Smith, M., Gaffney, O. *et al.* (2013). Policy: sustainable development goals for people and planet. *Nature*, **495**, 305–307.

Hansen, J., Sato, M. and Ruedy, R. (2012). Perception of climate change. *Proceedings of the National Academy of Sciences*, **109**, 14726–14727.

Heal, G. (2000). *Nature and the Marketplace: Capturing the Value of Ecosystem Services*. Washington DC: Island Press.

Huntjens, P., Pahl-Wostl, C. and Grin, J. (2010). Climate change adaptation in European river basins. *Regional Environmental Change*, **10**, 263–284.

Intergovernmental Panel on Climate Change (2012). Managing the risks of extreme events and disasters to advance climate change adaptation. A special report of working groups I and II of the Intergovernmental Panel on Climate Change, ed. Field, C. B., Barros, V., Stocker, T. F., Qin, D., Dokken, D. J., Ebi, K. L., Mastrandrea, M. D., Mach, K. J., Plattner, G.-K., Allen, S. K., Tignor, M. & Midgley, P. M. Cambridge: Cambridge University Press.

Keys, P. W., van der Ent, R. J., Gordon, L. J. *et al.* (2012). Analyzing precipitationsheds to understand the vulnerability of rainfall dependent regions. *Biogeosciences*, **9**, 733–746.

Kummu, M., Ward, P. J., de Moel, H. and Varis, O. (2010). Is physical water scarcity a new phenomenon? Global assessment of water shortage over the last two millennia. *Environmental Research Letters*, **5**, 034006.

L'vovich, M. I. (1979). *World Water Resources and their Future*. Washington DC: American Geophysical Union.

Lagi, M., Bar-Yam, Y., Bertrand, K. Z. and Bar-Yam, Y. (2012). *Economics of Food Prices and Crises*. New England Complex Systems Institute, Cambridge, MA. Available at: http://necsi.edu/publications/food/.

Lenton, T. M., Held, H., Kriegler, E. *et al.* (2008). Tipping elements in the Earth's climate system. *Proceedings of the National Academy of Sciences*, **105**, 1786–1793.

Millennium Ecosystem Assessment (2005). *Ecosystems and Human Well-being: Synthesis*. Washington DC: Island Press.

Molden, D. (ed.) (2007). *Water for Food, Water for Life: A Comprehensive Assessment of Water Management in Agriculture*. London: Earthscan.

Molden, D., Oweis, T. Y., Steduto, P. *et al.* (2007). Pathways for increasing agricultural water productivity. In *Water for Food, Water for Life: A Comprehensive Assessment of Water Management in Agriculture*, ed. Molden, D. London: Earthscan, pp. 278–310.

National Academy of Sciences of United States of America (2009). PNAS tipping elements in Earth systems special feature. *Proceedings of the National Academy of Sciences*, **106**.

National Research Council (1991). *Opportunities in the Hydrologic Sciences*. Washington DC: National Academies Press.

O'Brien, K. (2009). Do values subjectively define the limits to climate change adaptation? In *Adapting to Climate Change: Thresholds, Values, Governance*, ed. Adger, W. N., Lorenzoni, I. and O'Brien, K. Cambridge: Cambridge University Press, pp. 164–180.

Odum, E. P. (1989). *Ecology and Our Endangered Life-support Systems*. Sunderland, MA: Sinauer Associates.

Oki, T. and Kanae, S. (2004). Virtual water trade and world water resources. *Water Science and Technology*, **49**, 203.

Oki, T. and Kanae, S. (2006). Global hydrological cycles and world water resources. *Science*, **313**, 1068–1072.

Olsson, P., Folke, C. and Hughes, T. P. (2008). Navigating the transition to ecosystem-based management of the Great Barrier Reef, Australia. *Proceedings of the National Academy of Sciences*, **105**, 9489–9494.

Olsson, P. and Galaz, V. (2009). Transitions to adaptive approaches to water management and governance in Sweden. In *Water Policy Entrepreneurs. A Research Companion to Water Transitions around the Globe*, ed. Huitema, D. and Sander, M. Cheltenham, UK: Edward Elgar Publishing, pp. 304–324.

Ostrom, E. (2009). A general framework for analyzing sustainability of social-ecological systems. *Science*, **325**, 419–422.

Ostrom, E. (2010). Polycentric systems for coping with collective action and global environmental change. *Global Environmental Change*, **20**, 550–557.

Pahl-Wostl, C. (2007). Transitions towards adaptive management of water facing climate and global change. *Water Resources Management*, **21**, 49–62.

Pahl-Wostl, C., Jeffrey, P., Isendahl, N. and Brugnach, M. (2011). Maturing the new water management paradigm: progressing from aspiration to practice. *Water Resources Management*, **25**, 837–856.

Pahl-Wostl, C., Sendzimir, J., Jeffrey, P. *et al.* (2007). Managing change toward adaptive water management through social learning. *Ecology and Society*, **12**, 30.

Raudsepp-Hearne, C., Peterson, G. D. and Bennett, E. M. (2010). Ecosystem service bundles for analyzing tradeoffs in diverse landscapes. *Proceedings of the National Academy of Sciences*, **107**, 5242–5247.

Rimas, A. and Fraser, E. D. (2010). *Empires of Food: Feast, Famine, and the Rise and Fall of Civilizations*. New York: Free Press.

Rio Declaration (1992). *Rio Declaration on Environment and Development*. United Nations Environment Programme, Nairobi. Available at: http://www.unep.org/documents.multilingual/default.asp?documentid=78&articleid=1163.

Rockström, J. (1999). On-farm green water estimates as a tool for increased food production in water scarce regions. *Physics and Chemistry of the Earth, Part B: Hydrology, Oceans and Atmosphere*, **24**, 375–383.

Rockström, J., Falkenmark, M., Karlberg, L *et al.* (2009a). Future water availability for global food production: the potential of green water for increasing resilience to global change. *Water Resources Research*, **45**.

Rockström, J., Falkenmark, M., Lannerstad, M. and Karlberg, L. (2012). The planetary water drama: dual task of feeding humanity and curbing climate change. *Geophysical Research Letters*, **39**, L15401.

Rockström, J., Gordon, L., Folke, C., Falkenmark, M. and Engwall, M. (1999). Linkages among water vapor flows, food production, and terrestrial ecosystem services. *Conservation Ecology*, **3**, 5.

Rockström, J., Steffen, W., Noone, K. *et al.* (2009b). A safe operating space for humanity. *Nature*, **461**, 472–475.

Rost, S., Gerten, D., Hoff, H. *et al.* (2009). Global potential to increase crop production through water management in rainfed agriculture. *Environmental Research Letters*, **4**.

Schlüter, M., Hirsch, D. and Pahl-Wostl, C. (2010). Coping with change: responses of the Uzbek water management regime to socio-economic transition and global change. *Environmental Science & Policy*, **13**, 620–636.

Shiklomanov, I. A. and Rodda, J. C. (2004). *World Water Resources at the Beginning of the 21st Century. International Hydrology Series*. Cambridge: Cambridge University Press.

Snyder, P. K., Foley, J. A., Hitchman, M. H. and Delire, C. (2004). Analyzing the effects of complete tropical forest removal on the regional climate using a detailed three-dimensional energy budget: an application to Africa. *Journal of Geophysical Research: Atmospheres (1984–2012)*, **109**.

Steffen, W., Crutzen, P. J. and McNeill, J. R. (2007). The Anthropocene: are humans now overwhelming the great forces of nature. *Ambio*, **36**, 614–621.

Steffen, W. L., Sanderson, A., Tyson, P. D. *et al.* (2004). *Global Change and the Earth System: A Planet Under Pressure*. Berlin: Springer.

Stockholm Environment Institute (1997). *Comprehensive Assessment of the Freshwater Resources of the World.*

Stockholm: Stockholm Environment Institute and World Meteorological Organization.

Stockholm Memorandum (2011). Tipping the scales towards sustainability. Stockholm Resilience Centre, Stockholm. Available at: http://globalsymposium2011.org/wp-content/uploads/2011/07/memorandum-signed.pdf.

Sustainable Development Solutions Network (2013). *An Action Agenda for Sustainable Development.* Paris: Sustainable Development Solutions Network.

United Nations Department of Economic and Social Affairs (2011). *World Population Prospects: The 2010 Revision.* New York: United Nations Department of Economic and Social Affairs. Available at: http://esa.un.org/unpd/wpp/Excel-Data/population.htm (accessed 21 November 2012).

United Nations University-International Human Dimensions Programme and United Nations Environment Programme (2012). *Inclusive Wealth Report 2012: Measuring Progress toward Sustainability.* New York: United Nations University-International Human Dimensions Programme and United Nations Environment Programme. Available at: http://www.ihdp.unu.edu/article/iwr.

Voss, J.-P., Bauknecht, D. and Kemp, R. (2006). *Reflexive Governance for Sustainable Development.* Cheltenham, UK: Edward Elgar Publishing.

Walker, B., Barrett, S., Polasky, S. *et al.* (2009). Looming global-scale failures and missing institutions. *Science*, **325**, 1345–1346.

Walker, B. and Salt, D. (2006). *Resilience Thinking: Sustaining Ecosystems and People in a Changing World.* Washington DC: Island Press.

Walker, B. H., Anderies, J. M., Kinzig, A. P. and Ryan, P. (2006). Exploring resilience in social-ecological systems through comparative studies and theory development: introduction to the special issue. *Ecology and Society*, **11**, 12.

Westley, F., Olsson, P., Folke, C. *et al.* (2011). Tipping toward sustainability: emerging pathways of transformation. *AMBIO*, **40**, 762–780.

World Water Assessment Programme (2009). World Water Development Report 3: Water in a changing world. World Water Development Report. United Nations Educational, Scientific and Cultural Organization, Paris.

Young, O. R. (2011). Effectiveness of international environmental regimes: existing knowledge, cutting-edge themes, and research strategies. *Proceedings of the National Academy of Sciences*, **108**, 19853–19860.

Young, O. R., Berkhout, F., Gallopin, G. C. *et al.* (2006). The globalization of socio-ecological systems: an agenda for scientific research. *Global Environmental Change*, **16**, 304–316.

Glossary

	Chapter	
adaptive management	9	In general terms, adaptive management can be defined as a systematic process for improving management policies and practices through systematic learning from the outcomes of implemented management strategies and taking account of changes in external factors in a proactive manner. Adaptive management was introduced into ecosystem management a few decades ago to promote approaches that experimentally compare selected policies or practices by evaluating alternative hypotheses about the system being managed. Pahl-Wostl, C. (2007). Transitions towards adaptive management of water facing climate and global change. *Water Resources Management*, **21**, 49–62.
adaptive water management	8	Adaptive management approaches have become increasingly popular in water management. The emphasis on robust solutions and iterative learning processes is in stark contrast to the hitherto prevailing approach of searching for optimal solutions.
Anthromes anthropogenic biomes	2	Anthropogenic biomes are the global ecological patterns created by sustained direct human interactions with ecosystems. http://ecotope.org/anthromes/
Anthropocene	1	The scientific postulation that human pressures on the Earth System have reached a scale and pace that mean humanity (Anthros) has become a quasi-geological force for change at the planetary scale, i.e. that humanity constitutes its own new geological epoch. There is some scientific debate about the start of the Anthropocene era, e.g. whether it should be dated back to the start of human intensification on Earth, as early as the start of agriculture some 8000 years ago, or whether its onset coincides with the great acceleration of human enterprise in the mid-1950s, when human pressures on most environmental processes and parameters (e.g. emissions of greenhouse gases, freshwater use, deforestation, land degradation, air pollution and loss of biodiversity) started to increase abruptly and exponentially.
aridification	4	A process of the drying of landscapes and local to regional climates.
aridity index (P/PET)	6	A measure of how arid a specific region is, usually measured as the ratio of precipitation to potential evapotranspiration (P/PET).

AWM	6	Agricultural water management.
basin closure	4	Full allocation of all blue water resources in a basin so that any further allocation affects existing uses and there is no outflow to the sea. Falkenmark, M. and Molden, D. (2008). Wake up to realities of river basin closure. *International Journal of Water Resources Development*, **24**, 201–215.
biophysical system	1	The biological and physical processes, entities and interactions that constitute a system, such as a wetland, forest system, farmland or the biosphere.
blue water	1	The liquid water in rivers and aquifers. Blue water flows sustain aquatic ecosystems, irrigate agriculture and supply water to humans.
carbon sequestration	4	The process of capturing and the long-term storage of atmospheric carbon dioxide.
cascade	3	A series of reactions in which one process triggers another.
consumptive water use	4	Evaporative use of green water/soil moisture or blue water/surface water and groundwater, forming the moisture feedback to the atmosphere from which rain is regenerated.
decoupling	2	1. Spatial decoupling of production and consumption; 2. Decoupling of production and resource use.
disturbance regimes	1	A dynamic set of interacting disturbances, occurring with different magnitudes, duration, frequency and spatial distribution.
drylands	6	Arid, semi-arid, and dry sub-humid climate zones. The term is somewhat misleading, since it conveys the notorious image of dryness and low agricultural potential, which is not true for large parts of the world's semi-arid and dry sub-humid tropical regions.
dry spell	6	A short period of between a couple of days and a couple of weeks without any rain within a rainy season. Dry spells are not necessarily detectable when looking at cumulative seasonal rainfall records, but are a common reason for crop failure in the savannah zone.
Dublin Principles	8	The International Conference on Water and the Environment held in Dublin in 1992 (Dublin conference) adopted a statement that included four guiding principles for IWRM (Dublin principles): (1)Fresh water is a finite and vulnerable resource, essential to sustain life, development and the environment. (2) Water development and management should be based on a participatory approach, involving users, planners and policymakers at all levels. (3) Women play a central part in the provision, management and safeguarding of water. (4) Water has an economic value in all its competing uses, and should be recognised as an economic good.
ecosystem functions	1	The ecological processes that generate a function which has the potential to be an ecosystem service.
ecosystem services	1	The benefits humans derive, directly or indirectly, from ecosystem functions.

environmental flow	1	An ecologically acceptable flow regime, designed to maintain a river in an agreed or predetermined ecological state; which often represents a compromise between water resources development, on the one hand, and river ecology maintenance, on the other.
EU Water Framework Directive (WFD)	8	The WFD is an EU directive which entered into force in 2001. It commits EU member states to achieve good qualitative and quantitative states for their water bodies and introduces the basin scale, public participation and cost-effectiveness as guiding principles.
ex-situ systems	6	AWM interventions that divert runoff from external areas, such as a road or a degraded patch of land with limited infiltration capacity, to the field after heavy rainfall events.
external forces	3	Forces that act on a system from the outside.
external variables	3	Variables outside of the system.
feedback mechanism	1	A loop system, in which the system variables influence each other either in the same direction (positive or reinforcing feedback) or in the opposite direction (negative or dampening feedback).
fertility rates	2	The number of children per woman.
globalisation	2	The process of international integration, focused on trade and foreign direct investment, among other things.
great acceleration	2	The increase in human development after World War II: population doubled in 50 years and the global economy grew by a factor of 15. Fossil fuel consumption has grown 3.5 fold since 1960. From 1950 to 2000 the percentage of the global population living in urban areas increased from 30% to 50% and continues to grow. Steffen, W., Crutzen, P. J. and McNeill, J. R. (2007). The Anthropocene: are humans now overwhelming the great forces of nature? *Ambio*, **36**, 614–621.
green water	1	The rainfed soil moisture supporting plant growth. The water eventually returns to the atmosphere through evapotranspiration. Green water flows sustain terrestrial ecosystem services (rainfed food, forests, grazing lands and grasslands) and regulate the terrestrial moisture feedback that sustains the bulk of the rainfall over land areas on Earth.
Holocene	1	The interglacial geological epoch on Earth over the past 10 000 years. The Holocene is an extraordinarily stable equilibrium state of the planet, in geological terms, with average global temperatures varying by ±1°C as well as predictable temperatures, rainy seasons and precipitation levels. Science shows that the Holocene period has been a highly favourable state for the planet to support the development of the modern world as we know it.
human	2	Human drivers of change.
hysteresis	1	Hysteresis refers to the existence of different stable states under the same variables or parameters. It can be explained by *path dependency*, in which a threshold for the trajectory of 'A → B' is different to that for 'B → A'.

Term	Chapter	Definition
in-situ systems	6	AWM interventions that make better use of the rain that falls on a given parcel of land, e.g. conservation tillage or Zai pitting.
internal variables	3	Variables within a system.
IWRM	8	The most widely used definition of Integrated Water Resources Management (IWRM) is provided by the Global Water Partnership: 'IWRM is a process which promotes the coordinated development and management of water, land and related resources in order to maximise the resultant economic and social welfare in an equitable manner without compromising the sustainability of vital ecosystems'. The IWRM concept has been widely adopted by many countries as a guiding principle for national water policies.
land conversion	2	Conversion of natural land cover to human (in particular agricultural) land cover, including intensification of existing agricultural land use.
LPJmL (Lund–Potsdam–Jena Dynamic Global Vegetation Model with managed land)	5	The LPJmL model is designed to represent the key processes and their interaction in both natural vegetation and on managed lands (pastures, cropland). While the distributions of different types of natural vegetation (biomes) are genuine modelling results, the distribution of crops and pastures needs to be prescribed since it is not driven by biophysical processes but human activity. An important feature of the model is that the water cycle and the carbon cycle are explicitly linked by simultaneous water loss (transpiration) and CO_2 uptake (photosynthesis) through stomata. Key model outputs include variables of the carbon cycle (carbon stocks, carbon assimilation, respiration) and the water cycle (soil moisture, evaporation, runoff, river discharge), as well as crop yields and irrigation requirements on managed lands. The integration of ecosystems, agriculture and water cycle provides the basis for a wide range of assessments within and across different sectors.
meteorological droughts	6	When cumulative seasonal rainfall is not enough to produce a cereal crop. These are clearly detectable in rainfall records.
natural capital	9	The stock of environmental functions, from the local to the biosphere levels, that generate a flow of natural resources and ecosystem services of significance for social and economic development. Natural capital consists of non-renewable resources (such as ore and minerals, renewable resources and ecosystem services) – the latter two generated by life-supporting ecosystems interacting with water as the bloodstream of the biosphere.
net primary production (NPP)	2	The rate at which an ecosystem accumulates energy or biomass, excluding the energy it uses for the process of respiration.
nexus	9	A focus area with complex and critical interconnections which form a 'nexus' of human development in sustainability, governance, policy and management terms, such as the closely linked areas of energy, food and water where water is key for energy and food, and energy is key for water and food.

over-appropriation	1	Assignment of water resources beyond a predefined upper limit.
planetary boundary	1	Science-based definition of a safe level for control variables regulating critical environmental processes associated with the ability of the Earth System to remain in Holocene-like conditions. Moving beyond a planetary boundary for key environmental processes leads to a high risk of crossing a threshold with potentially irreversible and long-term implications for the ability of the biosphere to support human development. The planetary boundaries include climate change, ozone depletion, ocean acidification, land-use change, interference with the global nitrogen and phosphorus cycles (N and P), consumptive freshwater use, the rate of loss of biodiversity, chemical pollution and the release of new entities, and aerosol loading (air pollution).
poverty trap	3	A self-reinforcing mechanism that causes poverty to persist, reflecting a lack of options to develop or deal with change. Occurs in situations of low resource availability, low diversity and low connectedness to other systems, but can also happen in situations of failure of social institutions.
precipitationshed	4	The area from which a basin receives its precipitation through moisture recycling of evaporated water.
rainwater partitioning	4	The partitioning of rainwater at ground level into soil moisture/green water, as vapour flow returning to the atmosphere through evaporation and transpiration, and liquid water/blue water, as surface water or groundwater flowing through landscapes.
redundancy	3	Duplication of components in a system, some of which can protect the survival of the total system in case of a failure of single components.
regime shifts	3	Abrupt, persistent changes in the structure and function of a system; reorganisation from one dynamic system state (regime) to another.
reinforcing processes	3	A process that continually feeds back on itself to reinforce its own growth or collapse.
resilience	1	The capacity of a system to absorb disturbance and reorganise while undergoing change, which enables it to retain essentially the same function, structure feedbacks and identity. Resilience in this context is the capacity of a social–ecological system to continually change and adapt but remain within critical thresholds.
resilience building	9	A key strategy to sustain rainfall, e.g. by sustaining forested areas as critical zones for moisture feedback, and to maintain redundancy in landscapes through a high degree of biodiversity and a rich mosaic of different land-use types.
rigidity trap	3	When a system is characterised by low potential for change, high connectedness and high resilience; and where institutions are often highly connected, self-reinforcing and inflexible.
savannah zone	6	Semi-arid and dry sub-humid tropics where rainfed farming is a main livelihood source. There is a large convergence between these regions and the areas of the world with the highest poverty rates. These regions often have substantial untapped agro-hydrological potential.

savannisation	3	When a humid forest undergoes a regime shift to a savannah regime.
small-scale farming	6	Farming systems that are rainfed and non-mechanised, with limited use of external inputs and where at least part of the production is used for household consumption. This type of farming is often carried out at small spatial scales and is in many cases associated with traditional practices maintained by local communities.
social–ecological resilience	1	The resilience in coupled human–environment systems, e.g. in agriculture.
social–ecological system	1	An integrated system of people and nature with reciprocal feedback and interdependence; the concept emphasises the humans-in-nature perspective. Berkes, F. and Folke, C. (1998). *Linking Social and Ecological Systems: Management Practices and Social Mechanisms for Building Resilience.* Cambridge: Cambridge University Press.
stability landscape	3	The combination of the various basins, states with different degrees of stability that a system may occupy and the boundaries that separate them.
stewardship of landscapes	9	A suite of approaches whose goal is to sustain social–ecological systems, based on reducing vulnerability and enhancing adaptive capacity, resilience and transformability. Its goals are to respond to and shape change in social–ecological systems in order to sustain the supply and opportunities for the use of ecosystem and Earth System services by society. Chapin, F. S., Kofinas, G. P., Folke, C. and Chapin, M. C. (2009). *Principles of Ecosystem Stewardship: Resilience-based Natural Resource Management in a Changing World.* Heidelberg: Springer.
teleconnection	4	Cause-and-effect relationships over large distances, in particular through atmospheric transport and trade.
threshold critical transition	1	A level or amount of a controlling, often slowly changing, variable in which a change occurs to a critical feedback causing the system to self-organise along a different trajectory, i.e. towards a different attractor. Folke, C., Carpenter, S. R., Walker, B. *et al.* (2010). Resilience thinking: integrating resilience, adaptability and transformability. *Ecology and Society*, **15**, 20.
thresholds of particular concern (TPC)	9	Thresholds between multiple stable states of ecosystems, ranging from local lakes to regional systems such as rainfall regimes, subject to risk of large-scale regime shifts due to interactions among different drivers of change, and from processes of regional to global change not necessarily related to water.
tipping point	3	The point at which a regime shift that involves a hysteretic response takes place, i.e. when a slower, reversible change becomes either irreversible or relatively more difficult to reverse, often with dramatic consequences.

virtual water	1	The amount of water required to produce a commodity, in particular crops or livestock, which is saved by a country when substituting imports for local production.
virtual water flow	4	The amount of water used for the production of agricultural commodities and virtually associated when these commodities are traded.
water accounting	9	Similar to economic accounting, the quantification of water requirements in the generation of ecosystem functions and services, e.g. the amount of water required to produce food, uphold biodiversity and secure functioning landscapes.
water crowding	1	Population pressure on blue water availability expressed as people per million m^3 per year. Chronic water shortage occurs when water crowding exceeds 1000 people per million m^3 water/year.
water productivity (WP)	6	The amount of water required to produce a certain amount of food. Often measured in m^3 of water per tonne of food.
water resilience	1	The role of water for social–ecological resilience.
yield gap	6	The difference between farmers' actual and potential yields.

Index